METHODS IN MOLECULAR BIOLOGY™

Series Editor
John M. Walker
School of Life Sciences
University of Hertfordshire
Hatfield, Hertfordshire, AL10 9AB, UK

For further volumes:
http://www.springer.com/series/7651

Plant Mineral Nutrients

Methods and Protocols

Edited by

Frans J.M. Maathuis

Department of Biology, University of York, York, UK

Humana Press

Editor
Frans J.M. Maathuis
Department of Biology
University of York
York, UK

ISSN 1064-3745 ISSN 1940-6029 (electronic)
ISBN 978-1-4939-5973-0 ISBN 978-1-62703-152-3 (eBook)
DOI 10.1007/978-1-62703-152-3
Springer New York Heidelberg Dordrecht London

Printed on acid-free paper

Humana Press is a brand of Springer
Springer is part of Springer Science+Business Media (www.springer.com)

Preface

Plants require a range of essential macro and micronutrients for their growth and development. For terrestrial plants, all mineral nutrients derive from the soil. Although research into the roles and functions of such minerals stretches back more than 150 years, there are still many unresolved questions, even where the definition of "essential mineral" is concerned.

The study of plant mineral nutrition has both academic and applied aspects to it. The simple fact that the human diet is, directly or indirectly, plant based has obvious and profound implications in this respect. Today, research into plant mineral nutrition is more pertinent than ever in the face of a growing world population and the increasing need for sustainable agriculture.

The study of plant mineral research touches on many biological disciplines such as biophysical techniques to follow uptake and distribution of mineral ions, analytical methods to measure minerals in soil and tissue, whole plant physiology to assess growth and development in different conditions, and, more recently, the whole gambit of molecular approaches to characterize the relevant genes and proteins. Furthermore, it spans a large spatiotemporal range from subcellular to whole plants and from msec to months.

This volume contains a comprehensive collection of methodologies that are routinely used in plant mineral nutrition research. It describes easy-to-follow protocols that will allow the researcher to study the most relevant aspects of plant mineral nutrition, including growth parameters, ion contents and composition, soil analyses, flux measurements, and the use of public facilities for high-throughput analyses. As such this volume should be of great use to plant scientists at every level but particularly to plant physiologists, crop scientists, and horticulturalists.

York, UK *Frans J.M. Maathuis*

Contents

Contributors

M. Alexou • *National Agricultural Research Foundation (NAGREF), Forest Research Institute (FRI), Thessaloniki, Greece*

Veena S. Anil • *Department of (Agri) Biotechnology, College of Agriculture-Hassan, Subcampus of University of Agricultural Sciences, Bangalore, India*

Barbara Baldan • *Department of Biology, University of Padova, Padova, Italy*

Bettina Berger • *The Plant Accelerator, University of Adelaide, Urrbrae, SA, Australia*

Jayakumar Bose • *School of Agricultural Science, University of Tasmania, Hobart, TAS, Australia*

Marie Bourcy • *Institut des Sciences du Végétal, CNRS, Gif sur Yvette, France*

D.T. Britto • *Department of Biological Sciences, University of Toronto, Toronto, ON, Canada*

Peter Buchner • *Rothamsted Research, Harpenden, Hertfordshire, UK*

Viviane Cosson • *Institut des Sciences du Végétal, CNRS, Gif sur Yvette, France*

Tracey Cuin • *School of Agricultural Science University of Tasmania, Hobart, TAS, Australia; Biochimie et Physiologie Moléculaire des Plantes, Institut National de la Recherche Agronomique (INRA), Montpellier SupAgro, Université Montpellier II, Montpellier, France*

John M.C. Danku • *School of Biological Sciences, University of Aberdeen, Aberdeen, UK*

T.C. de Bang • *Plant and Soil Science Section, Department of Agriculture and Ecology, Faculty of Science, University of Copenhagen, Copenhagen, Denmark*

Martin D. de Jonge • *Australian Synchrotron, Clayton, VIC, Australia*

Bas de Regt • *The Plant Accelerator, University of Adelaide, Urrbrae, SA, Australia*

Eugene Diatloff • *INRA AgroParisTech, Inst Natl Rech Agron, Inst Jean Pierre Bourgin, UMR, Versailles, France*

Sylvie Dinant • *Institut Jean Pierre Bourgin (IJPB), INRA-AgroParisTech, Versailles, France*

Erica Donner • *Centre for Environmental Risk Assessment and Remediation, University of South Australia, Adelaide, SA, Australia; CRC Care, Salisbury, SA, Australia*

Wieland Fricke • *School of Biology and Environmental Science, University College Dublin (UCD), Dublin, Ireland*

Benjamin Gourion • *Institut des Sciences du Végétal, CNRS, Gif sur Yvette, France*

T.H. Hansen • *Plant and Soil Science Section, Department of Agriculture and Ecology, Faculty of Science, University of Copenhagen, Copenhagen, Denmark*

Malcolm J. Hawkesford • *Rothamsted Research, Harpenden, Hertfordshire, UK*

S. Husted • *Plant and Soil Science Section, Department of Agriculture and Ecology, Faculty of Science, University of Copenhagen, Copenhagen, Denmark*

KAVITHA P.G. • *Membrane Biophysics Group, National Centre for Biological Sciences, University of Agricultural Sciences-G.K.V.K Campus, Bangalore, India*

JULIA KEHR • *Centro de Biotecnología y Genómica de Plantas (CBGP), Campus de Montegancedo, Madrid, Spain*

PETER M. KOPITTKE • *CRC CARE, PO, Salisbury, SA, Australia; The University of Queensland, School of Agriculture and Food Sciences, St. Lucia, QLD, Australia*

H.J. KRONZUCKER • *Department of Biological Sciences, University of Toronto, Toronto, ON, Canada*

PAVAN KUMAR • *Membrane Biophysics Group, National Centre for Biological Sciences, University of Agricultural Sciences-G.K.V.K Campus, Bangalore, India*

SAM KURUVILLA • *Membrane Biophysics Group, National Centre for Biological Sciences, University of Agricultural Sciences-G.K.V.K Campus, Bangalore, India*

BRETT LAHNER • *Department of Horticulture & Landscape Architecture, Purdue University, West Lafayette, IN, USA*

K.H. LAURSEN • *Plant and Soil Science Section, Department of Agriculture and Ecology, Faculty of Science, University of Copenhagen, Copenhagen, Denmark*

ENZO LOMBI • *Centre for Environmental Risk Assessment and Remediation, University of South Australia, Adelaide, SA, Australia*

FRANS J.M. MAATHUIS • *Biology Department, University of York, York, UK*

M.K. MATHEW • *Membrane Biophysics Group, National Centre for Biological Sciences, University of Agricultural Sciences-G.K.V.K Campus, Bangalore, India*

ANTHONY J. MILLER • *Department of Disease and Stress Biology, John Innes Centre, Norwich Research Park, Norwich, UK*

ROBERTO MOSCATIELLO • *Department of Biology, University of Padova, Padova, Italy*

LORELLA NAVAZIO • *Dipartimento di Biologia, Università di Padova, Padova, Italy*

IAN NEWMAN • *School of Mathematics and Physics, University of Tasmania, Hobart, Australia*

SAROJ PARMAR • *Rothamsted Research, Hertfordshire, UK*

S. PEDAS • *Plant and Soil Science Section, Department of Agriculture and Ecology, Faculty of Science, University of Copenhagen, Copenhagen, Denmark*

A.D. PEUKE • *ADP International Plant Science Consulting, Gundelfingen, Germany*

DORINA PODAR • *Faculty of Biology and Geology, Department of Experimental Biology, Babes-Bolyai University, St. Cluj-Napoca, Cluj, Romania*

PASCAL RATET • *Institut des Sciences du Végétal, CNRS, Gif sur Yvette, France*

DAVID E. SALT • *School of Biological Sciences, University of Aberdeen, Aberdeen, UK*

J.K. SCHJOERRING • *Plant and Soil Science Section, Department of Agriculture and Ecology, Faculty of Science, University of Copenhagen, Copenhagen, Denmark*

MICHAEL SCHULTZE • *Biology Department, University of York, York, UK*

LANA SHABALA • *School of Agricultural Science, University of Tasmania, Hobart, Australia*

SERGEY SHABALA • *School of Agricultural Science, University of Tasmania, Hobart, Australia*

BO STENBERG • *Department of Soil and Environment, Swedish University of Agricultural Sciences, Skara, Sweden*

MARK TESTER • *Australian Centre for Plant Functional Genomics University of Adelaide, Urrbrae, SA, Australia*
RAPHAEL A. VISCARRA ROSSEL • *Department of Soil and Environment, Swedish University of Agricultural Sciences, Skara, Sweden*
JOHANNA WETTERLIND • *Department of Soil and Environment, Swedish University of Agricultural Sciences, Skara, Sweden*
ELENA YAKUBOVA • *Department of Horticulture & Landscape Architecture, Purdue University, West Lafayette, IN, USA*

Chapter 1

Roles and Functions of Plant Mineral Nutrients

Frans J.M. Maathuis and Eugene Diatloff

Abstract

Plants require macro- and micronutrients, each of which is essential for a plant to complete its life cycle. Adequate provision of nutrients impacts greatly on plant growth and as such is of crucial importance in the context of agriculture. Minerals are taken up by plant roots from the soil solution in ionic form which is mediated by specific transport proteins. Recently, important progress has been achieved in identifying transport and regulatory mechanisms for the uptake and distribution of nutrients. This and the main physiological roles of each nutrient will be discussed in this chapter.

Key words: Element composition, Macronutrient, Micronutrient, Mineral, Nutrient, Plant nutrient, Vacuole

1. Introduction

Apart from atmospheric oxygen and soil-derived water, plants typically require 14 essential nutrients. Six of these are needed in relatively large amounts and commonly referred to as "macronutrients" (1). The macronutrients comprise nitrogen (N), potassium (K), calcium (Ca), magnesium (Mg), phosphorous (P), and sulfur (S). The second group of "micronutrients" or "trace elements" is needed at much smaller concentrations in plant tissues and comprises chloride (Cl), copper (Cu), manganese (Mn), iron (Fe), zinc (Zn), cobalt (Co), molybdenum (Mo), and nickel (Ni). However, this list is a generalization and some authors might include minerals such as sodium (Na) and silicon (Si) (2).

In plant available form, mineral nutrients are usually present at low concentrations in the soil, and furthermore, their availability can fluctuate greatly in both space and time due to environmental factors such as weather and climate (e.g., precipitation, temperature, wind) and physicochemical properties such as erosion, soil

Frans J.M. Maathuis (ed.), *Plant Mineral Nutrients: Methods and Protocols*, Methods in Molecular Biology, vol. 953, DOI 10.1007/978-1-62703-152-3_1, © Springer Science+Business Media, LLC 2013

type, and soil pH. To sustain growth, plants therefore have had to develop adaptive and flexible mechanisms for the acquisition and distribution of nutrients and many of these processes show similarities in their basic aspects, irrespective of which nutrient is at issue. For example, uptake mechanisms at the root–soil boundary typically have multiple phases with varying affinities to accommodate uptake in the presence of different substrate supplies. Uptake of negative macronutrients (e.g., NO_3^-, PO_4^{3-}, SO_4^{2-}) typically requires energization which occurs via coupling of their transport to the proton (H^+) gradient. Uptake of micronutrients in most cases also requires H^+-coupled transport. On the other hand, positive macronutrients (K^+, Ca^{2+}, Mg^{2+}, NH_4^+) typically enter the plant via passive (ion channel) transport systems. Similarities in morphological adaptations are also obvious: root systems often develop in response to localized deficiency or abundance of nutrients with a proliferation of lateral roots or the development of cluster roots. Nutritional deficiencies for many nutrients cause exudation of compounds that help nutrient release from the soil, conversion into bioavailable forms and entry into the root symplast. Finally, within plant cells the central vacuole, particularly in shoot tissue, functions as a well buffered larder for many nutrients such as K^+, NO_3^-, SO_4^{2-}, Ca^{2+}, and PO_4^{3-}.

The above examples pertain to most mineral nutrients and it is therefore not surprising that many of the adaptive and molecular mechanisms recur when different nutrients are discussed. However, each nutrient has very specific and unique functions within cells and tissues. In the following sections, and in order of their abundance in a typical plant, we will discuss the availability of each essential nutrient inside and outside of the plant, mechanisms for their uptake and distribution, and the main biological functions.

2. Nitrogen

Availability: "Nitrogen" (N) derives from the Greek words nitron and genes, which together mean "saltpetre forming". Around 1.5% of plant dry weight is made up of N (Table 1) which ultimately derives from the atmosphere. Although the atmosphere contains large amounts of N, it is in the form of N_2 and the triple covalent bond between the two N atoms is extremely stable. Plants do not possess the enzymatic machinery to break this bond but, thankfully, some soil bacteria are capable of making nitrogenase, the enzyme that converts N_2 into ammonia (NH_4^+). NH_4^+ can be taken up by plants or is converted into NO_3^-, another important form in which N is assimilated by plants, by nitrifying bacteria. Some plants, particularly leguminous species such as alfalfa, beans, and peas, can form symbioses with the microorganisms that fix N. Roots of these

Table 1
Element composition of the earth's crust by weight (%) and the typical relative proportion of minerals found in plant tissue assuming N levels at 100%

		Plant tissue levels			
Earth's crust composition		**Macronutrients**		**Micronutrients**	
Oxygen	46.6	Nitrogen	100	Chlorine	0.05
Silicon	27.7	Potassium	50	Iron	0.03
Aluminium	8.1	Calcium	25	Boron	0.03
Iron	5.0	Magnesium	10	Manganese	0.02
Calcium	3.6	Phosphorous	8	Zinc	0.007
Sodium	2.8	Sulfur	5	Copper	0.002
Potassium	2.6			Nickel	0.0004
Magnesium	2.1			Molybdenum	0.0001
All others	1.5				

plants form nodules where N fixing bacteria of the *Rhizobium* genus exist in an oxygen free environment (necessary for the nitrogenase activity). In return for sugars that derive from plant mediated photosynthesis the bacteria pass NH_4^+ to the plant.

Plant uptake of N is predominantly as NH_4^+ or NO_3^- but the ratio between these two forms depends on soil conditions and plant species (2). Low pH and/or reducing soil conditions promote formation of NH_4^+ whereas high pH and aerobic surroundings favor formation of NO_3^-. Consequently plants that thrive in either habitat often show a propensity for the most abundant form of N. For example, paddy grown rice acquires most nitrogen in the form of NH_4^+, whereas comparable cereals such as barley and wheat that are grown in aerobic soils, prefer NO_3^- as N source. Not surprisingly, for most plants a mixture of NH_4^+ and NO_3^- provides the greatest growth stimulus.

In addition to the inorganic forms of N (NH_4^+ and NO_3^-) organic N compounds may be available, for example from plant litter or animal feces. These compounds include peptides, amino acids and nucleotides, and may require further breakdown before plant uptake is possible. Break down can be facilitated by microorganisms such as (mycorrhizal) fungi which often contribute significantly to plant N acquisition in areas that are nitrogen deficient (3, 4). However, there is growing evidence that these organic N compounds are important N sources to plants in their own right and that roots contain transporters to import amino acids directly from the soil (5). In spite of the widespread occurrence of inorganic and organic nitrogen, it is often rate limiting for plant growth and in agricultural settings large amounts of N fertilization occurs.

N uptake from the soil and distribution in the plant: NO_3^- is highly mobile in the soil while NH_4^+ is less so. Yet both ions reach roots via mass flow and diffusion. There are multiple transport systems involved in uptake and distribution of N, and this has been extensively studied in the model species *Arabidopsis* (for review see ref. 6). *Arabidopsis* takes up N primarily in the form of NO_3^-. Its roots contain both high (HATS) and low (LATS) affinity transport systems with affinities in the micromolar and millimolar range. The HATS has a constitutive (HATSc) and an inducible component (HATSi). NO_3^- transporting HATS and LATS function as H^+-coupled symporters and are encoded by genes from the NRT1 and NRT2 families respectively. The inducibility of HATSi by NO_3^- provides a regulatory mechanism that ensures increased uptake when substrate becomes available. This process is countered by other nutrients: NH_4^+ or addition of amino acids rapidly inhibits NO_3^- uptake, as does glutamine. Accumulation of the latter provides an inhibitory signal that reduces high-affinity uptake.

As stated above, NH_4^+ is often the prevalent form of inorganic N in non-aerobic environments. Plants grown in submerged environments such as rice have relatively large numbers of high- and low-affinity NH_4^+ transporters encoded by the AMT family. Two isoforms, AMT1;1 and 1;2, are present exclusively in rice roots, where the large majority of NH_4^+ is assimilated. Both *Arabidopsis* (7) and rice (8) AMT1;1 contain threonine residues which form a phosphorylation target. The phosphorylation status of this residue plays a critical role in the allosteric regulation of this ammonium transporter.

Both inorganic (NO_3^-) and organic (amino acids, peptides) N forms are transported throughout the plant. Xylem and phloem sap can contain considerable amounts of NO_3^- and amino acids whereas the phloem also contains many peptides. Some of these may have a signaling function but the majority derives from protein and nucleic acid turnover and constitutes an important source for N metabolism. Many amino acid and proton dependent oligopeptide transporters of the POT/PTR family are expressed in plants that are believed to mediate intercellular movement of organic nitrogen and contribute to overall nitrogen homeostasis (9). When in ample supply, a large proportion of NO_3^- is deposited in the (shoot) vacuoles where it is a significant contributor to turgor generation.

Assimilation and biological functions: N can assume a range of valencies but to be incorporated into organic compounds such as amino acids it needs to be reduced to its −3 valence state. In case N is taken up as NO_3^-, this reduction is achieved by the activity of two crucial enzymes, nitrate reductase and nitrite reductase. Nitrate reduction to ammonium can take place in both roots and shoots but is spatially separated between the cytoplasm (nitrate reductase activity) and plastids/chloroplasts (nitrite reductase activity) (6). The reduction of NO_3^- to NH_4^+ is highly endergonic and requires 15 ATP equivalents. This high energetic cost suggests NH_4^+ would be the preferable N source. However, due to its propensity to dissipate transmembrane gradients, NH_4^+ is far more toxic than NO_3^- which can be safely stored at high levels in vacuoles. Thus, NH_4^+ is rapidly assimilated into glutamine in chloroplasts and mitochondria to glutamine, the amino acid that forms the basis of all other N assimilation.

After carbon, N is typically the most abundant element in plant dry matter, indicative of its prominent role in all aspects of plant metabolism. Most of the plant N is fixed in amino acids and proteins (around 85%) and nucleic acids (around 5%) which leaves around 10% in inorganic and low molecular weight organic N compounds which are relatively mobile throughout the plant.

The foremost function of N is to provide amino groups in amino acids, the building blocks of every protein. N is also prolific in nucleotides, where it is incorporated in the ring structure of purine and pyrimidine bases. Nucleotides form the constituents of nucleic acids but also have many important functions in their own right such as in energy homeostasis, signaling, and protein regulation. In addition, N is essential in the biochemistry of many non-protein compounds such as coenzymes, photosynthetic pigments, secondary metabolites, polyamines, signaling molecules, and even some phospholipids.

The central role of N in protein synthesis means its nutrition is interlinked with that of other nutrients, particularly carbon and K^+. To maintain the overall C:N balance changes in CO_2 fixation are paralleled by those governing N reduction. NO_3^- translocation from root to shoot requires cations to maintain electroneutrality and this is normally achieved by K^+. Thus, adequate K^+ supply is crucial for N nutrition.

3. Potassium

Availability: K or "kalium" derives its name from the Latin word "alkali" which in turn came from Arabic meaning "plant ash". In most English speaking nations K is referred to as "potassium" from the Old Dutch word "potasch", and points to the old method of making potassium carbonate (K_2CO_3). K is one of the most abundant elements in the earth's crust (Table 1). However, a large proportion of soil K is not available to plants because in ionic form K^+ is tightly coordinated with oxygen atoms of soil minerals such as feldspars, illites, and vermiculites. Part of this electrostatically bound K^+ can be released through weathering of minerals and this released K^+ fraction is readily exchangeable with the third K^+ compartment, the soil solution. Within the soil solution, K^+ is extremely mobile and normally is between 0.1 and 1 mM K^+. Real K^+ deficiency is rare but to reach optimum plant production large amounts of K^+, in the form of potash fertilization, are often applied to arable land.

Uptake of K^+ and its distribution throughout the plant: Plant dry weight has around 1% K^+ (Table 1) requiring substantial net influx of this element. Similar to NO_3^-, K^+ uptake into plant roots is mediated by multiple systems with varying affinities (10). Many of these systems were characterized in the 1990s resulting in a model where passive transport through ion channels mediates low affinity (millimolar K_m for K^+ uptake) whereas active H^+-coupled cotransporters are responsible for high affinity (micromolar K_m for K^+ uptake) (10, 11). Several of these proteins have been cloned and identified at the gene and protein levels and functionally characterized. The *Arabidopsis AKT1* locus encodes a K^+ selective inward rectifying channel that is expressed in the root cortex. Reverse genetic studies with loss of function mutations in *AtAKT1* (12) have confirmed a role of this protein in low affinity K^+ uptake. However, this channel can also mediate some K^+ uptake in the high-affinity range if membrane potentials are sufficiently negative. *AKT1* transcript level is not affected by changes in ambient K^+ but K^+ deficiency is believed to provoke a Ca^{2+} signal which, via the Ca^{2+} sensors CBL1 and CBL9, activates the protein kinase CIPK23. CIPK23 phosphorylates AKT1 which increases K^+ uptake via this channel (13). In contrast, high-affinity K^+ uptake is primarily through carrier type transporters from the *KUP/HAK* gene family. In Arabidopsis roots, HAK5 is essential for high-affinity uptake (14) and its transcription is drastically induced by K^+ deficiency. Most of the K^+ that enters the root symplast ends up in shoot tissue, via long distance transport through the xylem. Loading of K^+ into the xylem is mediated by SKOR type channels that are expressed in the xylem parenchyma and release K^+ into the xylem (15). Retranslocation of K^+ from shoot to root is also considerable and may partly rely on the activity of another K^+ channel. This seemingly futile cycling of K^+ is believed to be important to maintain K^+ homeostasis and also necessary for electroneutrality of anions such as NO_3^- on their way to the shoot.

Subcellular K partitioning: In K^+ replete plant cells, most of the tissue K^+ is deposited in the vacuole to maintain turgor. In other conditions, vacuolar K^+ may need to be released, for example when cytoplasmic K^+ becomes deficient (16), when osmotic adjustment is necessary or when turgor-driven movement is required such as during stomatal closure. Vacuolar deposition and release are mediated by tonoplast transporters such as K^+:H^+ exchangers (e.g., from the CHX and NHX family) to drive vacuolar uptake (16) and cation channels such as TPC1 and TPK1 for K^+ release (17).

However, prolonged K^+ starvation can deplete vacuolar K^+ stocks to concentrations that are significantly lower than those in the cytoplasm (18) ruling out ion channel mediated transport. In such cases, HAK/KUP type systems may be important, several of which have been localized at the tonoplast (e.g., ref. 8).

Biological functions of K: K^+ is essential for all living organisms for a number of reasons, the most important being its capacity to activate a multitude of enzymes. Kinetic studies have shown that activation of many K^+ dependent enzymes occurs in the presence of 50–80 mM K^+, a value that is close to that measured in the cytoplasm (19). K^+ binds to enzymes via coordination with negative groups such as carboxyls, carbonyls, and hydroxyls. This interaction is very selective for K^+ and can not occur with other, chemically similar, ions such as Na^+ or Li^+. Processes and enzymes that have been found to rely on the presence of K^+ include pyruvate kinases involved in glycolysis, phosphofructokinase and ADP-glucose synthase that contribute to starch synthesis, and membrane transporters such as the vacuolar PPase. Protein synthesis at ribosomes is another key process that requires high concentrations of K^+.

K^+ is one of the most important contributors to cell turgor in plants. As such it participates in cell expansion and plant growth. Indeed, turgor pressure in K^+ deficient plants is often reduced leading to lower growth rates. Due to its role in turgor, movement of cells and organs also greatly depends on K^+; closure of the *Dionaea muscipula* (Venus flytrap) trapping leaves, the nyctinastic reorientation of leaves in response to diurnal rhythms and the change in stomatal aperture all constitute important processes where large amounts of K^+ are moved to cause changes in cell turgor which result in movement.

A further, biophysical rather than biochemical, function of K^+ is in the cell cytoplasm. The large mass:charge ratio of K^+ reduces its chaotropic effect; K^+ therefore has less tendency to sever hydrogen bonds between water molecules or between water molecules and macromolecules and as such can compensate the negative charge of many macromolecules without disturbing for example intricate hydrogen bonds. This role is not easily fulfilled by other cations and may explain the relatively high (~100 mM) cytosolic K^+ concentrations that are found in all living cells.

4. Calcium

Availability: Calcium (Ca) (from the Latin word "calcis" meaning lime) is even more abundant in the lithosphere than K with a mean value of about 3.5% Ca (Table 1). Soils derived from limestone or chalk are high in $CaCO_3$ (calcite) and can contain 50% Ca. Other prevalent forms of Ca include dolomite ($CaMg(CO_3)_2$), apatite

($CaHPO_4$) and gypsum ($CaSO_4$). Ca deficient soils are rare but a low soil pH (e.g., through acid rain precipitation) combined with severe weathering and leaching may lead to deficiency. Consequently acid soils generally have less Ca and high pH soils have more. In most soils, Ca is tightly adsorbed to colloids but a small fraction can exchange with the soil solution. In this liquid phase of the soil much of the "free" Ca forms insoluble complexes with other minerals such as phosphates and sulfates. Excess Ca levels in the soil are also rare but in extreme cases can reduce a plant's uptake of other nutrients such as phosphorus, potassium, and magnesium, resulting in deficiencies of these nutrients.

Ca uptake from the soil and distribution in plants: Ca is taken up as a divalent ion (Ca^{2+}) and around 0.5% of plant dry matter is Ca^{2+} (Table 1). In contrast to animals, plants do not appear to have Ca^{2+} selective ion channels. Instead, plants use fairly nonselective Ca^{2+}-permeable ion channels located at the root system to take up Ca^{2+} from the soil (see ref. 20 for review). Such channels are part of a large, ill-defined group that varies in selectivity and gating properties. For example, many of these nonselective channels are permeant to Ca^{2+} but also to other divalent cations and to monovalent cations (20). Gating of Ca^{2+} permeable channel can be under control of the membrane potential or ligands such cyclic nucleotides and glutamate (20). Since there are several large gene families that encode Ca^{2+} permeant nonselective cation channels, the identity of the specific gene(s)/protein(s) that mediates Ca^{2+} uptake is not known but there are likely to be multiple pathways for Ca^{2+} uptake.

The high affinity of Ca^{2+} for many biological constituents that carry negative charge, makes Ca^{2+} relatively immobile within plant tissues. This means it cannot be remobilised from older tissues and can have adverse consequences for plant growth particularly for fast growing plant organs and crops in agriculture. Poor delivery of Ca^{2+} via long distance pathways such as the xylem to leaf tips and fruits can cause diseases in fruits and vegetables such as "black heart" in celery, "blossom end rot" in tomatoes, or "bitter-pit" in apples. No specific proteins have been identified that are responsible for Ca^{2+} xylem loading and a proportion of the Ca^{2+} in the xylem may arrive entirely via the apoplast (21). In either case Ca^{2+} translocation is intricately linked to transpiration and conditions of high humidity, drought, or salinity can cause calcium deficiency because they decrease transpiration and/or water uptake by the plant. Within tissues, most Ca^{2+} will reside outside of the cytoplasm, particularly in the apoplast and endocompartments such as the ER and vacuole where concentrations may be in the millimolar region. Vacuolar sequestration is carried out by members of the CAX H^+:Ca^{2+} antiport family and by Ca^{2+}-ATPases from the ECA/ACA family whereas release from endocompartments is through nonselective ion channels (20, 22).

Biological functions of Ca: Functions of Ca^{2+} are based on its high affinity for negative groups and its extremely low cytoplasmic concentration (~200 nM). The latter is necessary to avoid precipitation of insoluble Ca-salts such as Ca-phosphates, Ca-sulfates, and Ca-oxalates. Ca^{2+} readily complexes with the negative phosphate and carboxyl groups of phospholipids, proteins and sugars: For example\, cross-linking of glycans and pectins in plant cell walls occurs via electrostatic coordination by Ca^{2+}. In this way Ca^{2+} acts as a "glue" and confers structure and rigidity to cell walls. Limited Ca^{2+} supply can therefore weaken cell walls and for example increase risk of invasion by pathogens which often penetrate through cell walls. Similarly, in membranes Ca^{2+} coordinates with phosphate groups from phospholipids (and possibly also between phospholipid phosphates and carboxyls of membrane proteins) and in this way stabilizes the lipid bilayer. This complexation occurs predominantly at the external face of the plasmamembrane, requiring relatively high levels of apoplastic Ca^{2+}. Removal of apoplastic Ca^{2+} therefore compromises membrane integrity and can cause electrolyte loss.

In addition to these structural functions of Ca^{2+}, it is an important signaling compound. Since Ca^{2+} readily forms insoluble salts, the free Ca^{2+} concentration in the cytoplasm is kept extremely low. In contrast, the apoplast and intracellular compartments such as the ER and vacuole can contain Ca^{2+} in the millimolar range and this makes Ca^{2+} an ideal secondary messenger. A wide range of biotic and abiotic stimuli has been shown to evoke rapid changes in cytosolic free Ca^{2+} in plants, including mechanical stimuli, pathogen attack, temperature shock, drought and changes in nutrient status (22). The changes in cytoplasmic Ca^{2+} are relayed to downstream components to evoke relevant responses such as changes in protein activity of alteration of gene transcription.

5. Magnesium

Availability: At around 2%, magnesium (Mg) is the eighth most abundant element in the earth's crust (Table 1). Its name derives from the Greek "Magnesia region" where talc was mined and as with Ca, Mg is found in a range of minerals such as dolomite ($CaCO_3 \cdot MgCO_3$), carnallite ($KCl \cdot MgCl_2 \cdot 6H_2O$), and magnesite (MgO). Plants take up Mg as a divalent cation (Mg^{2+}) the availability of which varies with soil organic matter content, pH and the presence of other cations. Mg^{2+} has a large hydrated ionic radius and electrostatic interaction with soil particles is therefore weaker than for most other ions. Thus, in many minerals Mg^{2+} is readily exchanged by H^+ and a low soil pH can therefore lead Mg^{2+} displacement from mineral exchange sites. The released Mg^{2+} is relatively mobile in the soil and at risk of leaching out of the top soil, making it inaccessible to plant roots. Leaching is particularly problematic in sandy soils with a low cation exchange capacity.

Mg uptake from the soil and distribution in plants: Magnesium is abundant in plant cells with typical values of around 0.5% DW (Table 1; (23)) In planta it occurs as Mg^{2+} and this is also the form in which Mg is taken up from the soil solution. Uptake of Mg^{2+} at the root:soil boundary is believed to be mediated by specific membrane transporters of the so-called MGT family. Overexpression studies with MGTs have shown increased uptake capacity and improved growth in Mg^{2+} deficiency conditions (24). The cytoplasmic Mg^{2+} is probably in the order of 0.5 mM and millimolar concentrations are also prevalent in the vacuolar compartment. In vacuoles, Mg^{2+} contributes to turgor generation and charge balancing of anions and its transport into this organelle is mediated by Mg^{2+}:H^+ antiporters. The exact pathway for vacuolar Mg^{2+} release remains to be identified but could include non selective tonoplast channels such as the slow vacuolar (SV) channel. Mg^{2+} is essential for photosynthesis (see below) and the highest tissue Mg^{2+} is usually measured in shoots. However, the proteins that are involved in this long distance movement are unknown.

Biological functions of Mg: Mg^{2+} plays a vital role in plants in the process of capturing light energy. Indeed, Mg^{2+} is essential for photosynthesis in several ways but primarily as the metal in the porphyrin cofactor of chlorophyll. As such it is an indispensable component of the reaction centers that make up the thylakoid photosystems. However, in addition to this structural role in chlorophyll, Mg^{2+} is also involved in stromal reactions during photosynthesis: in the stroma Mg^{2+} dissipates the electrical component of the proton motive force that is generated by light perception and as such allows the proton motive force to consist mainly of a trans pH gradient. In the stroma, Mg^{2+} also acts as a cofactor for ribuloso-1,6-bisphosphate carboxylase (RuBP carboxylase), the first enzyme of the Calvin cycle and for fructose-1,6-bisphosphatase an enzyme that regulates the partitioning between starch synthesis and export of triosephosphates to the cytosol. A third way in which Mg^{2+} impacts on photosynthesis is during biosynthesis and degradation of chlorophylls. In these processes, Mg^{2+} functions as metal cofactor for the enzymes (Mg-dependent chelatases and dechelatases) that mediate chlorophyll turnover (2). Not surprisingly then that 15–30% of total plant Mg^{2+} is found in chloroplasts (2).

Apart from the important functions in photosynthesis, Mg^{2+} can act as a cofactor or promoter for numerous enzymes throughout the plant. This includes many enzymes involved in transcription and translation, metabolic enzymes such as phosphatases, kinases, carboxylases, and synthases, and many transport enzymes such as plasma membrane H^+-ATPases and vacuolar pyrophosphatases. Many of the enzymes responsible for formation and break down of nucleic acids such as polymerases, DNases, and RNases require Mg^{2+}. Thus, gene transcription, gene translation and therefore cell division and protein synthesis all critically depend on adequate levels of cytoplasmic Mg^{2+}.

Outside of chlorophyll, the structural roles of Mg^{2+} are primarily in the stabilization of nucleic acids. In DNA, Mg^{2+} attaches to the major groove via hydrogen bonds, thereby increasing the DNA melting temperature and hence stability (25). In RNA, Mg^{2+} is essential for establishing and maintaining secondary structure, for example in tRNAs to coordinate and stabilize hairpin and other structural configurations (25).

6. Phosphorus

Availability: The word phosphorus (P) comes from the Greek for "light carrier". It does not occur in elemental form but is widely distributed in many minerals mainly as phosphates. However, most (> 90%) of the soil P is firmly fixed in minerals and cannot be used by plants. Another part of insoluble P, the "labile fraction", exchanges with the soil solution and the inorganic P (Pi) released from this fraction can be taken up by plants. The form in which Pi is found in the soil solution is pH dependent but in most soils (with a pH of around 5) it is as $H_2PO_4^-$ and this is the form in which plants take up Pi. The release of Pi in soils is extremely slow and P deficiency is therefore widespread in agriculture. Phosphate rock, a finite source material which is partially made of apatite, is an important commercial source of P and often applied as fertilizer to augment P availability. The low P availability means that in agricultural settings, P is replenished with large amounts of P fertilizer derived from phosphate rock.

Pi uptake from the soil and distribution in plants: Soils typically contain very low concentrations of Pi (~0.1–1 μM). Cellular levels of P are usually in the millimolar range so roots must have potent uptake mechanisms. Several Pi transporters of the Pht family have been characterized with Pht1 isoforms predominantly found in root tissues. These membrane proteins include transporters with high- and low affinity and are energized via H^+ symport (26). To increase Pi release from insoluble minerals calcium phosphates, some plants can form cluster roots. These dense clusters of fine lateral roots increase the soil volume hat can be exploited but also extrude large amounts of chelators (e.g., citric acid and malic acid) to dissolve the minerals. The limited supply of P in many environments has also led to symbiotic adaptations to increase plant P nutrition. The main example of this is mycorrhizas which are established by more than 90% of plant species. In many cases, the hyphae of mycorrhizal fungi deliver the majority of plant P albeit at a considerable cost to the plant in the form of reduced carbon that is delivered to the non-autotrophic fungus (27).

As with most other nutrients, excess supply of P leads to vacuolar accumulation with concentrations reaching millimolar levels. The transport steps involved are unknown but may involve anion channels. Sequestration of P in the vacuole is particularly important during seed development. In this reproductive growth phase large amounts of P and other minerals are stored in complex form such as myo-inositol salts and phytate. These compounds are stored in globoid inclusion bodies within seed protein storage vacuoles and hence require two transmembrane transport steps, for which no details are available.

Biological functions of P: In cells, P can occur in inorganic or organic form. The soluble forms of inorganic P are Pi (orthophosphate or $H_2PO_4^-$) or as PP (pyrophosphate). As a constituent of organic molecules, P can be bound via esterification reactions to many side groups such as hydroxyl groups of sugars and alcohols. Alternatively, Pi binds to other phosphate groups via pyrophosphate bonds. Formation and disruption of these ester and pyrophosphate bonds is one of the central mechanisms in cellular energy homoeostasis. For example formation and disruption of the pyrophosphate bond in the central energy intermediary ATP is accompanied by a change in free energy of around 50 kJ/mol. Similar energy rich phosphonucleotides are UTP, CTP and GTP. ATP, UTP, CTP and GTP form major sinks for cellular P and are central players in a multitude of metabolic pathways which include nucleic acid synthesis, sugar metabolism and phospholipid. UTP is also a building block of sucrose, starch and cellulose formation whereas CTP acts as energy rich compound during phospholipid formation. In nucleic acids Pi acts as a bridge between each nucleotide base by coupling C3 and C5 of two adjacent riboses via esterification. Another important area where P plays structural roles is in membrane lipids where Pi forms the link between the lipophilic fatty acid and the hydrophilic choline part of the lipid. The negative charge on the phosphate group makes this part of the lipid strongly hydrophilic and therefore helps a proper orientation in the membrane. The negative charges are compensated by electrostatic binding of divalent cations, in particular of Ca^{2+}.

In addition to structural roles, P is important for the regulation of protein activity. Reversible protein phosphorylation, which is mediated by kinases and phosphatases, is one of the most prominent mechanisms for the modulation of protein activity. Phosphorylation typically involves transfer of the terminal Pi from a nucleotide (e.g., ATP or GTP) to Ser or Thr residues of proteins. These posttranslational regulation mechanisms tend to be very specific, are strictly controlled in space and time and rely on the activity of particular kinases or phosphatases.

7. Sulfur

Availability: The name sulfur (S) almost certainly stems from the Arabic "sufra" for yellow the color in which elemental S occurs. In nature, sulfur can be found as the pure element but also in mineral form as sulfides and sulfates. In aerobic conditions, inorganic S is present mainly as sulfate (SO_4^{2-}) and this is also the form in which plants take up most S. However, the anaerobic and reducing environments created by flooding leads to production of sulfides such as FeS, FeS_2, and H_2S. Uniquely, plants can also extract S from the atmosphere, for example the SO_4^{2-} that precipitates in the form of acid rain. S toxicity is rare but can occur in saline soils with high levels of SO_4^{2-} salts.

S uptake from the soil and distribution in plants: H^+-coupled transport mechanisms reside in the root to take up SO_4^{2-} from the soil solution. This function is primarily carried out by members of the "Sultr" family which are located in the epidermal and cortical plasma membranes and have substrate affinities of around 10 μM. Transcription of Sultr transporters is induced or derepressed when S becomes deficient. The exact nature of the signal that generates this regulation remains to be revealed but it has been suggested that phloem glutathione levels signal the roots to either up- or downregulate uptake capacity (28). SO_4^{2-} is highly mobile and rapidly loaded into the xylem, possible by other members of the Sultr family, for transport throughout the plant. In the shoot, the much of the SO_4^{2-} is reduced before incorporation into organic compounds occurs. Surplus S is deposited in vacuoles as SO_4^{2-}. S assimilation involves reduction of SO_4^{2-} to SO_3^{2-} (sulfite) and subsequently to S^{2-} (sulfide). The reduced S^{2-} is incorporated into the amino acid cysteine with the help of acetyl CoA and the serine derivate, acetyl-serine. Cysteine thus forms a metabolic hub from which other cellular sulfur compounds are generated such as methionine and glutathione (Glu). Since this process takes place mostly in the shoot, roots must derive most reduced S via the phloem. Indeed, high levels of organic S can be found in phloem sap, particularly in the form of the tripeptide glutathione (Glu-Cys-Gly). Oligopeptide transporters are likely candidates for the distribution of glutathione and related compounds in the vascular system and between organs in general.

Biological functions: The main destination of S in plants is in proteins. Both the amino acids cysteine and methionine contain reduced S in the form of a sulfhydryl (–SH or thiol) group. Such groups are extremely important in the formation of protein structures and in the modulation of protein activity since they can undergo reversible oxidation/reduction and hence the formation/disruption of a covalent –S–S– bond. The formation and disruption of these S bridges impacts on tertiary and quaternary protein structure and therefore protein activity.

A further structural role of S can be seen in sulfolipids; lipids that contain sulfate rather than phosphate groups. Some sulfolipids are always present in chloroplast thylakoids where they may be contribute to the stabilization of photosystem components. In other membrane no or very little sulfolipids are detected normally but phosphorous deficiency can increase this fraction. Other environmental conditions, e.g., salinity, can also impact on the ratio between phospho- and sulfolipids, suggesting that in some circumstances raised levels of sulfolipids may confer some adaptive advantage (29).

The easy transition of different redox states of S is also exploited in redox homeostasis and the protection of cells against reactive oxygen species (ROS) and (heavy) metals. For example, glutathione not only serves as a mobile carrier of reduced S but it is an important reductant in the detoxification of reactive oxygen species (ROS) (30). Similarly, the glutathione-based phytochelatins (PCs), with a general structure of (Glu-Cys)n-Gly where $n=2$–5, have a high affinity for (heavy) metals and arsenic. Binding of the latter to the multiple –SH groups of PCs renders metals and metalloid inert. Not surprisingly, there is a strong correlation between heavy metal stress and sulfate uptake.

8. Chlorine

The word "chlorine" (Cl) originates from the Greek for "pale green". For plant physiologists, chloride (Cl^-) is usually associated with salt, salinity and toxicity. However, Cl^- was shown to be essential to plants in 1954 (cited in 31), but because Cl^- is generally ubiquitous in soils, Cl^- applications were thought to have little value in improving plant growth. Recently, some wheat varieties have been shown to exhibit Cl^- deficiency symptoms when grown on sandy soils with very low Cl^- levels and yields were increased by Cl^- application and chloride application has also helped suppress certain diseases. Some components of Cl^- uptake and distribution have been characterized such as members of the ClC chloride channel family which includes both Cl^- channels and Cl^-/NO_3^- antiporters, and CCC cation chloride transporters (32).

Chloride is best considered an inert anion which is involved in osmotic adjustment and cation neutralization roles which have important biochemical and biophysical consequences. An important function for Cl^- is as a counterion for K^+ (see above section on Potassium) and as such it contributes to turgor, for example in guard cells. Not surprisingly, loss of leaf turgor is one symptom of Cl^- deficiency. For the above functions, other anions can replace Cl^- but chloride is essential in its role as cofactor in oxygen evolution complex of photosystem II (33).

9. Iron

The symbol Fe is from the Latin *ferrum*, meaning "iron". Iron (Fe) is the second most abundant element in the earth's crust after aluminium; however, the solubility of Fe is very low in aerobic and neutral conditions (34). A few billion years ago, photosynthetic microorganisms appeared on earth and produced oxygen. This resulted in the oxidation of soluble Fe^{2+} (ferrous) into insoluble Fe^{3+} (ferric). Consequently, plants evolved mechanisms to acquire Fe from insoluble ferric oxides. There are two main mechanisms that plants use (35): Strategy I, found in dicots and nongraminaceous monocots, solubilization of Fe^{3+} by reducing it to Fe^{2+}, and then absorption of Fe^{2+} across the plasma membrane; Strategy II, used by graminaceous monocots, consists of the excretion of high-affinity Fe chelators (phytosiderophores) which bind and solubilize Fe^{3+}. The Fe-chelator complex is then transported across the plasma membrane. The Fe^{2+} influx transporters and phytosiderophore efflux and influx transporters have now been cloned and characterized (36).

Eusèbe Gris, a French professor of Chemistry, demonstrated in the 1840s that in an iron-free medium, leaves were chlorotic and plant growth was stunted. However, growth quickly resumed if a few drops of an iron salt solution were added to the medium, or if a dilute iron salt solution was applied to the leaves. This pointed at an important role of Fe in chloroplast development and/or function. Indeed, once absorbed into the plant, Fe is distributed as Fe-citrate, Fe-phytosiderophore and Fe-nicotianamine complexes and mainly deposited in the chloroplast where generally 80% of the total Fe in leaves in located. In photosynthetically active tissue, iron is present in a large range of enzymes involved in redox systems and also in chlorophyll synthesis. Consequently, Fe deficiency leads to rapid symptoms of iron chlorosis (interveinal yellowing).

10. Boron

The name boron (B) originates from the Arabic word "buraq" and the Persian word "burah" which are names for the mineral borax. In 1923, Warington showed that B was essential for broad bean and later Sommer and Lipman (1926) showed that B was also essential for nonlegumes (cited in 31). However, it is only recently that the biomolecules with which B interacts have been discovered; Boron has not yet been found to be an enzyme constituent but it forms very stable complexes with organic compounds such as polymeric sugars which are abundant in cell walls. Therefore, B is strongly complexed in cell walls (particularly with pectic polysaccharides) where over 80% of B resides under normal growing condition.

This amount rises to 95–98% in B deficient plants suggesting no or little involvement of B in other plant functions. Since B is cell wall associated, B is generally considered phloem immobile and B deficiency results in disorders in the structural development of organs and whole plants. However, recent evidence suggests that plants that contain sugar alcohols such as sorbitol show B retranslocation as B-sugar alcohol complexes (37).

In aqueous solution at pH < 7, B occurs mainly as undissociated electroneutral boric acid ($B(OH)_3$), which dissociates to $B(OH)_4^-$ at higher pH values. Boric acid is absorbed by plant roots via a subtype of aquaporins, the NIPs (nodulin-26-like intrinsic proteins) (38). Boric acid is then loaded into the xylem by boric acid efflux transporters such as BOR1 in *Arabidopsis* (38). The same type of system is also involved in B distribution within shoots and may provide tolerance to high B by mediating efflux of B from the cytoplasm into the apoplast. Boron toxicity is widespread globally. For example, over 5 million ha of crop growing areas in South Australia suffer from B excess, and it has been proposed to grow wheat and barley cultivars in such areas with increased BOR activity.

11. Manganese

Manganese (Mn) derives from the Greek *Magnesia* and Spanish *negra*. In 1912, Gabriel Bertrand, the Director of the Pasteur Institute, gave a lecture in New York to a standing ovation from international chemists describing his research on Mn biochemistry and the positive crop yield responses of 25–50% when Mn was applied as a fertilizer to Mn deficient soils. During 1922–1923, McHague using carefully prepared solution and sand cultures showed Mn to be indispensible for the growth of several plant species (31) while in the same decade Samuel and Piper in Australia discovered that the so called "grey speck disease" of oats, commonly reported in Europe and Australia, was caused by Mn deficiency (39).

Manganese exists in plants in several oxidation states, and therefore plays important roles in redox processes and in electron transfer reactions. Manganese is a constituent of essential metalloenzymes, including the antioxidant enzyme Mn^{2+}-dependent superoxide dismutase (SOD) and in the four Mn atom containing oxygen evolving complex that catalyzes water oxidation in photosystem II (2). Of the micronutrients, Mn appears to be the most important in the development of resistance in plants to both root and foliar diseases of fungal origin, through its involvement in phenol metabolism and lignin biosynthesis (2).

High affinity Mn uptake by roots of *Arabidopsis* is conducted by NRAMP1 ($K_m \approx 30$ nM) (40) whereas other NRAMP isoforms are important for the transit of Mn through the vacuole prior to the import into chloroplasts of mesophyll cells. Several other proteins that may transport Mn have been identified in plants, such as members of the CAX, CDF, and P2A-type ATPase families (40).

12. Zinc

From the German "Zink" which is related to *Zinken* "prong, point". The essentiality of Zn (Zn) to plants was established in 1926 (cited in 31). Unlike Fe, Mn and Cu, Zn is not subject to valency changes and exists in plants only as Zn^{2+}. This non-redox property of Zn makes it ideally suited as metal cofactor in transcription factors and other enzymes involved in DNA metabolism, as the use of redox-active metal ions for these tasks could lead to radical reactions and nucleic acid damage.

Only a few Zn-containing enzymes are known in higher plants such as alcohol dehyrogenase, Cu-Zn-SOD, carbonic anhydrase, and RNA polymerase. However, a very large number of enzymes is activated by Zn (41). Zinc deficiency is common world-wide and particularly affects calcareous, heavy clay, alluvial, peaty and sandy *soils.* In most of these, plant available Zn is very low but Zn efficient crop varieties that have been bred for such soils show improved root growth and enhanced Zn uptake, utilization and compartmentalization (42).

13. Copper

Copper (Cu) is derived from the Latin cuprum, itself a contraction of Latin for "brass of Cyprus". Copper was shown to be essential for plants in 1931 (cited in ref. 31) and is a redox-active element with roles in photosynthesis, respiration, C and N metabolism, lignification and protection against oxidative stress. Copper is a cofactor for plastocyanin, copper/zinc superoxide dismutase (Cu/ZnSOD), cytochrome-c oxidase, and the ethylene receptors for the apoplastic oxidases: ascorbate oxidase, diamine oxidase, and polyphenol oxidase (43). Copper deficiency causes decreased plant growth and results in distinct phenotypes such as chlorosis in young leaves, curled leaf margins, impaired fruit seed formation, defects in the development, and viability of pollen and seeds. Elevated Cu concentrations are toxic because free copper ions can generate reactive oxygen species (e.g., superoxide and hydroxyl radicals) which damage DNA, proteins and lipids.

Multiple members of the COPT (copper transporter) protein family have been shown to be important in Cu uptake and distribution and therefore overall Cu homeostasis. In *Arabidopsis* there are five members of this COPT family, COPT1 is the best physiologically characterized, mediates the copper uptake into the plant but also seems to be involved in the distribution of copper in the whole organism by catalyzing copper import into cells of different tissues. The physiological function of COPT2, 3 and 4 remains unclear, but COPT5 appears to be vacuolar expressed and mobilizes Cu to the cytoplasm, a task that is most critical under Cu deficiency (44). Some metal-transporting P-type ATPases can move Cu into organelles (43).

Copper is generally toxic to microorganisms and therefore has been used as the basis for many fungicide and bactericide formulations. Recently, it has been shown that some pathogenic bacteria have a sophisticated strategy to overcome rice defense mechanisms by secreting proteins into the plant that co-opt rice COPT transporters to remove Cu from the xylem allowing the bacteria to proliferate in this otherwise hostile compartment (45).

14. Nickel

From the German, *Kupfernickel*, meaning "copper-colored ore"; this referred to the ore niccolite from which it was obtained. Nickel (Ni) is the most recent element shown to be essential to plants in 1987 by Brown et al. (cited in 31). The authors showed that the viability of Ni deficient barley seed could not be restored by soaking the seeds in a solution containing Ni, demonstrating that Ni is essential for normal seed development and therefore, for competing the life cycle of barley. Nickel is chemically related to Fe and therefore can be transported by iron transporters such as the *Arabidopsis* IRT1 (46). A specific transporter for Ni has yet to be discovered in plants. Once inside the plant, Ni is found in urease (and urease accessory proteins), an enzyme that catalyzes the hydrolysis of urea into carbon dioxide and ammonia, and hence is important in N metabolism (47). This destination appears to be the only role of Ni in higher plants but in lower plants Ni may play additional roles and can substitute for Zn and Fe as a cofactor in some enzymes. Although Ni deficiency has been recorded in the field in many crop plants, generally there is more concern about Ni toxicity due to metal processing, land application of sludges, and the use of certain fertilizers. In addition to anthropogenic pollution, Ni may also accumulate naturally in soils formed from serpentine minerals which often contain high Ni concentrations. Consequently, researchers are also interested in Ni accumulators and hyper-accumulators for potential bioremediation (47) and as natural products to relieve Ni deficiency. Nickel tolerance in hyper-accumulators is achieved through the complexation of Ni by organic acids such as citric acid and also through the strong complexation with the amino acid histidine (47). It is interesting to note, that the latter mechanism is the basis for protein purification in molecular biology using poly-histidine tagged proteins and Ni purification columns.

15. Molybdenum

In Greek, *molybdos* means *lead-like*, as it forms a natural chemical compound molybdenite that looks like lead. In 1939, Molybdenum (Mo) was shown to be essential for tomatoes in carefully conducted solution culture experiments (cited in 31). In 1942, the effects of Mo nutrition were found to be more than a laboratory curiosity, when Mo deficiency was first recorded in the field and experiments showed that Mo fertilization could greatly improve plant growth (39). In the soil, soluble Mo is present as the molybdate anion (MoO_4^{2-}), at concentrations of around 1 mg/kg. Molybdate is absorbed by roots through a molybdate-specific transporter (MOT1 Molybdate transporter, type 1) with a K_m of approx 7–20 nM (48). MOT1 appears to be very unique, as at present it is the only characterized Mo transporter in eukaryotes and was detected in less than 40% of Mo-utilizing organisms. Once absorbed, Mo is incorporated into enzymes such as nitrate reductase (nitrate reduction to nitrite), sulphite oxidase (oxidizes toxic sulphite into sulfate), xanthine dehydrogenase (purine degradation in nitrogen metabolism), nitrogenase (nitrogen fixation), and aldehyde oxidase (synthesis of abscisic acid, auxin, glucosinolates). Consequently, during Mo deficiency, there is altered expression of genes involved in sulfur and nitrogen metabolism, stress responses, signal transduction, and the levels of amino acids and many other carbon and nitrogen compounds. Though required at extremely low levels, the importance of Mo to plants cannot be underestimated: Stout (1972) calculated that one atom of Mo can bring together 100 million carbon atoms in carbohydrates and proteins! (cited in ref. 39).

16. Conclusions

Although research into the roles and functions of plant minerals stretches back more than 150 years this is still a major area of research for a number of reasons. In some cases, the exact role of nutrients is still not clear (e.g., for boron and cobalt). However, the fact that the human diet is, directly or indirectly, plant-based means that research into plant mineral nutrition is more pertinent than ever in face of a growing world population and the increasing need for environmentally and sustainable agriculture. The latter will become more feasible if we are able to optimize crop growth with a minimum of input in the form of fertilizer. Furthermore, the optimization of crop growth and crop nutritional properties also requires profound understanding of the genetic, regulatory and transport mechanisms that underlie the uptake translocation and partitioning of minerals and nutrients. To assist in these studies, this volume contains a comprehensive collection of methodologies that are routinely used in plant mineral nutrition research.

References

1. Maathuis FJM (2009) Physiological functions of mineral macronutrients. Curr Opin Plant Biol 12:250–258
2. Marschner H (2012) Mineral nutrition of higher plants, 3rd edn. Academic, London, UK
3. Read DJ (1996) The structure and function of the ericoid mycorrhizal root. Ann Bot 77:365–374
4. Sokolovski SG, Meharg AA, Maathuis FJM (2002) *Calluna vulgaris* root cells show increased capacity for amino acid uptake when colonised with the mycorrhizal fungus *Hymenoscyphus ericae*. New Phytol 155:525–530
5. Jamtgard S, Nasholm T, Huss-Danell K (2008) Characteristics of amino acid uptake in barley. Plant Soil 302:221–231
6. Orsel M, Filleur S, Fraisier V, Daniel-Vedele F (2002) Nitrate transport in plants: which gene and which control? J Exp Bot 53:825–833
7. Loque D, Lalonde S, Looger LL, von Wiren N, Frommer WB (2007) A cytosolic trans-activation domain essential for ammonium uptake. Nature 446:195–198
8. Whiteman SA, Serazetdinova L, Jones AME, Sanders D, Rathjen J, Peck SC, Maathuis FJM (2008) Identification of novel proteins and phosphorylation sites in a tonoplast enriched membrane fraction of *Arabidopsis thaliana*. Proteomics 8:3536–3547
9. Stacey MG, Osawa H, Patel A, Gassmann W, Stacey G (2006) Expression analyses of *Arabidopsis* oligopeptide transporters during seed germination, vegetative growth and reproduction. Planta 223:291–305
10. Maathuis FJM, Sanders D (1994) Mechanism of high affinity potassium uptake in roots of *Arabidopsis thaliana*. Proc Natl Acad Sci USA 91:9272–9276
11. Maathuis FJM, Sanders D (1995) Contrasting roles in ion transport of two K^+-channel types in root cells of *Arabidopsis thaliana*. Planta 197:456–464
12. Hirsch RE, Lewis BD, Spalding EP, Sussman MR (1998) A role for the AKT1 potassium channel in plant nutrition. Science 280:918–921
13. Xu J, Li HD, Chen LQ, Wang Y, Liu LL, He L, Wu WH (2006) A protein kinase, interacting with two calcineurin B-like proteins, regulates K^+ transporter AKT1 in *Arabidopsis*. Cell 125:1347–1360
14. Armengaud P, Breitling R, Amtmann A (2004) The potassium-dependent transcriptome of *Arabidopsis* reveals a prominent role of jasmonic acid in nutrient signaling. Plant Physiol 36:2556–2576
15. Gaymard F, Pilot G, Lacombe B, Bouchez D, Bruneau D, Boucherez J, Michaux-Ferriere M, Thibaud J, Sentenac H (1998) Identification and disruption of a plant Shaker-like outward channel involved in K^+ release into the xylem sap. Cell 94:647–655
16. Pardo JM, Cubero B, Leidi EO, Quintero FJ (2006) Alkali cation exchangers: roles in cellular homeostasis and stress tolerance. J Exp Bot 57:1181–1199
17. Gobert A, Isayenkov S, Voelker C, Czempinski K, Maathuis FJM (2007) The two-pore channel TPK1 gene encodes the vacuolar K^+ conductance and plays a role in K^+ homeostasis. Proc Natl Acad Sci USA 104:10726–10731
18. Walker DJ, Leigh RA, Miller AJ (1996) Potassium homeostasis in vacuolate plant cells. Proc Natl Acad Sci USA 93:10510–10514
19. Britto DT, Kronzucker HJ (2008) Cellular mechanisms of potassium transport in plants. Physiol Plant 133:637–650
20. Demidchik V, Maathuis FJM (2007) Physiological roles of nonselective cation channels in plants: from salt stress to signalling and development. New Phytol 175:387–404
21. White PJ, Broadley MR (2003) Calcium in plants. Ann Bot 92:487–511

22. Dodd AN, Kudla J, Sanders D (2010) The language of calcium signaling. Annu Rev Plant Biol 61:593–620
23. Broadley MR, Bowen HC, Cotterill HL, Hammond JP, Meacham MC, Mead A, White PJ (2004) Phylogenetic variation in the shoot mineral concentration of angiosperms. J Exp Bot 55:321–336
24. Berezin I, Mizrachy-Dagry T, Brook E, Mizrahi K, Elazar M, Zhuo SP, Saul-Tcherkas V, Shaul O (2008) Overexpression of AtMHX in tobacco causes increased sensitivity to Mg^{2+}, Zn^{2+}, and Cd^{2+} ions, induction of V-ATPase expression, and a reduction in plant size. Plant Cell Rep 27:939–949
25. Robinson H, Gao YG, Sanishvili R, Joachimiak A, Wang AHJ (2000) Hexahydrated magnesium ions bind in the deep major groove and at the outer mouth of A-form nucleic acid duplexes. Nucleic Acids Res 28:1760–1766
26. Schachtman DP, Reid JD, Ayling SM (1998) Phosphorus uptake by plants: from soil to cell. Plant Physiol 116:447–453
27. Bucher M (2007) Functional biology of plant phosphate uptake at root and mycorrhiza interfaces. New Phytol 173:11–26
28. Rouached H, Wirtz M, Alary R, Hell R, Arpat AB, Davidian JC, Fourcroy P, Berthomieu P (2008) Differential regulation of the expression of two high-affinity sulfate transporters, SULTR1.1 and SULTR1.2, in *Arabidopsis*. Plant Physiol 147:897–911
29. Ramania B, Zornb H, Papenbrock J (2004) Quantification and fatty acid profiles of sulfolipids in two halophytes and a glycophyte grown under different salt concentrations. Z Naturforsch C 59:835–842
30. Clemens S (2006) Toxic metal accumulation, responses to exposure and mechanisms of tolerance in plants. Biochimie 88:1707–1719
31. Epstein E (2000) The discovery of the essential elements. In: Kung SD, Yang SF (eds) Discoveries in plant biology, vol 3. World Scientific Publishing Co. Pte. Ltd, Singapore, pp 1–16
32. Teakle NL, Tyerman SD (2010) Mechanisms of Cl^- transport contributing to salt tolerance. Plant Cell Environ 33(4):566–589
33. Kawakami K, Umena Y, Kamiya N, Shen JR (2009) Location of chloride and its possible functions in oxygen-evolving photosystem II revealed by X-ray crystallography. Proc Natl Acad USA 106(21):8567–8572
34. Lindsay WL (1979) Chemical equilibria in soils. Wiley, New York
35. Marschner H, Romheld V (1994) Strategies of plants for acquisition of iron. Plant Soil 165(2):261–274
36. Nozoye T, Nagasaka S, Kobayashi T, Takahashi M, Sato Y, Uozumi N et al (2011) Phytosiderophore efflux transporters are crucial for iron acquisition in graminaceous plants. J Biol Chem 286(7):5446–5454
37. Lehto T, Ruuhola T, Dell B (2010) Boron in forest trees and forest ecosystems. Forest Ecol Manag 260(12):2053–2069
38. Miwa K, Fujiwara T (2010) Boron transport in plants: co-ordinated regulation of transporters. Ann Bot 105(7):1103–1108
39. Donald CM, Prescott JA (1975) Trace elements in Australian crop and pasture production, 1924–1974. In: Nicholas DJ, Egan AR (eds) Trace elements in soil-plant-animal systems. Academic, New York, pp 7–37
40. Cailliatte R, Schikora A, Briat JF, Mari S, Curie C (2010) High-affinity manganese uptake by the metal transporter NRAMP1 is essential for Arabidopsis growth in low manganese conditions. Plant Cell 22(3):904–917
41. Broadley MR, White PJ, Hammond JP, Zelko I, Lux A (2007) Zinc in plants. New Phytol 173(4):677–702
42. Rengel Z (1999) Physiological mechanisms underlying differential nutrient efficiency of crop genotypes. In: Rengel Z (ed) Mineral nutrition of crops: fundamental mechanisms and implications. Food Products Press, New York, pp 227–265
43. Yruela I (2009) Copper in plants: acquisition, transport and interactions. Funct Plant Biol 36(5):409–430
44. Klaumann S, Nickolaus SD, Furst SH, Starck S, Schneider S, Neuhaus HE et al (2011) The tonoplast copper transporter COPT5 acts as an exporter and is required for interorgan allocation of copper in Arabidopsis thaliana. New Phytol 192(2):393–404
45. Yuan M, Chu ZH, Li XH, Xu CG, Wang SP (2010) The bacterial pathogen *Xanthomonas oryzae* overcomes rice defenses by regulating host copper redistribution. Plant Cell 22(9): 3164–3176
46. Nishida S, Tsuzuki C, Kato A, Aisu A, Yoshida J, Mizuno T (2011) AtIRT1, the primary iron uptake transporter in the root, mediates excess nickel accumulation in Arabidopsis thaliana. Plant Cell Physiol 52(8):1433–1442
47. Küpper H, Kroneck PMH (2007) Nickel in the environment and its role in the metabolism of plants and cyanobacteria. Metal Ions Life Sci 2:31–62
48. Tomatsu H, Takano J, Takahashi H, Watanabe-Takahashi A, Shibagaki N, Fujiwara T (2007) An Arabidopsis thaliana high-affinity molybdate transporter required for efficient uptake of molybdate from soil. Proc Natl Acad Sci USA 104(47):18807–18812

Chapter 2

Plant Growth and Cultivation

Dorina Podar

Abstract

There is a variety of methods used for growing plants indoor for laboratory research. In most cases plant research requires germination and growth of plants. Often, people have adapted plant cultivation protocols to the conditions and materials at hand in their own laboratory and growth facilities. Here I will provide a guide for growing some of the most frequently used plant species for research, i.e., *Arabidopsis thaliana*, barley (*Hordeum vulgare*) and rice (*Oryza sativa*). However, the methods presented can be used for other plant species as well, especially if they are related to the above-mentioned species. The presented methods include growing plants in soil, hydroponics, and in vitro on plates. This guide is intended as a starting point for those who are just beginning to work on any of the above-mentioned plant species. Methods presented are to be taken as suggestive and modification can be made according to the conditions existing in the host laboratory.

Key words: Plant growth, *Arabidopsis*, *Thlaspi*, Barley, Rice, Hydroponics, Soil growth, In vitro growth, Seeds sterilization

1. Introduction

Plants require very little for their growth in comparison to animals. As long as they have a substrate with even a small amount of mineral elements, water, and light plants will grow. Growing plants outdoor is subject to environmental conditions which vary greatly in space and time. For example, soil fertility and climatic conditions can rapidly change according the seasons and or geographical position. However, in order to ensure repeatability and uniformity across experiments, specific growth conditions need to be defined and controlled for research purposes. Over time, much interest and resources have therefore been dedicated to optimize the growing of plants indoor, under well-defined and controlled conditions. The optimum type and volume of soil, nutrients, and water dosages, light conditions and photoperiod required by different plant species were studied. Subsequently, hydroponic systems were introduced

Frans J.M. Maathuis (ed.), *Plant Mineral Nutrients: Methods and Protocols*, Methods in Molecular Biology, vol. 953, DOI 10.1007/978-1-62703-152-3_2, © Springer Science+Business Media, LLC 2013

with nutrient solutions that offer more uniform mineral composition and are easily adjusted. This has the advantage that the root systems develop in direct contact with the required minerals which are already in their ionic form and uniformly distributed in the substrate. Thus, macro- and micronutrients recipes formulated for the exact needs of each plant species have been developed. The level of control can be further increased by using in vitro plant cultures, especially when grown in dedicated growth cabinets or chambers where light regime, humidity, and CO_2 concentrations can be easily manipulated. In this chapter, I will give detailed protocols for growing some of the most often used plant species for research purposes. All protocols are based on indoor growth but various methods will be discussed such as growth on soil, in hydroponics, and in vitro cultures. The intention is to provide researchers who are starting to work with *Arabidopsis*, barley (*Hordeum vulgare*), or rice (*Oryza sativa*) a beginners guide. Growth conditions presented here can also be customized to suit other plant species and can also be tailored according to the conditions that exist in local growth facilities.

2. Materials

2.1. Growth Facilities

Growing plants indoor requires growth facilities in the form of, e.g., growth chamber, growth cabinet or greenhouse equipped with lighting and temperature control.

2.2. Sowing and Plant Growth on Soil

1. Plant seeds or seedling.
2. Soil/compost. In general, soil/compost is mixed with vermiculite, perlite, or polystyrene pellets in different ratios to prevent the soil from compacting and becoming too heavy after repeated watering and also to reduce the growth of fungi. Some people autoclave the soil before use. However, this practice is not recommended as it can affect the properties of the soil. Pesticides can be used to treat the soil prior to sowing to reduce fungal growth and development of insect larvae (see ref. 8). Soil/compost treated with pesticides is prepared at least 2–4 days in advance of sowing (see ref. 8). If the soil already contains fertilizers no further addition is required, at least, in the early stages. For suggestions regarding the type of soils used for *Arabidopsis*, barley, and rice, see Notes 1–3.
3. Trays/pots. There is a larger variety of trays/pots that are used and their sizes depend on the plant species they are used for. Pots are usually available at any gardening center. Trays/pots should be provided with drainage holes. If trays/pots are to be reused they should be thoroughly washed with detergent, or

even soaked in commercial bleach for about an hour in order to make sure that no fungi, seeds, or soil particles from previous experiments have been left. For suggestions of trays/pots used for growing *Arabidopsis*, barley, and rice, see Notes 1–3.

4. Lighting and temperature control. See Notes 1–3 for recommendations regarding light intensities and temperatures required by *Arabidopsis*, barley, and rice.
5. Water. To ensure repeatability between experiments distilled and deionized water (dIH_2O) is recommended to be used for watering the plants. If mineral composition of plants is also a matter of investigation, then the use of dIH_2O is a must. To ensure a healthy growth, fertilizers can be used.
6. Fertilizer. See Notes 1–3 for recommendations on fertilizers to be used for *Arabidopsis*, barley, and rice.
7. Pesticides. Pesticides are not a must, but if you are working in an institution with large facilities for growing plants, they will be used anyway to prevent plant damages and spread of any pest that is present. For scarid fly and greenfly control, compost can be treated with Intercept 70WG (Levington Horticulture, Ipswich, UK) before use. To control thrips, their mite predator *Amblyseius cucumeris*, can be applied to the leaves as larvae in a sawdust medium. Further thrip control can be provided by spraying with Conserve (Dow AgroSciences, Indianapolis, USA) every 2 months and with Dynamec (Fargro Ltd, Littlehampton, UK). To control mealy bug plants can be sprayed every 2 weeks with Provardo (Bayer Garden, Cambridge, UK). Fungal infection can be controlled by using sulfur burners. To provide a broad spectrum pest control plants can be sprayed weekly with Agri 50E (Fargro Ltd, Littlehampton, UK). The active ingredient present in the above-mentioned pesticides can be found in Table 1.

Table 1
List of pesticides used for treating the soil used for plant growth. Active ingredients and providers are also listed

Pesticide	Active ingredient	Manufacturer
Intercept 70WG	Imidacloprid 70 % w/w	Levington Horticulture, Scotts
Agri 50E	Dodecylphenol ethoxylate 27–33 % Tetrahydrofurfuryl alcohol 13.5–16.5 %	Fargro
Conserve	Spinosad 120 g/L (11.6 % w/w)	Fargro
Dynamec	Abamectin (1.8 %w/v) also contains hexan-l-ol	Fargro
Provado	Thiacloprid 0.150 g/L	Bayer

8. Labels and a permanent marker for labeling.
9. Waxed paper bags to collect the seeds.
10. Insulating tape to seal seed bags.
11. Fine mesh sieve to separate seeds from other plant material if dealing with small seeds such as *Arabidopsis.*
12. White paper for handling the seeds.
13. Support for larger plants such as barley and rice to prevent them from tangling. Optional.
14. ARACONs—harvesting devices used for *Arabidopsis* seeds. Their use prevents cross contamination between adjacent plants (e.g., Arasystem, Belgium; Lehle Seeds, Texas, USA).

2.3. Plant Growth in Hydroponics

1. Plant material. Both seeds and seedlings can be used for hydroponics.
2. Containers used for hydroponics are typically made of plastic. In general containers with lids are used in order to prevent evaporation of the solution, but other tops can also be used as long as the volume of the hydroponics is maintained constant by replacing the evaporated solution. In general, opaque containers are recommended as they reduce the growth of algae and confer appropriate conditions for roots growth. Usually, dark containers (brown, black) are used for this purpose. Dark containers are suitable for growth chamber and cabinets. However, when plants are grown in the greenhouse, white or light color nontransparent containers are recommended as they prevent over heating of the solution by reflecting the sunlight. In general, any plastic container can be used for hydroponics and there are numerous methods to adapt them for hydroponics. For example, transparent containers can be made opaque by either painting them with vinyl paint, or covering them with black insulating tape or aluminum foil. When using aluminum foil care should be taken in order to avoid its contact with the solution and thus prevent aluminum contamination. Containers can be washed with detergent every time the solution is changed, but often they are washed during an experiment only if required, i.e., if formation of algae or deposition of salts on the walls of the containers occurs.

   With respect to the size of the containers there is no uniformity in the literature. However, there are a few things that need to be considered when choosing a container. If the plants are to be analyzed separately and especially if the roots are to be analyzed (for biometric measurements or for mineral or biochemical analysis) then a deeper container should be used to allow the root system of each plant to develop without getting entangled. In this case, also the spacing between the plants should be considered.

A method for growing *Arabidopsis* in hydroponics using Magenta GA7 boxes and without the need of aeration (see ref. 4) was described by Arteca and Arteca (1). Indeed, the Magenta vessels work well. However, the downside of using these boxes is that they can only sustain growth of a very limited number of plants (about 4 plants). In many experimental contexts, (e.g., growth of wild-type plants is compared with that of mutant plants) a larger number of plants per box would be preferable. Therefore, people tend to use larger containers to be able to grow a larger number of plants within the same container. This ensures that both types of plants that are compared are exposed to exactly the same growth conditions.

3. Holders for plants. For holding the plants polyurethane foam, rockwool, or cotton wool cylinders of about 3 cm height can be used. Precut polyurethane foam stoppers of different diameters can be acquired from Fisher (http://www.fishersci.com). Rockwool can be bought from any gardening center as it is a very common material used for hydroponics or aquacultures. It is made of mineral fibers and has a water- and air-holding capacity of up to 80 and 17 %, respectively. To some people, rockwool can be a skin irritant. Rockwool can be difficult to cut when dry, so wetting the material may help. Before placing the seeds or the seedling within the rockwool cylinders they should be well soaked with nutrient solution to allow the pH of the rockwool to adjust to that of the solution.
4. Aeration system for the nutrient solution. Nutrient solutions should be well aerated. Any aquarium air pump or any commercial aeration pump from a gardening center and any rubber tubes for tubing can be used.
5. Reagents for nutrient solution. Reagents used to prepare the nutrient solution are to be of high purity (e.g., Fluka, Sigma, Supelco from Sigma-Aldrich http://www.sigmaaldrich.com) and solutions should be prepared in distilled and deionized (dIH_2O) water. Nutrient solutions used for plant growth are, in general, modified Hoagland's solution (2) and many variations exist (3). Stock solutions of nutrients (100 or 1,000× concentrated) can be prepared in advance and can be stored at 4°C. Below are given some examples of nutrient solutions used for different plants. For all the nutrient solution presented below, the concentration of the stock solution, the volume of the stock solution required for 1 L of hydroponic solution and the concentration of nutrients in the growth solution is given.
   (a) Nutrient solution for *Arabidopsis*.
   Gibeaut's solution (4) (Table 2). This is a modified Hoagland's solution (2) containing one-third of the concentration of macronutrients of Hoagland and full strength concentrations for the micronutrients. Reduced

Table 2
Nutrient solution used for *Arabidopsis* hydroponics (after Gibeaut et al. (4)). The micronutrients Si and Fe are prepared as separate stock solutions whereas the other micronutrients are mixed together in a single stock solution

	Stock solution concentration	Volume of stock solution to be added to 1 L final solution (ml/L)	Concentration in the hydroponics
Macronutrients			
$Ca(NO_3)_2 \times 4H_2O$	1 M	1.50	1.50 mM
KNO_3	1 M	1.25	1.25 mM
$Mg(SO_4) \times 7H_2O$	1 M	0.75	0.75 mM
KH_2PO_4	1 M	0.50	0.50 mM
Micronutrients			
$Na_2O_3Si \times 9H_2O$	0.1 M	1.00	0.1 mM
Fe (Sprint 330)[a]	0.072 M	1.00	0.072 mM
	One single stock solution containing		
KCl	50 mM	1.00	50 µM
$MnSO_4 \times H_2O$	10 mM		10 µM
$CuSO_4 \times 5H_2O$	1.5 mM		1.5 µM
$ZnSO_4 \times 7H_2O$	2 mM		2 µM
H_3BO_3	50 mM		50 µM
$(NH_4)_6Mo_7O_{24}$	0.075 mM		0.075 µM

[a]Fe (Sprint 330) is a commercial Fe product containing 10 % DTPA-Fe (DTPA = pentetic acid or *diethylene triamine pentaacetic acid*)

strength for the macronutrients prevents osmotic shock, whereas full strength of micronutrients prevents depletion. Gibeaut et al. (4) have added silicon (Si) into the medium because this is an important element for cell walls which in turn play a vital role in defense against pathogens. Another alternative is presented in Table 3 and is a modified solution after Arteca and Arteca (1). The pH of the hydroponic solution for *Arabidopsis* should be adjusted to 5.7–6.5 using 1 M sodium hydroxide (NaOH) or potassium hydroxide (KOH) or with 1 N hydrochloric acid (HCl).

(b) Nutrient solution for *Thlaspi caerulescens*.
An example of nutrient solution is presented in Table 4 and is also a modified Hoagland's solution. The pH of the hydroponic solution for *Thlaspi caerulescens* is similar to the *Arabidopsis* slightly acidic and it should be adjusted to 5.5–6.0 using 1 M NaOH or KOH or with 1 N HCl.

(c) Nutrient solution for barley.
Several hydroponic solutions for growing barley have been published. One example is given in Table 5. An alternative would be the nutrient solution used by Gries et al. (5).

Table 3
Nutrient solution used for *Arabidopsis* hydroponics modified after Arteca & Arteca (1)

	Stock solution concentration	Volume of stock solution to be added to 1 L final solution (ml/L)	Concentration in the hydroponics
Macronutrients			
KNO_3	1.25 M	1.00	1.25 mM
$Ca(NO_3)_2 \times 4H_2O$	0.50 M	1.00	0.50 mM
$MgSO_4 \times 7H_2O$	0.50 M	1.00	0.50 mM
KH_2PO_4	0.625 M	1.00	0.625 mM
Micronutrients			
FeNaEDTA	42.5 mM	1.00	42.5 μM
	One single stock solution containing		
$CuSO_4 \times 5H_2O$	160 μM	1.00	0.16 μM
$ZnSO_4 \times 7H_2O$	380 μM		0.38 μM
$MnSO_4 \times H_2O$	1.8 mM		1.8 μM
H_3BO_3	45 mM		45 μM
$(NH_4)_6Mo_7O_{24} \times 4H_2O$	15 μM		0.015 μM
$CoCl_2$	10 μM		0.010 μM

Table 4
Nutrient solution used for *Thlaspi caerulescens* hydroponics

	Stock solution concentration	Volume of stock solution to be added to 1 L final solution (ml/L)	Concentration in the hydroponics
Macronutrients			
KNO_3	0.5 M	1.00	0.5 mM
$Ca(NO_3)_2 \times 4H_2O$	0.4 M	1.00	0.4 mM
$MgSO_4 \times 7H_2O$	0.2 M	1.00	0.2 mM
KH_2PO_4	0.1 M	1.00	0.1 mM
Micronutrients			
FeNaEDTA	20 mM	1.00	20 μM
	One single stock solution containing		
$CuSO_4 \times 5H_2O$	0.2 mM	1.00	0.2 μM
$ZnSO_4 \times 7H_2O$	12 mM		12 μM
$MnSO_4 \times H_2O$	2 mM		2 μM
H_3BO_3	10 mM		10 μM
$Na_2MoO_4 \times 2H_2O$	0.1 mM		0.1 μM

Table 5
Nutrient solution used for spring barley (*Hordeum vulgare* cv. Golden Promise) hydroponics

	Stock solution concentration	Volume of stock solution to be added to 1 L final solution (ml/L)	Concentration in the hydroponics
Macronutrients			
KH_2PO_4	0.4 M	1.00	0.4 mM
K_2SO_4	0.4 M	1.00	0.4 mM
$MgSO_4 \times 7H_2O$	0.6 M	1.00	0.6 mM
NH_4NO_3	1.0 M	1.00	1 mM
$Ca(NO_3)_2 \times 4H_2O$	2.0 M	1.00	2 mM
Micronutrients			
$FeCl_3 \times 6H_2O$ or FeNaEDTA	75 mM	1.00	75 μM
	One single stock solution containing		
$MnCl_2 \times 4H_2O$	7 mM	1.00	7 μM
$ZnCl_2$	3 mM		3 μM
$CuSO_4 \times 5H_2O$	800 μM		0.8 μM
H_3BO_3	1.6 mM		1.6 μM
$Na_2MoO_4 \times 2H_2O$	0.83 mM		0.83 μM

The pH of the hydroponic solution for barley should be adjusted to 5.5–6.0 using 1 M NaOH or KOH or 1 N HCl.

(d) Nutrient solution for rice.
One example of nutrient solutions used for growing rice in hydroponics is given in Table 6. An alternative nutrient solution that works well for rice is the one used for *Arabidopsis* and described in Table 3. The pH of the hydroponic solution for rice should be adjusted to 5.5–6.0 using 1 M NaOH or KOH or 1 N HCl (6, 7).

6. Concentrated 1 N hydrochloric acid (HCl).
7. 1 M solution of sodium hydroxide (NaOH) or potassium hydroxide (KOH). (Dissolve 2 g of NaOH in 50 ml dIH_2O or 2,8 g KOH in 50 ml dIH_2O).
8. Bottles and dark bottles for preparation and storage of stock solutions.
9. Tweezers for handling larger seeds or seedlings.
10. Knife or sharp scalpel to cut the plant holders.
11. pH meter to check the pH of the hydroponic solution.
12. Magnetic stirrer to mix the nutrient solution when prepared.
13. Fridge or cold room to store the stock solution.
14. Labels and permanent marker for labeling.

Table 6
Nutrient solution used for rice (*Oryza sativa*) hydroponics (after Miyamoto et al. (16))

	Stock solution concentration	Volume of stock solution to be added to 1 L final solution (ml/L)	Concentration in the hydroponics
Macronutrients			
$(NH_4)_2SO_4$	0.09 M	1.00	0.09 mM
KH_2PO_4	0.05 M	1.00	0.05 mM
K_2SO_4	0.03 M	1.00	0.03 mM
$MgSO_4 \times 7H_2O$	0.07 M	1.00	0.07 mM
KNO_3	0.05 M	1.00	0.05 mM
$Ca(NO_3)_2 \times 4H_2O$	0.06 M	1.00	0.06 mM
Micronutrients			
Fe-EDTA	0.11 M	1.00	0.11 mM
	One single stock solution containing		
$MnSO_4$	1.8 mM	1.00	1.8 μM
$ZnSO_4 \times 7H_2O$	0.3 mM		0.3 μM
$CuSO_4 \times 5H_2O$	0.3 mM		0.3 μM
H_3BO_3	4.6 mM		4.6 μM

2.4. Sowing and Growing on Plates

1. Plates (sterile Petri dishes: round or square, the latest being used especially if plants are to be grown vertically, either for root measurements or to ease the removal of the seedlings for transfer to a different substrate), e.g., Square 100 × 100 mm (Sterilin Ltd, UK) or 120 × 120 mm (Greiner—Sigma-Aldrich; Sarstedt, Sarstedt AG & Co., Germany).
2. Reagents for preparation of media. Media used for in vitro culturing are based on Murashige and Skoog (1962) formula (6), thus named MS. It can nowadays be bought as a powder including all the minerals. Different variations exist in order to assist all needs.
   (a) ½ MS minimal medium (1 L) (half-strength MS, no sucrose (7)).
   - 1/2 MS basal salt mixture (without micronutrient and vitamins) (weigh half of the manufacturer's recommended amount for 1 L) (e.g., Sigma; Duchefa, The Netherlands, http://www.duchefa.com).
   - 0.8 % plant agar for horizontal growth and 1.5 % for vertical growth (e.g., Sigma; Duchefa).

   (b) 1× MS medium (1 L).
   - 1× MS basal salt mixture (without micronutrient and vitamins) (weigh the manufacturer's recommended amount for 1 L) (e.g., Sigma; Duchefa).

- 1× Gamborg's vitamins (weigh the manufacturer's recommended amount for 1 L, to be filter sterilized and added after autoclaving) (e.g., Sigma, Duchefa).
- 2.56 mM MES-KOH (e.g., Sigma) ((MES—KOH—Monohydrate 2-(N-morpholino)ethanesufonic acid) adjusted to pH 5.7 with 1 M KOH).
- 1 % sucrose (to be added after autoclaving from a filter-sterilized 20 % stock solution). (The percentage of added sucrose can be increased).
- 1.5 % phyto or plant agar (Duchefa).

3. dIH_2O (distilled and deionized water).
4. Filter-sterilized 20 % sucrose solution (Weigh 20 g sucrose (e.g., Formedium, UK, http://www.formedium.com) and add dIH_2O up to 100 ml volume. Add a magnet and place it onto a magnetic stirrer until dissolved. Within a laminar hood filter the solution through a sterile syringe-driven filter with 0.22 μm pore size (e.g., Millipore) into a sterile recipient (e.g., 2×50 ml Falcon-type tube or an autoclaved empty Duran bottle)).
5. Solutions for sterilization of seeds:
   (a) 70 % ethanol/ methanol or Mikrozid.
   (b) 96–100 % ethanol.
   (c) 30–50 % domestic bleach (e.g., Domestos) or 2.625 % sodium hypochlorite in final solution made in H_2O (0.02–0.05 % Triton X-100 or Tween 20 (Sigma) can be added to reduce surface tension and allow better surface contact).
   (d) Concentrated hydrochloric acid (HCl) (for Subheading "Sterilization of Large Numbers of *Arabidopsis* Seeds with Chlorine Gas").
   (e) Sterile dIH_2O (autoclaved water dIH_2O).
   (f) Sterile top agar (0.1 % agar prepared in dIH_2O and autoclaved) (Duchefa).
6. Duran bottle to autoclave the media and solutions.
7. Duran bottle for the liquid waste (make sure you label it appropriately) and a recipient to dispense the pipette tips.
8. 1.5 ml Eppendorf-like tubes or, if large numbers of seeds are to be handling, 15 ml Falcon-like tubes.
9. Racks for the tubes.
10. Labels and a permanent marker for labeling.
11. White paper for handling the seeds.
12. Tweezers for handling larger seeds or seedlings, or toothpicks (cocktail sticks) for small seeds.
13. Scissors.

14. 200 ml glass beaker (for Subheading "Sterilization of Large Numbers of *Arabidopsis* Seeds with Chlorine Gas").
15. Plastic box or lid large enough to cover the 200 ml beaker and the rack with seeds (for Subheading "Sterilization of Large Numbers of *Arabidopsis* Seeds with Chlorine Gas").
16. Pipettes (for up to 1,000 μl).
17. Sterile pipette tips (autoclaved tips; the box of tips should only be opened inside the sterile laminar hood).
18. Micropore tape (3 M) to seal the plates.
19. Timer.
20. Autoclave.
21. Laminar flow hood, possibly also fume hood (for Subheading "Sterilization of Large Numbers of *Arabidopsis* Seeds with Chlorine Gas").
22. Table-top centrifuge for 1.5 ml Eppendorf-like tubes or centrifuge for 15 ml Falcon-like tubes.

3. Methods

3.1. Sowing and Growing in Soil

Seeds can be either placed directly onto soil and thus germinated in pots or they can be first germinated and then seedlings can be transferred to soil. There is no need to sterilize the seeds as long as they are not to be use for in vitro culturing.

3.1.1. *Arabidopsis thaliana* (This Method Also Works for Other Small Seed Species Such As *Thlaspi caerulescens** and *Nicotiana sp.*)

1. If the seeds are to be germinated directly onto soil in single cells/pots, then one seed should be placed in the center. *Arabidopsis* seeds are very small and sometimes it is difficult to place only one seed onto soil. If more than one seed is placed and germinated, just remove the extra seedlings.
2. To ease the sowing the soil should be water-saturated, and water should be seen in the tray as well. Also, water can be sprayed directly onto the surface of the soil.
3. Seeds can be placed on white paper and single seeds can be taken using a wet wooden tooth pick (cocktail stick) and placed onto the surface of the soil. Another method to place seeds onto soil is to first put them into a 0.1 % agar (e.g., plant agar Duchefa) solution and then pipette them. Seeds require light to germinate, therefore there is no need to cover them with a lot of soil or push them deep into the soil.
4. After sowing, trays should be covered to maintain moisture either with tops (if the trays are provided with incubation tops, transparent plastic tops) or wrapped in transparent foil. If using the incubation lids then the vents should be closed.

5. If seeds have not been stratified prior to placing them onto soil, then the trays should be placed at 4°C for 48 h before moving them to growth cabinets or greenhouse.
6. If seeds are to be first germinated onto wet filter paper or in vitro, then 10–20 day old seedlings (usually two to four leaves) are transferred to the soil. A small hole needs to be made in the center of the pot/cell where seedling should be gently placed using tweezers. The roots of the seedling should be placed entirely in soil. As *Arabidopsis* seedling are prone to dehydration the soil should be saturated with water and the trays/pots should be covered with the incubation lids with vents closed.
7. After germination, plants should be watered as required. In general, the top soil should be slightly dry before watering. Depending on the conditions in the greenhouse or growth cabinet, administration of water is required every 5–7 days. Both water logging and complete dryness should be avoided.
8. If plants are grown for seeds, in order to obtain a larger number, the first bolt can be simply removed by hand. This will stimulate more bolts to form.
9. To collect the seeds, stop watering the plants once the siliques have developed and turn brown.
10. Bag the entire inflorescence/siliques of a plant by introducing it into a waxed paper bag. Do not remove the plant from the soil, but seal the bag to the stem using insulating tape. Wait for the siliques to dry as this will allow the seeds to mature. When the plants are dead, about 7–10 days after bagging, seeds can be collected by cutting off the inflorescence pedicel with scissors. Optionally, for collecting and avoiding cross-contamination of *Arabidopsis* seeds ARACONs can be used (http://www.arabidopsis.com/main/cat/arasystem/!as02.html).
11. To clean the seeds from the rest of the plant material, cut off a corner of the bag and poor its content onto a fine mesh placed above a piece of white paper. For a first rough cleaning a tea sieve can be used followed by sieving through a finer meshed sieve.
12. Once cleaned, seeds can be transferred into vials (e.g., 1.5 ml Eppendorf-like tubes or Falcon-like tubes depending on the amount of seed). Vials should be labeled properly, sealed with parafilm and stored in dry conditions at 2–8°C, usually in dark. For more information on storage conditions of *Arabidopsis* see ref. (8).
13. *Winter annuals or biennials plants such as *Thlaspi* sp. also require a period of cold (about 4°C) in order to flower, a process called vernalization. Therefore, 6–8 weeks old plants should be moved to 4°C for 10 weeks, 8/16 h photoperiod. Before moving to the cold, plants should have developed a

large rosette of leaves as they should have acquired enough carbohydrates in order to survive vernalization. During vernalization plants should be watered once per week, just to make sure the soil does not dry out. Also, plants should not be fertilized before vernalization. Fertilizers should be added together with water only once the plants have been vernalized and moved to long days (16/8 h day/night light regime) and temperatures above 20°C. For *Thlaspi* bolting will start 7–10 days later (9).

3.1.2. Barley and Rice

1. If seeds are to be germinated directly in soil, then place one in the center of each pot at about 2.5 cm depth and make sure you cover it with soil.
2. Alternatively, germination of seeds can be achieved on moist filter paper or wet vermiculate. Wet vermiculate would give the advantage of a solid support and therefore seedlings can grow as they would in soil for up to 2 weeks. Use dIH_2O as this will have the least effect on seedlings that are to be used in downstream experiment.
3. Germination can be performed in germinators, plastic boxes or Petri dishes at room temperature (<22°C) for barley and at 28°C for rice. At the time of placing the seeds, the filter paper or the vermiculate should be moist/wet for barley so that the seeds do not end up floating. For rice more water can be added. Cover the germination container to maintain moisture and place it in the dark for 3–5 days. Water can be added during this time if the paper or vermiculate gets dry.
4. Once the coleoptile has emerged, uncover the container and place it in the light (e.g., growth chamber, greenhouse or even window frame if seedlings are to be used only for molecular analysis such as RNA or gDNA extraction). Once the first leaf has emerged and seedlings are 5–10 cm in length, they can be transferred to soil or hydroponics.
5. Plants grown in soil should be watered as required. There is no consensus because this is highly dependent on the type of soil, volume of the pot, number of plants per pot, developmental stage, air-humidity, and temperature. Just avoid water-logging the soil and keep in mind that in plastic pots the soil tends to dry at the surface, but can be very moist deeper within.
6. For barley and rice, after 6–7 weeks of growing, plants require feeding, especially if they are to be grown to maturity and seeds are required. In general, fertilizers are added from the 6th to 7th week, once per week together with water. The type of fertilizer used varies a lot. One possibility is to use a general purpose fertilizer (e.g., "Phostrogen" NPK 14:10:27 that includes also trace elements (PBI Home and Garden Ltd.,

Hertfordshire. UK)). Another option would be "Osmocote Plus" (NPK-Mg, 5:5:11–1.2) also added weekly together with the water. Murray et al. (10) used Aquasol liquid fertilizer (NPK 23:4:18) once per week (5 g/L).

7. Support the plants to prevent tangling when needed.
8. Seed heads should be harvested when they turn brown.
9. Store seeds in waxed paper bags or just paper bags at room temperature in dry conditions.

3.2. Sowing and Growing in Hydroponics

3.2.1. Preparation of Nutrient Solution

1. To ease the preparation of hydroponics, stock solutions of nutrients are prepared beforehand and certain volumes of these solutions are then added to prepare the final hydroponic solution in which plants are grown. In general, macronutrients are prepared as individual stock solutions while micronutrients are mixed together in a single stock solution. However, each microelement should be dissolved separately in water and only then mixed together. It is possible that for $ZnCl_2$ you might need to add a small drop of concentrated hydrochloric acid (HCl) until the solution is clear. Iron (Fe) is also usually prepared separately and, as it is sensitive to light, the stock solution should be kept either in a dark bottle or covered with aluminum foil. It is also recommended that the stock solutions are kept at 4°C to prevent algae formation. If you see any aggregates floating in the solution discard it and prepare a new one. The pH of the hydroponic solution should be adjusted to slightly acidic using 1 M sodium hydroxide (NaOH) or potassium hydroxide (KOH) if a Na-free solution is needed. The recipient in which the hydroponic solution is prepared can be placed on a magnetic stirrer while the pH is adjusted as required for each plant species.
2. Examples of nutrient solutions used for some of the most commonly used plant species are listed under Subheading 2.3.
3. Once the nutrient solution is prepared proceed to sowing (see Subheading 3.2.2).
4. Then place the hydroponics into the growth chamber/cabinet or greenhouse.
5. Aerate the solution continuously via a needle, injecting compressed air into the solution. However, for *Arabidopsis*, continuous aeration should be avoided as it can inhibit root growth due to agitation (4, 11).
6. Some protocols recommend changing the solution every 3–4 days; however others change the nutrient solution once per week. When changing solution for the first time, check the pH and the level of solution and adjust the changing times according to needs and duration of the experiment.

3.2.2. Preparation of Plant Material and Holders

1. Cut the holder cylinders (foam, rockwool or cotton) by making an incision along their length up to the center. This cut will allow you to place the seedling in the center of the holder and then will allow the plant to develop its root down to the nutrient solution.
2. Before sowing the seeds or transferring the seedlings, wet the holders by soaking them into nutrient solution.
3. Seeds can be sown directly onto wet holding material. Place two to three seeds in the center of the holders. After germination seedlings are thinned to one per plug of holding material.
4. If seedlings are used, seeds need to be germinated and seedlings to be grown to a certain developmental stage. In many cases, for small seeds such as *Arabidopsis*, germination is performed in vitro on half-strength Murashige and Skoog (MS) medium and 10–20 days old seedlings are then transferred to hydroponics. For sowing and getting seedlings in vitro for *Arabidopsis* see Subheadings "Sterilization of Limited Numbers of Small Seeds Such As *Arabidopsis*" and 3.3. For larger seeds such as from cereals, seeds can be germinated on wet substrate (see Subheading 3.1.2) and seedlings, developed to one leaf stage, can be transferred to hydroponics.
5. Place one seedling in the center of the holder and place the holder into one of the holes in the lid/top of the hydroponic container, making sure that the root is coming out from the holder and is reaching the nutrient solution.
6. To avoid algae growth onto the holder you can sprinkle some soil on top of them. Otherwise, just wash the holders with detergent or replace them every time or every second time you change the nutrient solution.
7. Place the containers with the hydroponic solution and the plants in the growth chamber/cabinet, greenhouse.

3.3. Sowing and Growing on Plates

3.3.1. Preparation of Media and Plates

1. Weigh the required amount of reagents needed to prepare the medium and place them in a Duran bottle. Sucrose can be added as a feed source for seeds and plants. However if seeds are healthy they should have enough stored resources to germinate and once the first leaves appear the seedlings are autotrophic. The addition of sucrose can also increase the susceptibility of the media and, in consequence, of the plants to fungal infection. However, on the other hand, sucrose can help to ensure uniformity of germination and of early development of the seedlings. Two examples of media (without and with sucrose) that can be used for general purpose growth of *Arabidopsis* and other plant species are given under Subheading 2.3. The agar or phytagel should be added only after the pH has been adjusted.

2. Add 900 ml of dIH_2O and agitate until the reagents added before are dissolved. completely.
3. Adjust pH to 5.6–6.0 with 1 M KOH.
4. Add the agar or phytagel. Mix well.
5. Fill with water to 1 L. (If a certain volume of sucrose and/or vitamins is to be added after autoclaving take into consideration those volumes when adding the water so that, after their addition, the final volume will still be 1 L).
6. Autoclave for 20 min at 121°C.
7. Let the medium cool down and then pour it into plates inside the laminar hood. Plates should only be opened inside the laminar hood to keep them sterile.
8. Let plates dry for 30 min (some people recommend 45 min or even 1 h). If the experiment will involve comparing the plants grown in different batches, then time the drying and use the same time every time you pour the plates.
9. Plates can be stored upside down at 4°C for 1–2 weeks.
10. If plates are to be used immediately leave them covered (lids on) in the hood and proceed to surface sterilization of the seeds and then sowing.

3.3.2. Surface Sterilization of Seeds

1. Sterilization of seeds is required for any in vitro experiments in order to get rid of any pathogens or bacteria that can be present at the surface of the seeds.
2. All steps of sterilization should be conducted in a laminar flow hood, ideally supplied with a UV lamp. The surface of the flow hood should be cleaned, wiped with 70 % ethanol/methanol or Mikrozid. Optional: Place all the solutions and materials (except the seeds) in the hood and keep them under the UV light for 5–10 min.
3. All solutions placed on the seeds during the procedure are to be discarded into the autoclavable waste bottle and tips into the waste bin. It is important, especially when you work with transgenic seeds, that all the liquid and solid waste will be autoclaved, in order to avoid any accidental spread.
4. Make sure you swab you hands with ethanol before starting work in the laminar flow hood or wear gloves.

Sterilization of Limited Numbers of Small Seeds Such As *Arabidopsis*

1. Label the tubes for each batch of seeds you are using.
2. Transfer the required amount of seeds from the storage vial (tube) into a clean 1.5 ml Eppendorf-like tube. This can be done on a bench, in the laboratory. Do this on a white paper so that you can see any seeds that might be spilled. To ease the transfer into the Eppendorf-like tube you can use either a white weighing boat or a folded white piece of paper. Make sure that

there are no seeds spilled on the paper or weighing boat when you change to a different batch of seeds.

3. Take the tubes with the seeds into the laminar flow hood and make sure you have all the solutions and materials ready.

   The following steps are to be performed in sterile conditions inside the laminar flow hood, using only sterile materials and solutions.

4. Carefully open the first Eppendorf-like tubes with the seeds (avoid seeds jumping out due to static). Add slowly 1 ml of 96–100 % ethanol. Close the tube and mix by inverting it several times, so that all the seeds get in contact with ethanol.

   Do the same for each of the remaining tubes. Seeds should be maintained in ethanol for 1 min. If you have a large number of tubes, handle them in batches so that the seeds do not stay in ethanol longer than 2 min.

5. Remove the ethanol using a pipette. Once you get more experienced you can pour it out, but for beginners it is better to use a pipette. Discard the ethanol into the waste bottle and then the tip into the waste bin.

6. Add 1 ml of 30–50 % bleach (sodium hypochlorite) solution. Make sure the tubes are tightly closed and mix the seeds with the bleach solution by inverting the tubes. Keep the seeds in the bleach solution for 10 min. Do not exceed this time as it can have a detrimental effect on seed germination.

7. Take the tubes to a centrifuge and give them a quick spin so that the seeds are collected at the bottom of the tubes. Then take them back into the laminar flow.

8. Remove all the bleach solution from the tubes by using a pipette and discard it. Do it slowly and make sure you do not take out the seeds as well.

9. Add 1 ml of sterile dIH_2O into the tube. Close the tube and mix well by inversion. The aim is to remove any traces of the bleach.

10. Take the tubes to the centrifuge and spin them quickly to collect the seeds at the bottom.

11. Repeat the washing step one or two times.

12. At this stage, seeds can be either sown on plates or seeds can be left in the last washing with dIH_2O and placed the tubes at 4°C (in the fridge or cold room) for 2–3 days to be stratified. Seeds can be kept like this for up to 1 week.

13. If seeds are to be sown on plates immediately, proceed to (Subheading 3.3.3).

14. A slight modified version of sterilization would be to use: 70 % ethanol for 5 min, 50 % bleach supplemented with 0.02 % Triton X-100 (Sigma) for 5 min and four washes with dIH_2O.

Sterilization of Large Numbers of *Arabidopsis* Seeds with Chlorine Gas

If many batches of seeds are to be sterilized in one go, then chlorine gas can be used. This method saves the time required for handling each individual tube of seed. As chlorine is a toxic gas, the sterilization procedure should be performed in a fume hood rather than a sterile hood.

1. Place the seeds in 1.5 ml Eppendorf-like tubes. Tubes should be labeled by writing in pencil on sticky labels as pen writing directly on the tubes will be removed by the chlorine gas.
2. Add 100 ml of bleach in a 200 ml beaker.
3. Place the tubes with the seeds in a rack and open them.
4. Put the tube racks with the seeds and the beaker with the bleach in a box or under the lid in a fume hood.
5. Add 3 ml of concentrated HCl to the bleach. A small quantity of brown gas will evolve.
6. Immediately close the box and leave for about 4 h. Avoid accidental spillage of the bleach from the beaker. Longer times might be used, but exceeding 24 h can cause seed mortality.
7. When the time is up, open the box in the fume hood and leave it for 10 min for the gas to disperse.
8. Close the tubes and transfer them to a sterile hood.
9. Seeds can be placed directly onto MS media or can be resuspended in sterile 0.1 % agar for pipette sowing.
10. For sowing (plating) follow protocol in Subheading 3.3.3.

Sterilization of Large Seeds (e.g., Barley, Rice, Wheat)

1. Remove awns and take only the seeds. Dehusked or non-dehusked seeds can be used depending on the purpose of the experiment. However, if seeds are intended for transformation or if seedlings are to be kept for longer in vitro, then it is recommended to use dehusked seeds. This can be achieved manually.
2. Place the seeds in a Berzelius or Erlenmeyer (beaker). The size of the Berzelius/Erlenmeyer should be at least twice the volume of the seeds that are to be sterilized. At this stage a magnet can be added to help with agitation in the further steps.
3. Add enough 96–100 % ethanol to cover the seeds completely. Mix by agitating the beaker for a minute, or you can cover the Berzelius/ Erlenmeyer with aluminum foil and place it onto a magnetic stirrer for 1 min. An alternative is to place the seeds in a 50 ml Falcon-like tube (no more than half full) and rinse with 96–100 % ethanol for 1 min.
4. Instead of 96–100 % ethanol for 1 min, 70 % ethanol can be used for 3 min.
5. Take the Berzelius/Erlenmeyer back in the laminar flow hood and pour off the ethanol into a waste bottle.

6. Add 20 % bleach solution (or final 1 % sodium hypochlorite) containing 0.02–0.05 % Triton-X or Tween-20. Use about 1.5–2 times the volume of the seeds (e.g., if seeds occupy about 100 ml, then add 150–200 ml of bleach solution).
7. Cover the Berzelius/Erlenmeyer and place it onto a magnetic stirrer (if you had previously added the magnet) for 15–20 min. If no magnet was added, keep the Berzelius/Erlenmeyer in the laminar flow hood and agitate it occasionally. Do not exceed 20 min.
8. Take the Berzelius/Erlenmeyer back in the laminar flow hood and pour off the bleach solution into a waste bottle.
9. Add sterile dIH_2O. The volume of the water should be 1.5–2 times the volume occupied by the seeds. Agitate the Berzelius/ Erlenmeyer well under the laminar flow hood.
10. Discard the water by pouring it into the waste bottle and repeat the washing steps 5–8 times. The purpose of these washes is to get rid completely of the bleach, as this can later impede the germination of the seeds.
11. After the last wash leave a few drops of water onto the seeds to keep them moist until they are used. If they are not used immediately, cover the Berzelius/Erlenmeyer and place it at 4°C (in the fridge or cold room).

3.3.3 Sowing (Plating on Petri Dishes) and Growing

1. Sowing needs to be performed in the laminar hood and sterilized seeds are to be used.
2. Small seeds, like *Arabidopsis* or *Nicotiana* can be resuspended in 200–1,000 μl of sterile 0.1 % agar. The volume of the agar can be adjusted according to the volume of your seeds. To plate, take the seed suspension into a tip (1,000 μl tip) and transfer them individually onto the solid medium in the plates, by gently touching the surface of the medium with the tip.
3. Larger seeds, like from cereals, can be placed directly onto the solid medium using sterile tweezers.
4. Once sowing is done let the plates dry under the hood for about 5 min to evaporate the water placed onto the agar together with the seeds.
5. Seal the plates with micropore tape (3 M) and place them into a growth chamber to allow them to germinate. Micropore tape is a porous material that allows water to evaporate and therefore reduced water condensation is observed in comparison to parafilm.
6. Some seeds require stratification (e.g., *Arabidopsis*). Stratification can be performed by placing the sterile seeds at 4°C for 2–3 days before or after sowing. If stratification is done after sowing, the plates (Petri dishes) are covered with aluminum

foil and placed at 4°C (in the fridge or cold room) and only afterwards moved to the growth room.

7. Plates can be placed either horizontally or vertically. Vertical growth is usually preferred if roots are intended for further investigations. Vertical position prevents roots to enter into the agar. Thus the roots can be easily measured for length (e.g., for root phenotypes) or removed for further investigations. Seedlings intended for hydroponics can also be grown upright as they are easier to be removed from the plate and the root system has place to develop vertically.

4. Notes

Notes presented are with regard to growth conditions used for Arabidopsis, barley and rice. With respect to light regimes and temperatures used for plant growth some alternatives are listed here, though they should be considered as consultative. Also, if plants are to be grown only for the extraction/preparation of nucleic acids there is no need to strictly follow or adapt your laboratory/greenhouse protocol to the growth conditions presented in this section.

1. *Light intensity*. Approximate 100–200 μmol/m^2 s from fluorescent bulbs, although lower light intensities have also been used (e.g., 40–60 μmol/m^2 s (12)). Also, higher light intensities of 250 μmol/m^2 s provided by mercury vapor lamps have been reported (13).

   Photoperiod. Long days >12 h accelerate the reproductive cycle—plants grown under long day regimes produce a small biomass (few and small leaves) while development of inflorescences occurs early. One example is 16/8 h day/night regime at 23/18°C. Short days <12 h favor vegetative growth—a 10 h/14 h day/night regime provides a good compromise between sufficient shoot biomass production and pre-flowering growth (4).

   Temperature. 20–22°C day/17–18°C night. The temperatures required for *Arabidopsis* growth depends on the ecotype. In general, temperatures around 20–22°C during the day and 17–18°C during the night are fine for any of the ecotypes. However, *A. thaliana* Landsberg erecta grows better at day temperature around 23°C, whereas Columbia 0 (Col 0) shows stress symptoms at this temperature (4).

   Stratification of seeds. *Arabidopsis* seeds, depending on the ecotype, can germinate without stratification. However, stratification of seeds ensures a more uniform germination and therefore a better comparison of results between different lots of seeds that are used in an experiment. Stratification of seeds is performed at 4°C for 2–4 days depending on the ecotype. Three days stratification periods would suit any of the *Arabidopsis* ecotypes. Stratification can be performed before sowing by placing the seeds into a vial in water or after sowing by placing the plates (in vitro experiments) or the pots with soil at 4°C. Seeds can be kept at 4°C for up to a week.

Humidity. High humidity is required for *Arabidopsis* seeds to grow, e.g., 65 % (7), 75 % (4), 60/75 %day/night (13).

Soil. In general, any standard peat-based soil/compost would work for *Arabidopsis.* Here are two examples: GS90 (composition: peat, clay, coconut fiber, 2 g/L salt, 160 mg/L N, 190 mg/L P_2O_5, 230 mg/L K_2O, pH 6; supplied by Werner Tantau GmbH & Co. KG, Germany) (14) and Levington F2 seed and modular compost pH 6 (Green-tech, UK). Soil is mixed 1:1 v/v with vermiculate or perlite.

Trays/pots. The size of trays/pots used for *Arabidopsis* growth depends on the number of plants that is needed for the experiment and available space for growing. In general, people use flat cell trays if individual plants are required or large trays if plants are to be pooled. For trays with individual cells or pots a square of around 7 × 7 cm works well. However smaller cells/pots (36 cells trays) also give good results.

Containers for hydroponics. One example would be to use a 5 L container 35 × 35 × 20 cm, holding up to 36 plants, with a diameter of the hole of 1.5–2.0 cm and a distance between the holes of 4–5 cm.

2. *Light intensity.* Barley requires high light intensities and therefore the light quality can significantly influence the performance of the plants. 350–500 mmol/m^2 s photosynthetically active radiation (5, 15) is adequate.

   Photoperiod. 16/8 h light/dark photoperiod is used most of the time for barley.

   Temperature regime varies between experiments and may also depend on whether the conditions in the greenhouse or growth chamber can be adapted for barley or whether other plant species are to be grown. However, keeping the temperature in the low range 10–20°C would be beneficial. High temperatures should be avoided. If plants are to be grown for the whole life cycle then ~15°C/10°C day/night temperatures gives good results (15). Other day/night temperature regimes that can be used are: 20/15°C, 20/17°C, 21/18°C, 23/17°C (5) or 24/18°C.

   Humidity. Relative air-humidity varies between different published data, but in general is between 60 and 80 %.

   Soil type. There are different types of soil that can be used to grow barley, but here are three possibilities that all work well: (1) Standard sphagnum (or sphagnum mixed with polystyrene pellets in a 2:1 ratio), (2) Levington F2 seed and modular compost pH 6 (Green-tech, UK) mixed with perlite in a 3:1 ratio or (3) compost and perlite in a 4:1 ratio (10).

Pots size. A 2 L pot works well. However, smaller pots (15 × 15 × 15 or 13 × 13 × 13 cm square) can also be used, but if there are no space constraints, then larger pots would work better.

Containers for hydroponics. The volume of the containers varies largely between experiments. In general, if the plants are to be grown in hydroponic solution up to maturity and seeds are required 1 L hydroponic solution/ plant is recommended. Alternatively smaller containers can be use such as 5 L containers that can hold up to ten plants. Diameters of the holes should be 2.0 cm and a distance between the holes of at least 5 cm.

3. *Light intensity*. Rice, like barley, requires high light intensities and therefore the light quality can significantly influence the performance of the plants. 300–500 μmol/m^2 s photosynthetically active radiation (16, 17).

   Photoperiod.16/8 h light/dark photoperiod is used most of the time for rice. However 12/12 hday/night regime can also be used.

   Temperature regime varies between experiments. However, rice prefers higher temperatures, in general above 20°C. Some examples of day/night temperatures used are: 25/20°C (17) or 27/22°C (16).

   Humidity. Relative air-humidity varies between different published data, but in general is between 40 and 80 %.

   Soil type. There are different types of soils that can be used to grow rice; in general clay loam mineral soil should be avoided. Levington F2 seed and modular compost pH 6 (Green-tech, UK) mixed with vermiculite or perlite in a 1:1 v/v ratio works well for rice.

   Pots size. See barley (Subheading 4.1.2).

   Containers for hydroponics. The volume of the containers varies largely between experiments. In general, 0.4–1 L hydroponic solution/ plant is recommended (e.g., 20–24 plants/8–10 L container (16, 17)). Alternatively smaller containers can be use such as 3 L containers that can hold up to 16 plants. Diameter of the hole is 2.0 cm and a distance between the holes of at least 5 cm should be used.

Acknowledgments

The author would like to thank Dr. Anthony Dodd for his useful suggestions and discussions and Dr. Frans J.M. Maathuis for his help with editing. The author is supported by CNCSIS-UEFISCSU, PN II_ PD_424/2010.

References

1. Arteca RN, Arteca JM (2000) Physiol Plant 108:188–193
2. Hoagland DR, Arnon DI (1950) Calif Agric Exp Stn Circ 347:1–32
3. Tocquin P, Corbesier L, Havelange A, Pieltain A, Kurtem E, Bernier G, Perilleux C (2003) BMC Plant Biol 3:2
4. Gibeaut DM, Hulett J, Cramer GR, Seemann JR (1997) Plant Physiol 115:317–319
5. Gries D, Brunn S, Crowley DE, Parker DR (1995) Plant Soil 172:299–308
6. Murashige T, Skoog F (1962) Physiol Plant 15:25
7. Peiter E, Montanini B, Gobert A, Pedas P, Husted S, Maathuis FJM, Blaudez D, Chalot M, Sanders D (2007) Proc Natl Acad Sci 104:8532–8537
8. Hay FR, Mead A, Manger K, Wilson FJ (2003) J Exp Bot 54:993–1011
9. Peer WA, Mamoudian M, Lahner B, Reeves RD, Murphy AS, Salt DE (2003) New Phytol 159:421–430
10. Murray F, Brettell R, Matthews P, Bishop D, Jacobsen J (2004) Plant Cell Rep 22:397–402
11. Toda T, Koyama H, Hara T (1999) Biosci Biotechnol Biochem 63:210–212
12. Yamanaka T, Nakagawa Y, Mori K, Nakano M, Imamura T, Kataoka H, Terashima A, Iida K, Kojima I, Katagiri T, Shinozaki K, Iida H (2010) Plant Physiol 152:1284–1296
13. Arrivault S, Senger T, Krämer U (2006) Plant J 46:861–879
14. Koslowsky S, Riegler H, Bergmüller E, Zrenner R (2008) Plant Biotechnol J 6:281–294
15. Holme I, Brinch-Pedersen H, Lange M, Holm P (2006) Plant Cell Rep 25:1325–1335
16. Miyamoto N, Steudle E, Hirasawa T, Lafitte R (2001) J Exp Bot 52:1835–1846
17. Martínez-Atienza J, Jiang X, Garciadeblas B, Mendoza I, Zhu J-K, Pardo JM, Quintero FJ (2007) Plant Physiol 143:1001–1012

Chapter 3

Protocols for Growing Plant Symbioses; Mycorrhiza

Michael Schultze

Abstract

Arbuscular mycorrhizal symbiosis is receiving increased attention as a potential contributor to sustainable crop plant nutrition. This chapter details a set of protocols for plant growth to study the development and physiology of the arbuscular mycorrhizal symbiosis, and how to establish root organ cultures for the production of axenic inoculum.

Key words: Arbuscular mycorrhiza, Medicago, Glomeromycota, Glomus, Root organ culture, Hairy roots

1. Introduction

Arbuscular mycorrhizal (AM) symbiosis has the potential to contribute to crop productivity and hence global food security (1). This chapter describes a few basic protocols that should enable the beginner to establish AM symbiosis in the laboratory, and thus produce material for molecular biological studies, genetic analysis of plant mutants, testing nutrient exchange, etc. Growth conditions are given for the model plant *Medicago truncatula*. As a legume, *Medicago* has been used to study mycorrhizal symbiosis in relation to nitrogen-fixing plant-*Rhizobium* interactions. Since key regulatory pathways are shared between the two symbioses (2) a legume plant such as *Medicago* will give a good laboratory model for the analysis of structural and functional aspects of the mycorrhizal symbiosis. The protocols described here are likely adaptable to other plant species that undergo AM symbiosis.

Procedures are described for setting up AM symbiosis in the glass house and in vitro in root organ culture. Root organ culture has been shown to reflect fairly accurately many aspects of the symbiosis seen in intact plants (3). Thus this monoxenic system has been an invaluable tool for the study of AM interactions as it

Frans J.M. Maathuis (ed.), *Plant Mineral Nutrients: Methods and Protocols*, Methods in Molecular Biology, vol. 953,
DOI 10.1007/978-1-62703-152-3_3, © Springer Science+Business Media, LLC 2013

minimizes the number of confounding variables and provides defined conditions for studying, for example, nutrient flow or the effect of chemical signals on gene expression.

2. Materials

2.1. Modified R, MW and M Media (Table 1) (See Note 1)

1. To prepare stock solutions dissolve salts in 1 L of deionized water and keep in the refrigerator.

20× R major salts	g/L
$MgSO_4 \cdot 7H_2O$	5
KCl	1.5
KH_2PO_4	0.3
$Ca(NO_3)_2 \cdot 4H_2O$	9.5

20× MW major salts	g/L
$MgSO_4 \cdot 7H_2O$	15
$Na_2SO_4 \cdot 10H_2O$	9
KNO_3	1.6
KCl	1.3
KH_2PO_4	0.4
$Ca(NO_3)_2 \cdot 4H_2O$	6

20× M major salts	g/L
$MgSO_4 \cdot 7H_2O$	15
KNO_3	1.6
KCl	1.3
KH_2PO_4	0.1
$Ca(NO_3)_2 \cdot 4H_2O$	6

100× M minor salts	mg/L
NaFeEDTA	800
KI	10
$MnCl_2 \cdot 4H_2O$	40
$ZnSO_4 \cdot 7H_2O$	30
H_3BO_3	100
$CuSO_4 \cdot 5H_2O$	10
$Na_2MoO_4 \cdot 2H_2O$	2

200× M vitamins	mg/100 ml
Glycine	60
Thiamine·HCl	2
Pyridoxin·HCl	2
Nicotinic acid	10
Myo inositol	1,000

Table 1
Final concentrations in modified M (MW[a]) medium

	mg/L		μM	
$MgSO_4 \cdot 7H_2O$	750		3,043	
$Na_2SO_4 \cdot 10H_2O$	–	(450)	–	(1,397)
KNO_3	80		791	
KCl	65		872	
KH_2PO_4	5	(20)	37	(147)
$Ca(NO_3)_2 \cdot 4H_2O$	300		1,271	
NaFeEDTA	8		22	
KI	0.1		0.6	
$MnCl_2 \cdot 4H_2O$	0.4		2.0	
$ZnSO_4 \cdot 7H_2O$	0.3		1.0	
H_3BO_3	1		16	
$CuSO_4 \cdot 5H_2O$	0.1		0.4	
$Na_2MoO_4 \cdot 2H_2O$	0.02		0.08	
Glycine	3		40	
Thiamine·HCl	0.1		0.3	
Pyridoxine·HCl	0.1		0.5	
Nicotinic acid	0.5		4.0	
Myo inositol	50		277	
Sucrose	10,000	(30,000)		
Phytagel	3,000	–		
Difco Bacto Agar	–	(3,000)		
pH	5.5	5.5		

[a]For modified MW medium only deviations from M are shown

2. Modified Rorison (R) nutrient solution (modified from ref. 4). To 940 ml deionized water, add 50 ml 20× R major salts and 10 ml 100× M minor salts. Mix and adjust the pH to 6.5 with 0.1 M KOH (see Notes 2 and 3).
3. M medium (M + Suc). To 935 ml deionized water, add 50 ml 20× M major salts, 10 ml 100× M minor salts, 5 ml 200× M vitamins. Mix and adjust the pH to 5.5 with 0.1 M KOH. Add 10 g sucrose. With rapid stirring add 3 g Phytagel™ (Sigma P8169) (see Note 4). Autoclave at 121°C for 15 min. Let cool to 60°C before pouring plates (see Note 5).
4. M medium without sucrose (M – Suc). As above but omit sucrose.
5. MW medium (MW + Suc). To 935 ml deionized water, add 50 ml 20× MW major salts, 10 ml 100× M minor salts, 5 ml 200× M vitamins. Mix and adjust the pH to 5.5 with 0.1 M KOH. Add 30 g sucrose. With rapid stirring add 3 g Phytagel™ (Sigma P8169). Autoclave at 121°C for 15 min. Let cool to 60°C before pouring plates.
6. MW medium without sucrose (MW – Suc). As above but omit sucrose.
7. MW medium, carbenicillin. Dissolve 5 g carbenicillin in deionized water and fill to 10 ml. Filter-sterilize and store in 1 ml aliquots at –20°C. Before casting the medium, add carbenicillin to a final concentration of 500 mg/L.
8. LB medium: 10 g tryptone, 5 g yeast extract, 10 g NaCl, pH 7 with NaOH.
9. Water agar. Add 10 g Bacto agar to 1 L of deionized water. Autoclave.
10. Carbenicillin: 500 mg/mL. Filter-sterilize. Store at –20°C.
11. Citrate buffer for solubilization of phytagel. Add 10 mM citric acid solution to 10 mM sodium citrate until the pH is 6.0. The volume ratio is about 1:7. Filter-sterilize.
12. Sand/Terragreen substrate for growing plants in pot culture. Measure out 4 volumes coarse sand and 5 volumes calcined clay Terragreen™ (Oil-Dri Corporation) (see Note 6). Wash both substrates with plenty of tap water to remove soluble nutrients. Thoroughly mix the sand, Terragreen and 1 volume of commercial arbuscular mycorrhizal inoculum, such as Endorize granular AMF inoculum (http://www.agrauxine.com) (see Note 7). Fill the mix into compartmentalized seed trays (e.g., P40, Desch Plantpak, 50 ml per compartment). For one P40 tray 2 L of substrate mix are needed. Saturate with tap water before sowing.
13. Whatman filter paper no. 2, cut to 1 × 3 cm strips folded in the middle. Autoclave.

14. Two-compartment Petri plates (9 × 1.6 cm).
15. PVC food wrap (see Notes 8 and 9). To seal Petri plates individually, cut a roll of food wrap into 1.5 cm slices, using a sharp knife or scalpel. Do not use more than two layers around the edge of the plate.
16. Sieves of various mesh sizes (e.g., http://www.endecotts.com).

3. Methods

3.1. Establishing Mycorrhizal Symbiosis in Medicago truncatula Under Glass House Conditions

1. Scarify seeds gently with fine sand paper (see Note 10). Place the seeds in Petri plates on a double layer of wet kitchen towel (see Note 11). Leave in the dark at room temperature for 1 day.
2. Using a blue micropipette tip, make a 0.5 cm-deep planting hole in each compartment of a P40 tray filled with Sand/Terragreen/AMF inoculum. With fine forceps plant one germinated seed per compartment. Close the hole with gentle pressure (see Notes 12 and 13).
3. Grow plants in the glasshouse or growth cabinet at sufficiently high light intensity for 4–6 weeks (see Note 14). Water with deionized water as necessary. Avoid keeping the substrate too wet.
4. When plants have developed one or two trifoliate leaves (about 2 weeks after sowing), water once a week with nutrient solution containing a reduced concentration of phosphate (e.g., R solution) (see Note 15).

3.2. Testing the Level of Colonization

1. A day or so before testing the plants reduce water supply to facilitate the removal of the plants from the container. Separate the strips of the P40 trays, incise the rims of each compartment with a pair of scissors. Carefully remove the plant, wash in a large vessel with tap water, and place the plants in containers with fresh tap water (see Note 16).
2. Cut small sections of different parts of the root system and place into vials containing 10% KOH for clearing (see Notes 17 and 18).
3. Root samples may be left for several days in 10% KOH and clearing may then suffice, depending on the plant species and thickness of the root. To accelerate clearing, heat the vials up to 90°C for a variable length of time. Optimal temperature and duration need to be found for each plant species and for a given species grown under different conditions. Young roots of *M. truncatula* are cleared sufficiently after 3–4 days at room temperature, or 20 min at 90°C.

4. Remove KOH and wash root samples three times with deionized water.
5. Add staining solution (5% pen ink, 5% acetic acid) and shake gently for 10–20 min (see refs. 5, 6) (see Note 19).
6. Remove staining solution, rinse with 1% acetic acid, and then shake in 1% acetic acid for 20 min (see Note 20).
7. Roots may be inspected at low magnification in Petri dishes, or at high magnification mounted in glycerol on microscope slides. Percent root length colonization can be estimated as described (7, 8).

3.3. Isolation of Fungal Spores from Soil

Modified from refs. (9, 10).

1. To 100 ml of soil add 1 L of water (see Notes 21 and 22). Stir for 10 min using a magnetic stirrer.
2. Leave for a few seconds to allow heavier particles to sediment.
3. Decant the suspension onto a stack of sieves with the top one 1–2 mm, the bottom about 40 μm mesh size, e.g., 1,000, 500, 250, 100, 40 μm (see Note 23).
4. Wash with plenty of water.
5. Recover spores of different sizes (large, medium, small) from the last three sieves: With the help of a wash bottle collect solid material with spores in a beaker. Distribute to 2 or more 50 ml centrifuge tubes.
6. Centrifuge at 1,000 × *g* for 4 min.
7. Decant the supernatant.
8. Add 30 ml of a 45% (w/v) sucrose solution and centrifuge for 2 min.
9. Immediately pour the supernatant onto a 40 μm sieve and wash with plenty of water (see Note 24).
10. Wash spores from sieve into a Petri dish and examine under a dissecting microscope.

3.4. Multiplication of Crude Inoculum in Pot Culture

1. Collect spores of similar morphology and inoculate young leek (*Allium porrum*) seedlings grown on washed Terragreen/sand. Take up five to ten spores with a pipette and inoculate by dipping the pipette tip 2–3 cm into the substrate next to the leek seedling.
2. Grow the pot culture for about 6 months.
3. Use the pot culture to isolate spores following the wet sieving protocol above. Alternatively, collect colonized root segments as inoculum.

3.5. Surface Sterilization of Fungal Inoculum

Modified from refs. (3, 11).

1. Pipette 20 ml of 2% chloramine T solution into a 50 ml polyethylene centrifuge tube. Add two drops of Tween 20. Add spores and incubate at 4°C with gentle shaking for 15 min.
2. Centrifuge at 2,000 × *g* for 1 min and remove supernatant by decanting or aspiration. Repeat the incubation with fresh chloramine T solution.
3. Wash three times in cold sterile deionized water.
4. Add 20 ml of a solution of 0.02% streptomycin sulfate and 0.01% gentamicin sulfate. Shake gently for at least 20 min. Centrifuge and wash three times with sterile deionized water.
5. Spread spores onto plates with 3% phytagel, 0.1% $MgSO_4 \cdot 7H_2O$ (about 50–100 spores per plate). Store at 4°C.

3.6. Carrot Root Organ Culture

Modified from refs. (11, 12).

1. Wash and peel a carrot. Cut 5 mm slices. Soak for 10 s in 95% ethanol, then treat with 1/4 strength commercial bleach for 10 min. Wash with sterile deionized water and place upside down (distal side up) on a 1% water agar plate.
2. Using an inoculation loop, streak freshly grown *A. rhizogenes* onto the carrot slices.
3. Incubate in the dark at 28°C for about 3 weeks.
4. Under a laminar flow hood remove roots emanating from the carrot slices and transfer to Petri plates with modified MW medium, containing a suitable antibiotic, e.g., 500 mg/L carbenicillin.
5. Subculture root pieces every 3 weeks in the presence of antibiotic until no more bacterial growth is observed when samples of root pieces are transferred to antibiotic-free MW medium. At least two subculture events will be required.
6. Maintain the root culture by transferring two root segments including their apexes to fresh plates every 4–6 weeks.

3.7. Producing AMF Inoculum in Root Organ Culture

Modified from refs. (11–15).

1. Fill one half of a two-compartment Petri plate with 20 ml M medium, the other half with M medium lacking sucrose. Before the medium solidifies place a Whatman filter strip with the fold over the central ridge (see Note 25).
2. Remove a 4 cm root piece (with apex) from a carrot root culture and transfer it to the M + Suc side.
3. Place five to ten surface-sterilized spores near the apex of the root fragment. Seal plates with plastic film, either individually or as a stack.

4. Incubate the plates in the dark at 24–28°C for 3 months until hyphae have covered the distal compartment containing M – Suc medium (see Note 26). Depending on the fungal species, up to 30,000 spores may have formed in this hyphal compartment (see Note 27). Regularly inspect the plates and trim off roots that have grown into the hyphal compartment.
5. To maintain and multiply the cultures, cut a piece of colonized root with a scalpel or pair of scissors and transfer it to a new two-compartment plate on the M + Suc side. Place an actively growing apex of a transformed carrot root in the vicinity (see Notes 28 and 29).

3.8. Harvesting Spores from Root Organ Culture

After ref. (16).

1. Remove gel from the hyphal compartment of the Petri plate and distribute it to two 50 ml centrifuge tubes. Store at 4°C until use.
2. Add 40 ml citrate buffer pH 6 and shake gently until the phytagel is dissolved (see Note 30).
3. Centrifuge at 2,000 × *g*, decant supernatant, wash twice with sterile deionized water and resuspend spores in an appropriate volume of deionized water. Vortex vigorously to separate clusters of spores. Determine the number of spores per ml using a disposable counting chamber.

3.9. Medicago Root Organ Culture

Modified from ref. (17)

1. Scarify seeds gently with fine sand paper.
2. Gently shake in household bleach for 10 min. Wash five times with autoclaved tap water.
3. Place the seeds in Petri plates on a double layer of wet sterilized kitchen towel. Leave in the dark at room temperature for 1–2 days.
4. From germinated seedlings, cut off the root tip with a scalpel. Dip the cut end of the seedling into a lawn of *Agrobacterium rhizogenes* grown on LB agar plates (see Note 31).
5. Place seedlings on a Petri plate containing MW medium without sucrose. Incubate plates at an angle of 45° in a growth room at 20°C for 2–3 weeks. Transformed "hairy" roots appear earliest 1 week after inoculation directly at the section, whereas normal roots grow out earlier above the section.
6. Excise hairy roots from the young seedlings and transfer to MW medium (with sucrose), supplemented with 400 mg/L augmentin. Incubate in the dark for 2–3 weeks and repeat the subculturing twice. Maintain the root culture by transferring two root segments including their apexes to fresh plates on MW medium.

7. To study symbiotic interactions with AM fungi, transfer contaminant-free root pieces (4 cm long with apex) to a two-compartment Petri dish on M + Suc medium (see Note 32).
8. Place a defined number of spores from carrot root organ culture in the vicinity of the root apex.
9. Inspect development of the hyphae at regular intervals. Compare roots derived from different genotypes or transformed with different constructs.

4. Notes

1. The composition of MW and M medium is based on ref. (12), with modifications to the microelements as reported by refs. (14, 18, 19). A further tenfold reduction in Mn^{2+} may be needed to alleviate inhibition of spore germination for some species (see ref. 20).
2. R medium: When used as fertilizer in open pot culture, there is no need to autoclave.
3. When the growth substrate contains phosphorus, KH_2PO_4 may be reduced or omitted.
4. Phytagel™ is a trademark of gellan gum, a polysaccharide gelling agent that forms a gel at low concentrations in the presence of cations. Other brands, such as Gel-Gro™ (MP Biomedicals 150180) have been used as well.
5. Some authors have added buffers, e.g., 10 mM MES pH 6.5, to stabilize the pH during the first few weeks of culture (e.g., see refs. 19, 21).
6. Similar products, such as Turface can be used as well.
7. Agrauxine sells several AMF species in the form of granular inoculum, but the cheapest option is "Endorize mix", which has worked very well, for example, in large scale screening projects to identify mutants defective in AM symbiosis.
8. For plant tissue culture applications, PVC food wrap has been shown to be superior to Parafilm® M, a sealing film commonly used in laboratories (22). Parafilm was reported to release toxic compounds (23). Compared to PVC, Parafilm also has a much lower permeability to gases, such as CO_2 and ethylene (24).
9. Non-PVC food wrap made of low density polyethylene (LDPE) allows even better gas exchange than PVC, but is much less self adhesive and therefore cannot be used in stripes to seal the edge of Petri plates. However, when wrapping a stack of plates together, this type of food wrap may be the best option.

10. A few scratches on the seed coat will be sufficient to facilitate imbibition of the seeds.
11. Fold two sheets of kitchen towel. Using a pencil, draw a circle around a Petri plate placed on the towel. Cut out a disc slightly smaller than the diameter of the plate. Place the disc inside the plate, add tap water to moisten the paper and decant surplus water.
12. Depending on the kind of analysis to be performed the size of the compartments can be changed. Growth in individual pots is also possible. If the symbiotic proficiency of individual plants is to be studied, e.g., in mutant screens, it is essential to grow no more than one plant in a compartment. AM fungi emanating from wild-type or nurse plants may successfully colonize mutant plants (see refs. 25, 26), hence making the mutants appear as wild types.
13. The protocol described for *Medicago* is likely to work with other plant species. However, compartment size and nutrient regimes might need modification.
14. Standard glasshouse conditions with supplemental light are sufficient in most cases.
15. For different plant species the ratios of nutrients may be altered to achieve optimal growth. Other nutrient solutions may be used as long as phosphate concentrations are low. In case plants grow slowly there is a risk of phosphate accumulation and hence suppression of colonization.
16. Plants can be left in water for a few hours before and after testing and collection of root samples.
17. If plants are to be repotted to produce seed, it is essential to sample only a proportion of the root system.
18. Clearing will be slow at room temperature. This will give the opportunity to process the samples at a later stage if necessary.
19. For a detailed review of the different staining techniques used to visualize AMF colonization, see ref. (6).
20. Destaining works best if the volume is sufficiently large. It is convenient to process the samples in 50 ml polypropylene tubes and decant the solutions.
21. The optimal volumes may depend on the soil type.
22. To facilitate isolation of spores and colonized root pieces, it may be helpful to plant leek at the collection site and then harvest the leek.
23. You may try different sizes that are available in the laboratory.
24. Make sure that spores are in contact with the sucrose solution for as short a time as possible to avoid osmotic damage.

25. The filter bridge (15) provides a permanent liquid film connecting the two compartments and therefore facilitates migration of hyphae from the root compartment to the hyphal compartment.
26. Hairy root carrot cultures may display negative geotropism. Therefore, plates may be incubated upside down to facilitate contact of roots with the medium. Likewise, negative geotropism is displayed by some AM fungi, such as *Gigaspora* (see ref. 27). Therefore, to maximize colonization success, plates may be placed vertically or at an angle.
27. Up to 75,000 spores of *Glomus intraradices* were obtained per plate, when after 3 months glucose was added to the root compartment and M medium minerals replenished in the hyphal compartment, followed by incubation for a further 2 months (28). This way spore inoculum can be increased substantially in a period of time that is shorter compared to starting a fresh coculture. It is also possible to harvest the entire hyphal compartment after 3 months, add glucose to the root compartment and replace the M medium in the hyphal compartment. Two further harvests 5 and 7 months after the start of the culture can be obtained (28).
28. Alternatively, using a cork borer a gel plug, about 1 cm in diameter, can be removed from a hyphal compartment of a 3 month old culture and inserted into the root compartment of a new plate from which a plug of the same size has been cut out. Again, place an actively growing apex of a transformed carrot root in the vicinity of the plug.
29. The monoxenic root organ culture may be propagated over several generations by adding colonized root pieces of a previous culture to uncolonized carrot roots. Depending on the fungal species, infectivity may be gradually lost with each generation (29), while in other cases infectivity was maintained for 30 generations (30). To cope with the risk of loss of infectivity and to make sure that viable fungal inoculum is available for experiments at all times, it is wise to start new monoxenic cultures at regular intervals, using spores or root segments from pot cultures.
30. Less time is needed when the mixture is shaken at 30°C rather than at room temperature (16).
31. The same procedure for generating *Medicago* hairy roots can be used to introduce transgenes, e.g., for demonstrating gene function after reintroduction of a wild-type copy into mutant background, or gene silencing constructs to study the effect on mycorrhizal symbiosis.
32. Nutrient transport studies can be carried out using this system. In this case the hyphal compartment is filled with liquid

medium (31) and a paper bridge is not used. Once hyphae have covered the entire distal compartment, the medium is carefully removed and replaced with fresh solution with appropriate nutrient composition, with the possibility to include labeled compounds.

References

1. Smith FA, Smith SE (2011) What is the significance of the arbuscular mycorrhizal colonisation of many economically important crop plants? Plant Soil 348:63–79
2. Ercolin F, Reinhardt D (2011) Successful joint ventures of plants: arbuscular mycorrhiza and beyond. Trends Plant Sci 16:356–362
3. Fortin JA, Bécard G, Declerck S, Dalpé Y, St-Arnaud M, Coughlan AP (2002) Arbuscular mycorrhiza on root-organ cultures. Can J Bot 80:1–20
4. Booth RE, Mackey JML, Rorison IH, Spencer RE, Gupta PL, Hunt R (1993) ISP germination and rooting environments: sand, compost and solution culture. In: Hendry GAF, Grime JP (eds) Methods in comparative plant ecology: a laboratory manual. Chapman Hall, London, pp 19–23
5. Vierheilig H, Coughlan AP, Wyss U, Piché Y (1998) Ink and vinegar, a simple staining technique for arbuscular-mycorrhizal fungi. Appl Environ Microbiol 64:5004–5007
6. Vierheilig H, Schweiger P, Brundrett M (2005) An overview of methods for the detection and observation of arbuscular mycorrhizal fungi in roots. Physiol Plant 125:393–404
7. Giovannetti M, Mosse B (1980) Evaluation of techniques for measuring vesicular arbuscular mycorrhizal infection in roots. New Phytol 84:489–500
8. McGonigle TP, Miller MH, Evans DG, Fairchild GL, Swan JA (1990) A new method which gives an objective-measure of colonization of roots by vesicular-arbuscular mycorrhizal fungi. New Phytol 115:495–501
9. Gerdemann JW, Nicolson TH (1963) Spores of mycorrhizal *Endogone* species extracted from soil by wet sieving and decanting. Trans Br Mycol Soc 46:235–244
10. McKenney MC, Lindsey DL (1987) Improved method for quantifying endomycorrhizal fungi spores from soil. Mycologia 79:779–782
11. Bécard G, Piché Y (1992) Establishment of vesicular-arbuscular mycorrhiza in root organ-culture: review and proposed methodology. Methods Microbiol 24:89–108
12. Bécard G, Fortin JA (1988) Early events of vesicular arbuscular mycorrhiza formation on Ri T-DNA transformed roots. New Phytol 108:211–218
13. St-Arnaud M, Hamel C, Vimard B, Caron M, Fortin JA (1996) Enhanced hyphal growth and spore production of the arbuscular mycorrhizal fungus *Glomus intraradices* in an in vitro system in the absence of host roots. Mycol Res 100:328–332
14. Declerck S, Strullu DG, Plenchette C (1996) In vitro mass-production of the arbuscular mycorrhizal fungus, *Glomus versiforme*, associated with Ri T-DNA transformed carrot roots. Mycol Res 100:1237–1242
15. Dalpé Y, Seguin S (2010) A "paper bridge" system to improve in-vitro propagation of arbuscular mycorrhizal fungi. Botany 88: 617–620
16. Doner LW, Bécard G (1991) Solubilization of gellan gels by chelation of cations. Biotechnol Tech 5:25–28
17. Boisson-Dernier A, Chabaud M, Garcia F, Bécard G, Rosenberg C, Barker DG (2001) *Agrobacterium rhizogenes*-transformed roots of *Medicago truncatula* for the study of nitrogen-fixing and endomycorrhizal symbiotic associations. Mol Plant Microbe Interact 14: 695–700
18. Mugnier J, Mosse B (1987) Vesicular-arbuscular mycorrhizal infection in transformed root-inducing T-DNA roots grown axenically. Phytopathology 77:1045–1050
19. Karandashov V, Kuzovkina I, Hawkins HJ, George E (2000) Growth and sporulation of the arbuscular mycorrhizal fungus *Glomus caledonium* in dual culture with transformed carrot roots. Mycorrhiza 10:23–28
20. Hepper CM (1979) Germination and growth of *Glomus caledonius* spores: the effects of inhibitors and nutrients. Soil Biol Biochem 11:269–277
21. Hildebrandt U, Janetta K, Bothe H (2002) Towards growth of arbuscular mycorrhizal fungi independent of a plant host. Appl Environ Microbiol 68:1919–1924

22. Ribeiro APO, Picoli EAT, Lani ERG, Vendrame WA, Otoni WC (2009) The influence of flask sealing on in vitro morphogenesis of eggplant (*Solanum melongena* L.). Vitro Cell Dev Biol Plant 45:421–428
23. Selby C, McRoberts WC, Hamilton JTG, Harvey BMR (1996) The influence of culture vessel head-space volatiles on somatic embryo maturation in Sitka spruce (*Picea sitchensis* (Bong.) Carr.). Plant Growth Regul 20: 37–42
24. Marino G, Berardi G (2004) Different sealing materials for Petri dishes strongly affect shoot regeneration and development from leaf explants of quince 'BA 29'. Vitro Cell Dev Biol Plant 40:384–388
25. David-Schwartz R, Badani H, Smadar W, Levy AA, Galili G, Kapulnik Y (2001) Identification of a novel genetically controlled step in mycorrhizal colonization: plant resistance to infection by fungal spores but not extra-radical hyphae. Plant J 27:561–569
26. Murray JD, Muni RRD, Torres-Jerez I, Tang YH, Allen S, Andriankaja M, Li GM, Laxmi A, Cheng XF, Wen JQ, Vaughan D, Schultze M, Sun J, Charpentier M, Oldroyd G, Tadege M, Ratet P, Mysore KS, Chen RJ, Udvardi MK (2011) Vapyrin, a gene essential for intracellular progression of arbuscular mycorrhizal symbiosis, is also essential for infection by rhizobia in the nodule symbiosis of *Medicago truncatula*. Plant J 65:244–252
27. Watrud LS, Heithaus JJ, Jaworski EG (1978) Geotropism in endomycorrhizal fungus *Gigaspora margarita*. Mycologia 70:449–452
28. Douds DD (2002) Increased spore production by *Glomus intraradices* in the split-plate monoxenic culture system by repeated harvest, gel replacement, and resupply of glucose to the mycorrhiza. Mycorrhiza 12:163–167
29. Plenchette C, Declerck S, Diop TA, Strullu DG (1996) Infectivity of monoaxenic subcultures of the arbuscular mycorrhizal fungus *Glomus versiforme* associated with Ri-T-DNA-transformed carrot root. Appl Microbiol Biotechnol 46:545–548
30. Vimard B, St-Arnaud M, Furlan V, Fortin JA (1999) Colonization potential of in vitro-produced arbuscular mycorrhizal fungus spores compared with a root-segment inoculum from open pot culture. Mycorrhiza 8:335–338
31. Maldonado-Mendoza IE, Dewbre GR, Harrison MJ (2001) A phosphate transporter gene from the extra-radical mycelium of an arbuscular mycorrhizal fungus *Glomus intraradices* is regulated in response to phosphate in the environment. Mol Plant Microbe Interact 14:1140–1148

Chapter 4

Protocols for Growing Plant Symbioses; Rhizobia

Benjamin Gourion, Marie Bourcy, Viviane Cosson, and Pascal Ratet

Abstract

Legume plants are used as a protein source for human and animal nutrition. The high protein content of legume plants is achieved via the establishment of a root symbiosis with rhizobia that allows the reduction of atmospheric nitrogen. In recent years, *M. truncatula* has been used as a legume model in view of its small, diploid genome, self-fertility, and short life cycle, as well as availability of various genomic and genetic tools. The choice and use of this model legume plant in parallel with the other model legume *Lotus japonicus* for molecular studies has triggered extensive studies that have now identified the molecular actors corresponding to the first steps of the plant–bacterial interaction. The use of this plant as model in an increasing number of laboratories has resulted in the development of numerous protocols to study the establishment of the symbiosis. The media and growth conditions used in our laboratory to nodulate wild-type or transgenic plants as well as wild-type plants with transgenic hairy root system are described below.

Key words: *Medicago*, Legumes, *Rhizobium*, Symbiosis, In vitro culture, Greenhouse culture, Hairy root

1. Introduction

Different laboratories have developed several protocols or techniques for studying the legume–*Rhizobium* interaction and more particularly for *Medicago truncatula* inoculation using *Sinorhizobium meliloti*. Inoculation of *M. truncatula* roots with *S. meliloti* bacteria results in the formation of nitrogen-fixing root nodules. This nodulation is negatively regulated by the presence of combined nitrogen in the growth medium. Therefore, nodulation assays have to be performed in nitrogen-free or nitrogen-limiting conditions. Plants can be inoculated when grown in inert substrates (sand, perlite, vermiculite) in the green house, in aeroponic conditions, in semi-sterile conditions in pouches or in vitro in tubes or square Petri dishes using agar-based media ((1); http://www.isv.cnrs-gif.fr/embo01/manuels/pdf/module1.pdf). In contrast to alfalfa (*Medicago sativa*), *M. truncatula* is sensitive to the agar type

Frans J.M. Maathuis (ed.), *Plant Mineral Nutrients: Methods and Protocols*, Methods in Molecular Biology, vol. 953,
DOI 10.1007/978-1-62703-152-3_4,

used for in vitro experiments (2). Low efficiency of nodulation was reported in early experiments with *M. truncatula* grown on agar media. In addition, exposure of the root system to light had a detrimental effect on nodule development. These problems are overcome by the use of special brands of agar and by addition of the ethylene biosynthesis inhibitor AVG to agar plates. For greenhouse experiments, any substrate that allows good aeration of the root system can be used. We found that perlite, or a mixture of river sand with perlite, is a convenient substrate, as it can be easily washed off for analysis of the root system. Although nodulation is generally quite robust and can be obtained upon inoculation with a wide range of initial bacterial densities, we describe here some of the commonly used conditions in which successful nodulation for laboratory studies can be obtained.

2. Materials

2.1. Media

1. Nitrogen-rich nutrient solution for greenhouse culture.
   This nitrogen-rich solution is used when growing plants in soil without symbiotic conditions and is suitable for *Medicago* and pea plants. We also use it to initiate growth of the plants when they are later grown with nitrogen-limited solutions. The nitrogen-rich solution is prepared using the N.P.K. 18.6.26 Soluplant fertilizer (see Note 10) composed of 18%N (12% nitric and 6% ammonium), 6% P (phophoranhydride), 26%K (potassiumoxide) and contains 2% MgO as well as oligo elements (0.025% B, 0.005% Cu, 0.06% Fe, 0.025% Mn, 0.0025% Mo, 0.005% Zn). Other commercial fertilizers can be used if they have a similar NPK composition (see Note 1).

   To prepare 1 L of the nitrogen-rich solution add to 987 ml of tap water, 0.8 g of N.P.K. 18.6.26 fertilizer and 12.5 ml of nitric acid. Mix until complete dissolved. This solution is autoclaved before use.
2. Nitrogen-limited solution I.
   This nitrogen-limited medium can be used for nodulation assays both in aeroponic tanks and in combination with perlite, vermiculite, or sand with manual irrigation. Nodulation assays using sand, perlite, or vermiculite can be performed in the greenhouse. This medium is prepared from stock solutions as described in Tables 1 and 2.

   The stock solution (40×) for the nitrogen-limiting solution is prepared by mixing the A-B-C components in the order A1 to C in deionized water as indicated in Table 3.

   The nitrogen-limiting solution (0.125 mM NO_3) is prepared by dissolving 25 ml of the above 40× stock solution

Table 1
Components for the nitrogen-limiting stock solution (see Note 2 for NaFeEDTA). Dissolve each component (A1-A5-B) separately in deionized water to 1 L final volume

A–B components for the stock solution

Component	Chemicals	Amount for 1 L	Final concentration in each component
A1	KNO_3	20.2 g	200 mM
A2	KH_2PO_4	27.2 g	200 mM
A3	$CaCl_2 \cdot 2H_2O$	73.0 g	495 mM
A4	$MgSO_4 \cdot 7H_2O$	24.6 g	100 mM
A5	K_2SO_4	43.5 g	250 mM
B	NaFeEDTA	8.2 g	22 mM

Table 2
Composition of the C component for the nitrogen-limiting stock solution. Dissolve the minerals in the given order in deionized water to 1 L of final volume. Heating at 70°C increases solubility

C components for the stock solution

Chemicals	Amount for 1 L	Final concentration in component C
H_3BO_3	11 g	178 mM
$MnSO_4 \cdot H_2O$	6.2 g	36.7 mM
$ZnSO_4 \cdot 7H_2O$	1 g	3.5 mM
$Na_2MoO_4 \cdot 2H_2O$	1 g	4 mM
$CuSO_4 \cdot 5H_2O$	0.5 g	2 mM
H_2SO_4 (density 1.83)	0.5 ml	–

(Table 3) in 1 L of water. This solution is autoclaved before use in the symbiotic tests. The final concentrations of this solution are given in Table 4.

3. BNM (Buffered Nodulation Medium; (1)) solution used for in vitro symbiotic tests.
   This BNM solution does not contain any nitrogen for the plant and is relatively easy to prepare. In addition it allows efficient and rapid nodulation in vitro.

Table 3
Preparation of the 40× nitrogen-poor stock solution

Stock solution (40×)

Components	Amount for 1 L
A1	25 ml
A2	25 ml
A3	25 ml
A4	25 ml
A5	25 ml
B	25 ml
C	0.27 ml

Table 4
Composition of the nitrogen-poor solution for symbiotic tests

Nitrogen-limiting solution

Chemical names	Chemicals	Amount for 1 L	Final concentration in each component
Potassium nitrate	KNO_3	12.6 mg	0.125 mM
Di-hydrogen phosphate	KH_2PO_4	17 mg	0.125 mM
Calcium chloride	$CaCl_2 \cdot 2H_2O$	45.6 mg	0.310 mM
Magnesium sulfate	$MgSO_4 \cdot 7H_2O$	15.3 mg	0.062 mM
Potassium sulfate	K_2SO_4	27.2 mg	0.156 mM
Ferric complex of EDTA	$C_{10}H_{12}N_2O_8$ FeNa	5.12 mg	0.014 mM
Boric acid	H_3BO_3	0.075 mg	1 μM
Manganese sulfate	$MnSO_4 \cdot H_2O$	0.042 mg	0.25 μM
Zinc sulfate	$ZnSo_4 \cdot 7H_2O$	0.0067 mg	0.02 μM
Sodium molybdate	$Na_2MoO_4 \cdot 2H_2O$	0.0067 mg	0.027 μM
Copper sulfate	$CuSO_4 \cdot 5H_2O$	0.0034 mg	0.0135 μM
Sulfuric acid (density 1.83)	H_2SO_4	0.0034 μL	–

The preparation of the BNM stock solutions is described Tables 5, 6, 7 and 8.

The preparation of the BNM solution is described Table 9.

Table 5
Preparation of the 200× Nod major salts stock for the BNM solution

200× BNM major salts

Chemicals	Amount for 1 L	Final concentration in the stock solution
$MgSO_4 \cdot 7H_2O$	24.6 g	100 mM
KH_2PO_4	13.6 g	100 mM
H_2O	Make up to 1 L	–

Table 6
Preparation of the 200× Nod minor salts I stock for the BNM solution

200× BNM minor salts I

Chemicals	Amount for 1 L	Final concentration in the stock solution
$ZnSO_4 \cdot 7H_2O$	920 mg	3.2 mM
H_3BO_3	620 mg	10 mM
$MnSO_4 \cdot H_2O$	1.69 g	10 mM
H_2O	Make up to 1 L	–

Table 7
Preparation of the 200× Nod minor salts II stock for the BNM solution

200× BNM minor salts II

Chemicals	Amount for 1 L	Final concentration in the stock solution
$Na_2MoO_4 \cdot 2H_2O$	50 mg	0.2 mM
$CuSO_4 \cdot 5H_2O$	5 mg	0.02 mM
$CoCl_2 \cdot 6H_2O$	5 mg	0.02 mM
H_2O	QSP: 1 L	–

Table 8
Preparation of the 200× FeEDTA stock for the BNM solution

200× BNM Fe EDTA

Chemicals	Amount for 1 L	Final concentration in the stock solution
$Na_2EDTA \cdot 2H_2O$	3.73 g	10 mM
$FeSO_4 \cdot 7H_2O$	2.78 g	10 mM
H_2O	Make up to 1 L	–

Table 9
Preparation of the BNM medium. The BNM medium can be stored in bottles for 1 month at 4°C in the dark (see Note 3). BNM plates containing AVG can be stored for 2 weeks at 4°C in the dark

Chemicals or stocks	Amount for 1 L	Final concentration
$CaSO_4 \cdot 2H_2O$	344 mg	2 mM
MES buffer	390 mg	2 mM
200× BNM major salts	5 ml	–
200× BNM minor salts I	5 ml	–
200× BNM minor salts II	5 ml	–
BNM Fe-EDTA	5 ml	–
Adjust pH with KOH to 6.5		–
Kalys Agar	10 g	1%
AVG (0.5 mM)	1 ml	0.5 μM
H_2O	Make up to 1 L	–

4. Preparation of the Aminoethoxyvinylglycine (AVG) solution. Dissolve 25 mg of AVG powder (Sigma 06665) in 5.2 ml H_2O to prepare a 25 mM stock solution. This stock solution is stored in 200 μL aliquots at −20°C. The 0.5 mM AVG solution is prepared by diluting 200 μL of a 25 mM solution aliquot in 10 ml distilled water. This stock solution is filter-sterilized and stored at −20°C. The 0.5 mM stock solution is added to the medium just before pouring the plates (see Note 3).

5. Solidifying agent (see Note 4)
   Kalys agar HP 696-7470 (Kalys, 59100 Roubaix, France; http://www.kalys.com/): used for the in vitro nodulation tests.
6. TA or YTB (3), (for *Rhizobium* culture): For 1 L, add 10 g tryptone, 1 g yeast extract, 5 g NaCl, 1 ml $CaCl_2$ (1 M stock solution), 1 ml $MgCl_2$ (1 M stock solution), pH 7.0 (adjusted with NaOH). For solid medium, add 15 g/L Bacto agar before autoclaving.
7. YEB (4), (for *Agrobacterium* culture): For 1 L, add 5 g of Bacto beef extract, 1 g of Bacto yeast extract, 5 g of Peptone, 5 g of Saccharose, 2 ml of Magnesium sulfate (1 M stock solution), pH: 7.2. For solid medium, add 15 g/L Bacto agar before autoclave.

2.2. Plant Material

1. *M. truncatula* (Gaertn.) line R108-1 (c3) is described in ref. (5) and called R108 throughout this text (see Note 5).
2. *M. truncatula* (Gaertn.) line Jemalong (J5) is described in ref. (6) and called Jemalong throughout this text (see Note 5).

2.3. Rhizobium Strains

1. *Rhizobium meliloti* Rm41 (7) nodulates efficiently and fixes nitrogen in symbiosis with *M. sativa* and *M. truncatula* R108. The Rm41 strain does not fix nitrogen (fix$^-$) in combination with *M. truncatula* Jemalong.
2. *Rhizobium meliloti* Rm1021 (8) nodulates and fixes nitrogen in symbiosis with *M. sativa* and *M. truncatula* R108 and Jemalong. The Rm1021 strain is, however, known to be poorly efficient for nitrogen fixation in combination with *M. truncatula* Jemalong (9). Our experience indicates that it is also poorly efficient in combination with R108.

2.4. Agrobacterium Strains and T-DNA Vectors

1. *Agrobacterium rhizogenes* A4Tc24 strain (10) and Arqua 1 (11) are recommended for R108 transformation experiments (see Note 6). Conventional plant transformation binary vectors (see Note 7) are introduced in the *Agrobacterium* strain by triparental mating as previously described (12) or by electroporation.

2.5. Stock Solutions

2.5.1. Antibiotics Stocks (See Note 8)

1. Streptomycin (Sigma S6501): Stock 500 mg/ml in sterile water. Final concentration for media is 500 mg/L.
2. Kanamycin (Sigma: K4000): Stock 100 mg/ml in sterile water. Final concentration is 25 mg/L for plant media and 100 mg/L for bacterial media.

2.6. Other Supplies

1. Sterile forceps and razor blades.
2. Sand paper (commercial fine sand paper for hard material).
3. Bayrochlor solution: Bayrochlor; BAYROL GMBH, Germany (http://www.bayrol.de/). The active compound in

this product is sodium dichlorisocyanurate. Bayrochlor is an industrial disinfectant. Use a 7 g/L solution for seed sterilization (see Note 9).

4. Household plastic film.
5. In vitro growth chamber lighting: We use alternating Mazdafluor Prestiflux-HF Incandia: 4A TF″P″58 W/inc and Mazdafluor Blanc Industrie 33 (6J TF58W/BI) tubes in order to have an appropriate light spectrum (see Note 11).
6. 10 ml plastic syringes.
7. 20 μm filters adaptable to syringes.
8. Sterile water.
9. Sulfuric acid 96%.
10. Na-hypochlorite solution (6, 9 or 12°Cl).
11. Washed river sand used for construction (sand grains < 5 mm) (see Notes 12 and 13).
12. Perlite (reference 885 161; Puteaux SA, France (http://www.puteaux-sa.fr/)) (see Note 12).
13. 2 L plastic pots, Ø int 14.9 cm × H 17.4 cm (conteneur SH 532302, Puteaux SA, France (http://www.puteaux-sa.fr/)).
14. Plastic trays STR 45 (45 cm × 30 cm) BHR Bouillard France (http://www.bouillard.fr/index.html).

3. Methods

3.1. Preparation of Plant Materials for Symbiotic Tests

Symbiotic tests can be performed in vitro or in the greenhouse. In both cases, growth conditions are important for the success of the symbiotic test (see Notes 14 and 15). We describe below two protocols for seed scarification and sterilization. One uses a commercial compound (Bayrochlor), the other sulfuric acid. These two methods give similar results. We have used *M. truncatula* R108 or Jemalong (J5) with equal success under the conditions described below.

3.1.1. Seed Sterilization

1. Seeds are scarified with sand paper, sterilized for 30 min in a Bayrochlor solution, rinsed five times in sterile water, and placed on sterile 1% Kalys agar (weight-volume in water) or BNM medium in 12 mm square Petri dishes at 4°C in the dark for 2–3 days (see Note 16).
2. Seeds are simultaneously scarified and sterilized for 8 min in 96% sulfuric acid. All seeds must be covered by sulfuric acid. Immediately afterwards, several washes are done with large volumes of sterile water. Seeds are then placed on solid medium in square Petri dishes at 4°C in the dark for 2–3 days as indicated above.

3.1.2. Seed Germination

The plates containing the sterile seeds are either placed at room temperature in the dark for 10–16 h or overnight in the dark at 24°C, with the agar side up in order to allow the growing root tip to develop strait down toward the lid. This allows having seedling with strait roots not attached to the medium. Germinating seedlings can be used for symbiotic tests in greenhouse (nonsterile conditions) or in vitro (sterile conditions).

3.2. Preparation of Rhizobium Culture

For the symbiotic test the choice of the bacterial strain is important (see Note 17).

1. Two days before the inoculation, start a *Rhizobium* liquid culture by inoculating a freshly grown single colony of *Rhizobium* in 2 ml TA liquid medium supplemented with appropriate antibiotics if required (see Note 18). Incubate on a shaker (200 rpm), at 30°C, overnight.
2. The day before the inoculation of plants, a 200 ml flask containing 30 ml *Rhizobium* culture medium (with appropriate antibiotics for selection if needed) is inoculated with 1 ml of the 2 ml overnight pre-culture. This culture is incubated on a shaker (200 rpm) at 30°C over night. The OD_{600nm} of the culture should reach 0.6 the day of the inoculation.
3. Centrifuge the 30 ml culture at 3,000 × *g* for 20 min and gently resuspend the pellet in 50 ml sterile water (see Note 19). Adjust the OD_{600nm} to 0.1 in sterile water (see Note 18).
4. For the greenhouse tests, 10 ml of inoculum per pot will be necessary.
5. For the in vitro tests 50 µL of inoculum per plant will be necessary.
6. Alternatively, 15 ml of the diluted *Rhizobium* culture can be added directly to the plate containing the germinating seedlings before they are transferred to the BNM AVG plates for the symbiotic test. For this, after addition of the inoculums, plates are gently moved in order to allow contact of the growing roots with the bacterial culture. The seedlings are left in contact with the diluted bacterial culture for 1 h before transferring them to the test plates (see Note 20).

3.3. Greenhouse Symbiotic Tests

1. Prepare a Perlite-sand (3/1 v/v) mixture and moisten it with sterile water in order to avoid the dispersal of Perlite dust and to avoid that young seedlings dry out during planting.
2. The plastic pots are filled with the mixture of Perlite-sand up to 5 cm from the upper edge of the pot.
3. 10–12 germinated seeds are placed on the surface of the substrate and covered by 1–2 cm of the same Perlite-sand mixture and watered immediately with sterile water.

4. These pots are placed in the greenhouse (16 h day period, 60% relative humidity, 16°C at night and 22°C daytime temperature with additional light: 200 µE/m^2/s). The pots are placed on plastic trays (6 pots per tray) that are filled with sterile water. The trays avoid draining of the nutrient solution and washing out of the substrate during watering. They also avoid dispersal of the bacteria. The pots are covered with a transparent plastic cover or with plastic foil in order to keep the surface of the pot humid.
5. Once the cotyledons emerge from the substrate, the plantlets are watered once with nitrogen-rich nutrient solution (see Note 21) and the plastic foils or covers are removed.
6. From this time onwards the plants are watered alternatively with nitrogen-limited solution and water (see Notes 21 and 22).
7. Plants are inoculated with a diluted culture of *Rhizobium* (5 ml/pot) prepared as described in Subheading 3.3 as soon as the first leaf is fully expanded (see Note 18).
8. Formation of nodules can be observed after 3 weeks but we generally inspect plants 4–5 weeks post inoculation. At this stage the aerial part of the plant has normally developed three to five nodes.
9. In order to observe the root system of the nodulated plants the content of the pot is poured on a bench and the roots are gently moved to remove the excess of substrate (see Note 23). To further clean the root system, roots are dipped in a half full 10–15 L water container and again gently shaken. This procedure allows the sand and perlite to separate from the roots without damaging them.
10. The root system can then be easily examined by eye or by using a stereo microscope.

3.4. In Vitro Symbiotic Tests

1. Pour 12 cm square Petri dishes with BNM Kalys agar medium supplemented with AVG (see Note 3). Prepare enough plates for your experiments taking into account that you will place eight to ten seedlings per plate.
2. Place eight to ten germinating seedlings in a row at 3–4 cm distance from the top side of the Petri dish. The root tip of the seedlings should be placed toward the bottom side of the plate.
3. The plates are then sealed with household plastic film (see Note 24) on three sides, leaving 60–80% of the top side unsealed (see Note 25).
4. The bottom part of the plate where the roots are growing (60% of the plate) is covered with aluminum foil in order to protect the growing part of the root system from excessive light.

This means that only the part of the plant comprising the epicotyl and the aerial part is exposed to full light (see Note 26).

5. Plates are incubated vertically (or nearly vertically, see Note 27) in a plant growth room at 24°C.
6. When the roots have grown 5 cm long (this takes between one night and 30 h), each root is inoculated by placing 50 μL of the diluted bacterial solution on the root.
7. Plates are replaced in the growth chamber vertically.
8. Nodule primordia appear generally 5–8 days post inoculation if the bacterial strain is efficient. Plants can be kept for a month on this medium.
9. This growing system allows observing the development of the symbiotic organ during the experiment.
10. Optimally nitrogen fixing nodules will have developed 15 days post inoculation. The experiment is generally stopped after 4 weeks or earlier, depending on the experiment.

3.5. Nodulation of Hairy Roots Containing Chimeric Plants

The possibility to construct chimeric plants using the hairy root technology has facilitated the study of the interaction between *Rhizobium* and *Medicago*. This system is using wild-type or mutant plants with a transgenic root system induced by *Agrobacterium rhizogenes*. The system described below is derived from the protocol developed by Boisson-Dernier et al. (13). We found that the complete procedure (transformation and symbiotic tests) can be followed using the BNM medium described above. The kanamycin selection for transformed roots described in ref. (13) is working on this medium.

1. Seeds are sterilized by one of the two methods described in Subheading 3.1.1 above and placed for 2–3 days at 4°C in the dark on BNM medium. 1 day before the transformation plates are placed at 24°C overnight for germination as described in Subheading 3.2.
2. Three days before the transformation experiment, start an *Agrobacterium* liquid culture by inoculating a freshly grown single colony in 2 ml YEB liquid medium supplemented with the appropriate antibiotics for selection of the transformation vector. Shaker incubates (200 rpm) at 30°C. The day before the transformation (day -1), 100 μL of the liquid culture are plated on a YEB agar plate supplemented with appropriated antibiotics for the selection of the transformation vector. Incubate overnight at 30°C to obtain an *Agrobacterium* layer on the plate (see Note 28).
3. Seedling root tips (3–5 mm including the meristem region) are cut with a razor blade and the sectioned region of the root is dipped into the *Agrobacterium* layer (see Note 29). The *Agrobacterium* will attach to and infect the sectioned extremity of the root.

4. Place the inoculated seedlings on BNM plates supplemented with Kanamycin (25 mg/L) with roots oriented toward the lower part of the plate. Infected seedlings can be placed on two rows on a plate. Plates are incubated at an angle of 45° for 15–21 days in the growth chamber at 20°C (see Note 30). Transgenic roots will start to grow after 1 week on the selective medium.
5. When transgenic roots are 5–10 cm long, plants that have produced roots (not all plants produce hairy root) are transferred to new BNM plates without kanamycin and inoculated with 50 μL of a dilute culture of *Rhizobium* as described in Subheading 3.5. Place only eight to ten seedlings per plate as in Subheading 3.5, step 2. Plates are incubated vertically (or nearly vertically) in the growth room (24°C).
6. Nodules generally appear 10 days post inoculation if the bacterial strain is efficient. Plants can be kept for a month on this medium.
7. Nodules can be observed 10–15 days later.

4. Notes

1. Any commercial fertilizer used in horticulture with a NPK content similar to the Soluplant fertilizer can be used to prepare the nitrogen-rich solution. This solution generally contains an equivalent of 10 mM NO_3.
2. FeEDTA (Ethylenediaminetetraacetic acid ferric sodium salt; $C_{10}H_{12}N_2O_8$ FeNa) can be purchased from Sigma (ref: E6760).
3. Nodulation is also working without AVG but in order to get reproducible efficiency of nodulation we add it for in vitro tests. As AVG is thermo sensitive make sure that the medium is not too hot before adding it.
4. *M. truncatula* is very sensitive to the type of agar used for preparing the solid medium. We found that Kalys agar HP 696 works well for in vitro symbiotic tests. Other gelling agents can induce root growth arrest of the seedlings and complete inhibition of the symbiotic interaction (see ref. 2).
5. *M. truncatula* R108 and Jemalong seeds can be requested from Dr. JM Prosperi (http://www1.montpellier.inra.fr/BRC-MTR/mauguio/mauguio.php).
6. Some laboratories have reported differences in the efficiency of transformation, hairy root production and efficiency of nodulation of the transgenic roots between these two strains. We did not observe these differences under our testing conditions.

7. Any plant transformation binary vector that allows kanamycin selection *in planta* can be used for *Medicago* transformation experiments.
8. Antibiotics from other companies can be used. Sigma references are given as examples.
9. Other disinfectants can be used. We have also used commercial Chlorine tablets but for each new sterilizing compound the efficiency of sterilization and the possible negative effects on seed germination should be tested.
10. The NPK fertilizer can be purchased from: Duclos International (Soluplant), Lunel Viel, France, http://www.duclosinternational.eu/index.html or Fertilex International (Nutriplant suprême), Barcelona, Spain (http://www.fertilex.es/71240_en.html).
11. This light tube mixture is used in order to provide red light in addition to white light. *Medicago* plants grow better in the presence of red light.
12. The mixture of sand and perlite, in addition to providing a well-aerated substrate favoring the plant–bacteria interaction, does not stick to the root, which facilitates their cleaning. Other substrates like vermiculite are often tightly attached to small roots making their observation more tedious.
13. For reliable nodulation assays, sand (or other substrates) needs to be sterilized.
14. The greenhouse culture conditions are: 16 h day period, 60% relative humidity, 16°C at night and 22°C daytime temperature with additional light: 200 $\mu E/m^2/s$.
15. The growth chamber culture conditions are: 16 h day period, 60% relative humidity, 22°C temperature with 200 $\mu E/m^2/s$ light.
16. Placing the scarified and sterilized seeds for 2 days in the dark at 4°C allows them to germinate synchronously.
17. *Rhizobium* Rm41 and Rm1021 are both nodulating *M. truncatula* R108, but the *Rm41* is inducing faster nodule formation and seems more efficient for nitrogen fixation. *Rm1021* was traditionally used for symbiotic tests with *M. truncatula* Jemalong. Rm41 is fix$^-$ (no N_2 symbiotic fixation) on *M. truncatula* Jemalong.
18. This rhizobia solution if transgenic should be sterilized by adding 10% (v/v) Na-hypochlorite (9°) before discarding it.
19. The TA medium used to grow *Rhizobium* is a rich medium containing various compounds that can elicit defense reactions in the plant. This is reduced by resuspending the bacterial culture in water. Rhizobia are soil bacteria that can be suspended at low density in pure water without damage.

20. This method of inoculation allows having all the seedlings treated the same way to achieve uniform inoculation.
21. When starting watering plants with only the nitrogen-poor solution we sometime observed poor or nonreliable nodulation because plants start to starve before the establishment of the symbiosis. Addition of a nitrogen-rich solution at the beginning of the culture allows the plantlets to grow well at the start and will give a more reliable nodulation during the test. If the plants show nitrogen starvation symptoms during the experiment (red stems or leaves) they should be watered exclusively with the nitrogen-poor solution instead of alternating water and nitrogen-poor solution.
22. Watering should be done gently with a watering can the first week in order to avoid drying of the seedlings. At this stage the roots are not yet deep in the substrate that will dry more rapidly on the surface. When the plants have their first trifoliate leaf developed, watering is done in the tray in order not to water the leaves. The solution is then absorbed by capillary force. This avoids keeping the leaves moist and limits the risk of diseases. It is important to avoid keeping the substrate permanently wet. Do not fill the tray with solution or water as soon as it is empty, the surface of the perlite/sand mixture should dry before watering again.
23. Do not water during 2 days before observation of the root system. Sand and perlite separate easier from the root if the substrate is slightly dry.
24. It is important to use this household film rather than parafilm during in vitro experiments in order to allow gas exchanges between the plant material and the outside atmosphere.
25. Once inoculated, it is important for gas exchanges to leave the top part of the plate unsealed. We observed complete blockage of the plant and nodule development in some experiments when the plates are sealed completely.
26. It is well known that light is inhibiting root growth and nodule development. This effect is partly mediated by ethylene and reduced by the presence of AVG in the medium. Covering the part of the plate where the root system is developing by wrapping it with aluminum foil increases the nodulation efficiency.
27. Roots that grow inside the agar medium have a different morphology and nodulate differently that the roots growing on the surface of the medium. Placing the plates vertically avoids penetration of the roots into the agar medium and thus allows an easier observation.
28. The *Agrobacterium* culture must be freshly prepared for the experiment as cultures stored on plates for few days lose their virulence.

29. Avoid drying of the roots by closing the plate containing the germinating seedlings when not used. Laminar benches are drying the root seedlings rapidly and this will damage them and reduce the efficiency of transformation. If necessary add a small amount of sterile water to the plate containing the germinated seedlings.
30. As described in ref. (13) transformation works better at 20°C than at 24°C.

Acknowledgements

We would like to thank Dr. M. Schultze for careful reading of the manuscript.

References

1. Ehrhardt DW, Atkinson EM, Long SR (1992) Depolarization of alfalfa root hair membrane-potential by Rhizobium meliloti Nod factors. Science 256:998–1000
2. Kamaté K, Rodriguez-Llorente ID, Scholte M, Durand P, Ratet P, Kondorosi E, Kondorosi A, Trinh TH (2000) Transformation of floral organs with GFP in *Medicago truncatula*. Plant Cell Rep 19:647–653
3. Orosz L, Sváb Z, Kondorosi A, Sik T (1973) Genetic studies on rhizobiophage 16-3. I. Genes and functions on the chromosome. Mol Gen Genet 125:341–350
4. Van Larebeke N, Genetello C, Hernalsteens JP, De Picker A, Zaenen I, Messens E, Van Montagu M, Schell J (1977) Transfer of Ti plasmids between *Agrobacterium* strains by mobilisation with the conjugative plasmid RP4. Mol Gen Genet 152:119–124
5. Trinh TH, Ratet P, Kondorosi E, Durand P, Kamaté K, Bauer P, Kondorosi A (1998) Rapid and efficient transformation of diploid *Medicago truncatula* and *Medicago sativa* ssp. falcata in vitro lines improved in somatic embryogenesis. Plant Cell Rep 17:345–355
6. Sagan M, Morandi D, Tarenghi E, Duc G (1995) Selection of nodulation and mycorrhizal mutants in the model plant Medicago truncatula (Gaertn.) after γ-ray mutagenesis. Plant Sci 111:63–71
7. Kondorosi E, Banfalvi Z, Kondorosi A (1984) Physical and genetic analysis of a symbiotic region of *Rhizobium meliloti*: identification of nodulation genes. Mol Gen Genet 193: 45–452
8. Galibert F, Finan TM, Long SR, Pühler A, Abola P, Ampe F et al (2001) The composite genome of the legume symbiont *Sinorhizobium meliloti*. Science 29:668–672
9. Terpolilli JJ, O'Hara GW, Tiwari RP, Dilworth MJ, Howieson JG (2008) The model legume *Medicago truncatula* A17 is poorly matched for N_2 fixation with the sequenced microsymbiont *Sinorhizobium meliloti* 1021. New Phytol 179:62–66
10. Petit A, David C, Dahl GA, Ellis JG, Guyon P, Casse-Delbart F, Tempe J (1983) Further extension of the opine concept: plasmids in *Agrobacterium rhizogenes* cooperate for opine degradation. Mol Gen Genet 190:204–214
11. Quandt H-J, Pühler A, Broer I (1993) Transgenic root nodules of *Vicia hirsuta*: a fast and efficient system for the study of gene expression in indeterminate-type nodules. Mol Plant Microbe Interact 6:699–706
12. Ratet P, Schell J, de Bruijn FJ (1988) Mini-*Mulac* transposons with broad-host-range origins of conjugal transfer and replication designed for gene regulation studies in Rhizobiaceae. Gene 63:41–52
13. Boisson-Dernier A, Chabaud M, Garcia F, Bécard G, Rosenberg C, Barker DG (2001) Agrobacterium rhizogenes-transformed roots of *Medicago truncatula* for the study of nitrogen-fixing and endomycorrhizal symbiotic associations. Mol Plant Microbe Interact 14: 695–700

Chapter 5

Plant Cell Suspension Cultures

Roberto Moscatiello, Barbara Baldan, and Lorella Navazio

Abstract

Plant cell suspension cultures are widely used in plant biology as a convenient tool for the investigation of a wide range of phenomena, bypassing the structural complexity of the plant organism in toto. The homogeneity of an in vitro cell population, the large availability of material, the high rate of cell growth and the good reproducibility of conditions make suspension-cultured cells suitable for the analysis of complex physiological processes at the cellular and molecular levels. Moreover, plant cell cultures provide a valuable platform for the production of high-value secondary metabolites and other substances of commercial interest. Here we describe how to initiate and maintain plant cell cultures starting from explants obtained from in vitro germinated seedlings. Isolation of protoplasts from plant cell suspension cultures and regeneration of plants via organogenesis and somatic embryogenesis are also presented.

Key words: Plant explants, Culture medium, Plant hormones, Callus, Cell suspension, Protoplasts, Organogenesis, Somatic embryogenesis

1. Introduction

In vitro cultivation of plants under controlled conditions in hydroponic cultures has helped to define essential macro- and micronutrients and assess their minimal concentrations required for healthy plant growth and development. Increasing knowledge of plant mineral nutrition has paved the way for the setup of plant cell cultures, because it has allowed the formulation of culture media suitable to maintain a cell population growing and proliferating in vitro. Plant cell cultures require a sterile environment and a culture medium consisting of water, inorganic salts, sucrose, vitamins and plant hormones. A sterile environment, a fundamental condition to avoid contamination of the plant cell population by fungi, bacteria and viruses, is ensured by the use of laminar airflow hoods, sterile plastic and glass labware and autoclaved solutions.

Plant cell cultures are initiated by transferring different types of sterile plant explants to a nutrient medium containing plant growth

Frans J.M. Maathuis (ed.), *Plant Mineral Nutrients: Methods and Protocols*, Methods in Molecular Biology, vol. 953,
DOI 10.1007/978-1-62703-152-3_5, © Springer Science+Business Media, LLC 2013

regulators (1). The more common plant cell culture systems are represented by cultures of calli, clusters of dedifferentiated cells proliferating on the surface of solid culture media and cell suspension cultures, in which cells grow in liquid culture media under constant shaking. Cell cultures have to be subcultured periodically, with a frequency depending on the cell growth rate. Callus cultures, which grow at a slower rate compared with cell suspension cultures, allow for long-term maintenance of plant cell lines and provide the experimentalist with a source of new plant material in case of microorganism contamination of the cell suspension cultures. Alternatively, cryopreservation of plant cell suspensions is also possible (2).

The commercial availability of cell culture media consisting of mixtures of macro- and microelements plus vitamins, such as Murashige and Skoog medium (3) and Gamborg B5 medium (4), has enormously facilitated the setup of plant cell cultures and reproducibility of cell culture conditions all over the world.

Plant cells cultures exhibit two main remarkable differences with respect to animal cell cultures: (1) cultured plant cells proliferate in vitro indefinitely, whereas primary animal cell cultures usually undergo only a limited number of cell divisions; (2) they are totipotent, i.e., under appropriate hormonal conditions, they may regenerate the whole plant organism. Additional distinctive features between plant and mammalian cell cultures concern in vitro culture conditions, such as temperature (24°C vs. 37°C), source of organic carbon (sucrose vs. glucose), pH (5.5–5.8 vs. 7.2–7.4), light requirement (yes vs. no) and type of cell growth (suspension/calli on solid medium vs. suspension/monolayer).

Plant cell cultures allow for dissection of the complexity of a plant organism in its fundamental units, thus favoring an easier analysis of several plant physiological processes. Many differentiated plant tissues, when cultivated in vitro under proper culture conditions, dedifferentiate and regain the competence to express their whole genetic potential. The availability of a homogeneous cell population, rapidly proliferating in vitro and able to uniformly respond to a wide array of signals of both abiotic and biotic nature, can in some cases amplify the detection of responses sometimes limited to a specific tissue or cell type. For example, plant cell cultures have been shown to be a valuable tool to analyze calcium-based signaling pathways during plant-microbe interactions (5–8).

Historically, in vitro plant cultures have helped to uncover the role of some plant hormones (9–11). Nowadays, they are an essential intermediate step in plant genetic engineering, allowing the generation of transgenic plants endowed with new characters, starting from cultured cells or protoplasts (i.e., plant cells that have been deprived of the cell wall) that have been transfected with exogenous DNA.

Plant regeneration may be achieved mainly in two ways: organogenesis or somatic embryogenesis. Plant regeneration via organogenesis is based on the fact that small pieces of plant tissues can be induced to regenerate the whole plant. The formation of a desired organ is obtained by manipulating the concentration of plant growth regulators in the culture media. An optimized regeneration protocol is needed to obtain clonal plants expressing a particular phenotype or exogenous genes (12).

Somatic embryogenesis relies on the fact that somatic tissues, calli or cell suspension cultures are able to produce embryos. This process allows the regeneration in vitro of a whole new plant that is identical to the plant from which the explants or cells were derived. It was first described in carrot and since then protocols for production of somatic embryos have been reported in several species (13, 14). Somatic embryogenesis is an efficient method to obtain large numbers of clonal plantlets that have green cotyledons, hypocotyls, and radicles and are able to complete the regeneration of entire plants.

Recent fields of application of plant cell cultures are the large-scale production of secondary metabolites and other molecules of pharmaceutical interest, such as heterologous proteins (15). Diverse stress conditions, such as deprivation of essential nutrients, may often enhance the accumulation of a given chemical by in vitro cell cultures. It has recently been shown that the fatty acid content of some microalgae, important for biodiesel production, significantly increases in culture media deprived of either nitrogen or phosphorus (16). A better understanding of the different plant requirements in terms of mineral nutrients may facilitate the development of new biotechnological tools based on the use of plant cell cultures as a production platform for a wide range of valuable compounds.

2. Materials

2.1. Plant Material

Seeds of various plant species (e.g., *Arabidopsis thaliana* L., *Nicotiana tabacum* L., *Lotus japonicus* L., *Daucus carota* L.).

2.2. Laboratory Equipment and Materials

1. Laminar airflow cabinet equipped with a Bunsen burner.
2. Culture room at 24°C, with a photoperiod of 16-h light/8-h dark.
3. Orbital shaker at 80 rpm.
4. Oven.
5. Autoclave.
6. Fume hood.

7. Desiccator jar.
8. Magnetic stirrer.
9. pH meter.
10. Refrigerator.
11. Freezer.
12. Microwave.
13. Vacuum pump.
14. Analytical balance (accuracy 0.001 g).
15. Benchtop centrifuge and microcentrifuge.
16. Light and stereo microscopes.
17. Dissecting forceps and scalpel.
18. Sterile Petri dishes 100 × 20 mm.
19. Phytatray™ disposable plant cell culture vessels (Sigma).
20. Erlenmeyer culture flasks.
21. Beakers and bottles.
22. Sterile disposable and autoclavable vacuum filtration units.
23. Sterile 50- and 15-mL tubes, 2- and 1.5-mL microtubes.
24. Pipet-aid and sterile disposable 25-, 10-, 5-, 2-mL pipettes.
25. Adjustable micropipettes and sterile tips.
26. Parafilm.
27. Aluminum foil.
28. Circular filter papers (40-mm diameter).
29. 1-mL syringe.
30. Fine nylon mesh filters with 50- and 120-μm mesh openings.
31. Sterile syringe filter units 0.20 μm.
32. Cell counting chambers.
33. Microscope slides and cover glasses.

2.3. Chemicals, Culture Media, and Plant Hormones

1. Sterile distilled water.
2. Sodium hypochlorite.
3. Ethanol.
4. Triton X-100.
5. Concentrated HCl (12 N).
6. Gamborg B5 basal medium (micro- and macroelements including vitamins) (Duchefa, Sigma).
7. Murashige and Skoog basal medium (micro- and macroelements including vitamins) (Duchefa, Sigma).
8. Sucrose.

9. Mannitol.
10. 2-(N-Morpholino)ethanesulfonic acid (MES).
11. Agar (Duchefa).
12. 2,4-dichloro-phenoxyacetic acid (2,4-D).
13. 6-benzylaminopurine (BAP).
14. α-naphthaleneacetic acid (NAA).
15. Kinetin.
16. 3-indolbutyrric acid (IBA).
17. Cellulase "Onozuka" R-10 (Yakult Honsha Co., Ltd.).
18. Macerozyme® R-10 (Yakult Honsha Co., Ltd.).
19. Trypan blue.

3. Methods

3.1. Sterility Conditions

When working with in vitro cell cultures, care must be taken to maintain sterile conditions all the time. All appropriate manipulations have to be carried out in a laminar airflow cabinet equipped with a Bunsen burner, using sterile materials and aseptic techniques (see Note 1).

3.2. Seed Sterilization

Herewith three different protocols to surface-sterilize seeds are described. Use the one that ensures asepsis as well as an appropriate level of seed germination, depending on the plant species. For example, protocol A is commonly used for *Lotus japonicus* (17), protocol B for *Daucus carota* (18), and protocol C for *Arabidopsis thaliana* (19).

3.2.1. Protocol A

1. Transfer seeds to a sterile tube of appropriate volume (1.5-, 2-, 15-, or 50-mL), depending on the number and size of seeds.
2. Add to the seeds a solution containing 4% sodium hypochlorite, 0.1% Triton X-100. Use 1 mL of disinfection solution if 1.5–2 mL tubes are used, or 10–40 mL if 15–50-mL tubes are used.
3. Incubate for 20 min on a shaker.
4. Carefully remove the disinfection solution and extensively rinse the seeds six times with sterile distilled water.
5. Proceed immediately with seed germination (see Subheading 3.4) or keep the sterilized seeds for 24–48 h at 4°C if vernalization is required. Some seeds require an overnight incubation in H_2O for proper imbibition before germination.

3.2.2. Protocol B

1. See step 1 of Subheading 3.2.1.
2. Add to the seeds 75% ethanol. Use 1 mL of disinfection solution if 1.5–2 mL tubes are used, or 10–40 mL if 15–50-mL tubes are used.
3. Incubate for 5–10 min on a shaker (see Note 2).
4. Remove the first solution.
5. Add 15% sodium hypochlorite. Use 1 mL of disinfection solution if 1.5–2 mL tubes are used, or 10–40 mL if 15–50-mL tubes are used.
6. Incubate for 5–15 min on a shaker (see Note 2).
7. Remove the bleach and wash three times with sterile water.
8. See step 5 of Subheading 3.2.1.

3.2.3. Protocol C (Vapor-Phase)

1. See step 1 of Subheading 3.2.1.
2. Place open tubes with seeds inside a desiccator jar.
3. Place the desiccator in a fume hood.
4. Place a 250-mL beaker containing 100 mL 15% sodium hypochlorite into the desiccator jar.
5. Immediately prior to sealing the jar, carefully add 3 mL concentrated HCl (12 N) to the bleach.
6. Seal the desiccator jar with parafilm and allow sterilization by chlorine fumes to continue for a time interval of 3–16 h (see Note 3).
7. Open the desiccator jar in a laminar flow cabinet.
8. Close the tube containing the sterilized seeds.
9. See step 5 of Subheading 3.2.1.

3.3. Preparation of Culture Media

The composition of some culture media commonly used for plant cell cultures in our laboratory is shown in Table 1. Such media are based on Gamborg B5 basal medium and Murashige and Skoog basal medium (micro- and macroelements including vitamins). The most suitable basal salt mixtures, sucrose concentration, and type/concentrations of plant hormones (usually auxins and cytokinins) need to be assessed carefully depending on the plant species.

3.3.1. Liquid Culture Media

1. Dissolve the appropriate basal salt mixture including vitamins (Gamborg B5 basal medium 3.2 g/L; Murashige and Skoog basal medium 4.4 g/L) in distilled water.
2. Add the proper amount of sucrose and stir on a magnetic stirrer.
3. Adjust the pH to the requested value with 1 N KOH and make up to volume with distilled water.
4. Sterilize in autoclave at 121°C for 20 min.

Table 1
Composition of some culture media commonly used in the laboratory of Plant Cell Biology, Department of Biology, University of Padova (Italy)

Plant species	Basal medium[a]	Sucrose (%)	2,4-D (mg/L)	BAP (mg/L)	NAA (mg/L)	Kinetin (mg/L)	pH
Arabidopsis thaliana	Murashige and Skoog	3	0.5	0.25	–	–	5.7
Lotus japonicus	Gamborg B5	2	2	–	–	–	5.5
Glycine max	Murashige and Skoog	0.5	–	–	1	0.2	5.7
Daucus carota	Gamborg B5	2	0.5	–	–	–	5.5
Nicotiana tabacum	Murashige and Skoog	3	0.5	0.25	–	–	5.8

[a]Micro- and macroelements including vitamins

5. Prepare the plant hormone stock solutions according to manufacturer's instructions and sterilize them by filtration through sterile 0.2-μm filter units. Store at 4°C.
6. Add the required plant hormones to the autoclaved (cooled down) culture medium.
7. Store at room temperature.

3.3.2. Solid Culture Media

1. Prepare a liquid culture medium (see Subheading 3.3.1) containing a twofold concentration of plant hormones (Solution A).
2. Prepare a solid culture medium containing a twofold concentration of agar, by adding 1.6% (w/v) agar directly into the bottle before sterilization in the autoclave (Solution B).
3. Pipette 10 mL Solution A and 10 mL Solution B in Petri dishes (100 × 20 mm) (see Note 4).
4. Carefully mix the medium components by gently moving the Petri dish in a cross-like motion, without lifting it.
5. Leave the Petri dish with the lid on top slightly inclined, until the medium solidifies (about 30 min).
6. If not used immediately, seal the Petri dishes with Parafilm and store at room temperature for a maximum of 1 month.

3.4. Seed Germination

1. Prepare a half-strength medium based on Gamborg B5 medium or Murashige and Skoog medium (micro- and macroelements including vitamins) supplemented with sucrose, by diluting the medium twice with sterile distilled water.
2. Add 0.8% agar and autoclave for 20 min at 121°C.

3. Transfer 20 mL of the half-strength agar medium to 100 × 20 mm Petri dishes.
4. Distribute the sterilized seeds on top of the solidified medium, by using a sterile pipette tip, capillary or sterile forceps.
5. Seal the Petri dishes with Parafilm.
6. Place the Petri dishes flat, lid side up, into the incubation room (24°C, 16 h light/8 h dark) until the seedlings show two to four leaves and an emerging root.

3.5. Callus Induction

1. Prepare the Callus Induction Medium (CIM): 3.2 g/L Gamborg B5 basal medium, 0.5 g/L MES, 2% sucrose, 0.7% agar, pH 5.7, 1 mg/L 2,4-D, 0.1 mg/L kinetin.
2. Transfer sterile seedlings with forceps from the solidified medium into a sterile Petri dish lid.
3. Dissect the seedlings using a sterile scalpel and forceps in order to obtain some explants (about 2-mm length). Different parts of a plant can be used as starting material, for example cotyledons, hypocotyls, or root segments.
4. Transfer the explants into Petri dishes containing CIM.
5. Seal the Petri dishes with Parafilm.
6. Place the Petri dishes flat (lid side up) into the culture room (24°C, 16 h light/8 h dark) until callus formation can be observed (about 3–4 weeks depending on plant species).
7. Transfer a fraction of each callus onto fresh CIM medium every 4 weeks for two to three times, and then subculture them onto the appropriate solid medium (e.g., basal medium containing the appropriate concentrations of sucrose and plant hormones, and supplemented with 0.8% agar) (Fig. 1) (see Note 5).

3.6. Initiation of Plant Cell Suspension Cultures

1. Pipette 10 mL of the appropriate liquid culture medium (basal salt medium containing sucrose and hormones, Table 1) into a sterile 50-mL Erlenmeyer culture flask (see Note 6).
2. Transfer a portion of the callus with sterile forceps into the Erlenmeyer flask containing the culture medium (see Note 7).
3. Gently grind the callus by using a sterile pipette.
4. Place the flask into the culture room on an orbital shaker (80 rpm) until an adequate fragmentation of the callus and cell density is reached (usually 1 week).

3.7. Determination of the Growth Curve of Plant Cell Suspension Cultures

It is often difficult to determine the cell concentration of a plant cell suspension culture by using a cell counting chamber, because plant cells usually grow as aggregates (clumps). The more common methods used to evaluate the growth of plant cell suspension cultures consist in measuring either the fresh weight or the dry weight of cells at given days after the onset of the suspension culture.

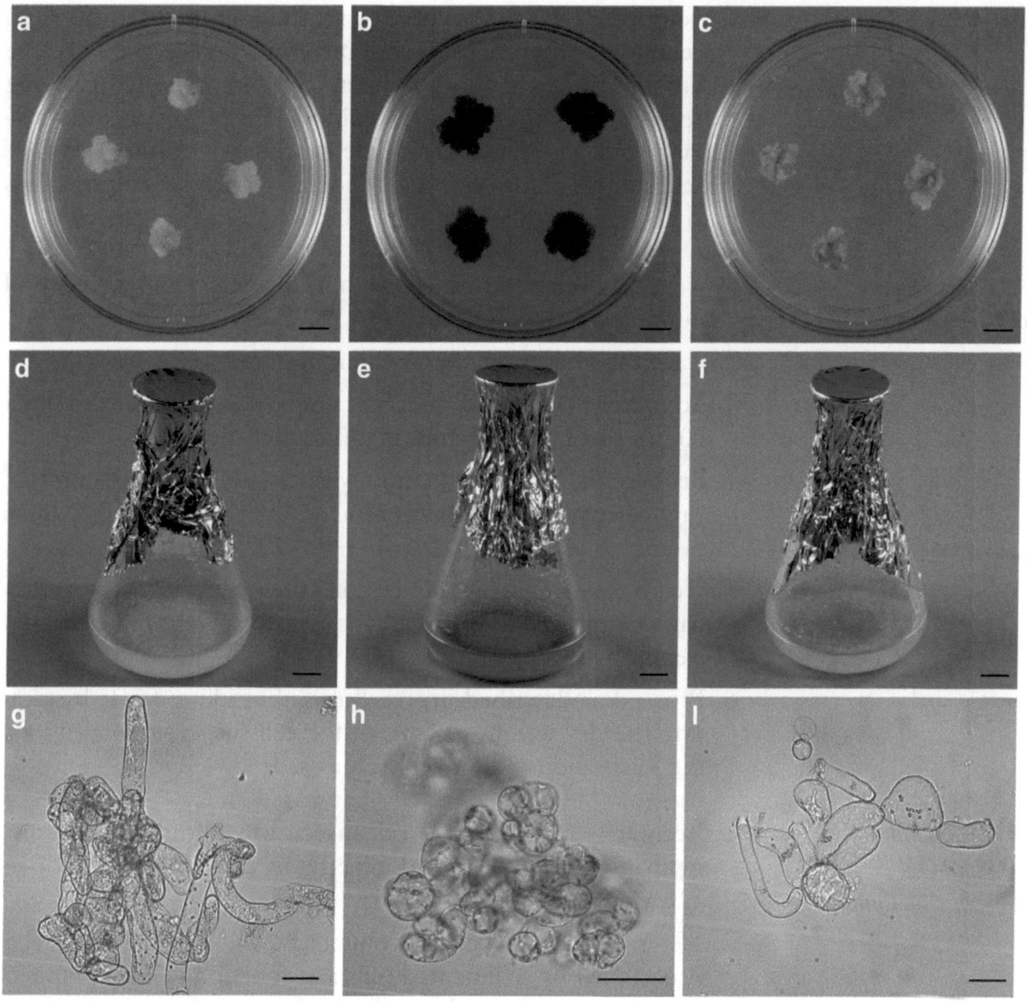

Fig. 1. Cell cultures of *Arabidopsis thaliana* L., *Glycine max* L, *Lotus japonicus* L. (**a–c**) Calli growing on solid culture media in Petri dishes. Bar, 1 cm. (**d–f**) Suspension-cultured cells, growing in liquid culture media in Erlenmeyer flasks. Bar, 1 cm. (**g–i**) Light microscopy images of plant cell suspension cultures. Bar, 50 μm. (**a**, **d**, **g**) *A. thaliana*; (**b**, **e**, **h**) *G. max*; (**c**, **f**, **i**) *L. japonicus*.

3.7.1. Measurement of the Fresh Weight

1. Weigh a 1.5-mL tube.
2. Resuspend the cell suspension culture and transfer 1 mL into the pre-weighed tube by using a precut blue tip (with a wider bore, obtained by removing the tip end with a sterile scalpel blade).
3. Centrifuge at full speed for 1 min.
4. Discard most of the supernatant.
5. Repeat centrifugation as above.
6. With a 1-mL syringe with a fine needle, remove all of the remaining culture medium.
7. Weigh the tube containing the cells and subtract the tare.

8. Determine the growth curve by plotting on the *y* axis the fresh weight per mL of cell suspension culture, and on the *x* axis the sampling time (day in which the cells are collected, starting from transfer into the fresh medium). The lag phase, exponential phase and stationary phase of the cell culture are thus determined. This is essential to identify when subculturing has to be carried out (Subheading 3.8).

3.7.2. Measurement of the Dry Weight

1. Weigh a circle filter paper (40-mm diameter) and a square of aluminum foil.
2. Place the filter paper on a vacuum filtration unit.
3. Accurately resuspend the cell suspension culture and filter 10 mL through the filtration unit under vacuum.
4. Recover the circle filter paper with the collected cells on top, wrap them in the aluminum foil and dry them at 60°C in an oven.
5. Weigh the sample (aluminum foil + filter paper + cells) for a few days, until a constant weight is reached. Determine the dry weight by subtracting the tare.
6. Proceed as described in step 8 of Subheading 3.7.1, by plotting the dry weight per mL of the cell suspension culture on the *y* axis and the time on the *x* axis.

3.8. Subculturing of Plant Cell Suspension Cultures

1. Accurately resuspend the cell suspension culture in late exponential phase with a 10-mL pipette, by pipetting up and down at least ten times (see Note 8).
2. Determine the packed cell volume (PCV), i.e., the volume that is occupied by sedimented cells as a function of the volume of the cell suspension culture. To do this, transfer a small volume of the cell suspension culture (e.g., 5 mL) into a 15-mL graduated centrifuge tube and centrifuge at 500 × *g* for 2 min.
3. To start a new subculture, inoculate 1 mL PCV into 20-mL fresh cell culture medium. To do this, transfer the appropriate volume of late exponential phase cell culture, corresponding to 1 mL PCV, into a 100-mL Erlenmeyer flask containing 20 mL culture medium (final volume) (see Note 9).
4. Place the flask into the culture room (24°C, 16 h light/8 h dark) on an orbital shaker (80 rpm) and leave it until the cell suspension culture again reaches the end of the exponential phase of the growth curve (Fig. 1).
5. Subculturing is carried out every 1–4 weeks, depending on the growth curve of the plant cell suspension culture (Subheading 3.7).

3.9. Isolation of Protoplasts from Plant Cell Suspension Cultures

1. Prepare Enzyme Solution: 0.75% (w/v) cellulase "Onozuka" R-10, 0.25% (w/v) macerozyme® R-10, 0.55 M mannitol in cell culture medium (without sucrose). Sterilize by filtration through sterile vacuum filtration units. Store at −20°C in 10-mL aliquots.
2. Prepare Protoplast Resuspension Buffer: 0.55 M mannitol in cell culture medium (without sucrose). Sterilize in autoclave. Store at −20°C in 15-mL aliquots.
3. Use mid-exponential phase suspension-cultured cells.
4. Incubate in the dark 2.5 mL packed cell volume (Subheading 3.8) with 4 volumes (10 mL) of Enzyme Solution for 2 h (e.g., *Arabidopsis*) or overnight (e.g., carrot) in a 50-mL tube kept flat on an orbital shaker in the incubation room in darkness (see Note 10).
5. Filtrate the cell suspension through a sterile 50-μm nylon mesh mounted on a 100-mL beaker (see Note 11).
6. Gently transfer the protoplast suspension from the beaker to a 15-mL tube.
7. Centrifuge at 60 × *g* for 5 min at room temperature and discard the supernatant.
8. Wash twice with 5 mL Protoplast Resuspension Buffer, each time gently resuspending the protoplasts and centrifuging as above (see Note 12).
9. Count the protoplasts under a light microscope by using a Bürker chamber (see Note 13).
10. Resuspend the protoplasts at the desired final concentration (e.g., 10^6 protoplasts/mL) (Fig. 2).

3.10. Plant Regeneration via Organogenesis

Herewith the protocol for regeneration via organogenesis in tobacco is described. Use sterile (from in vitro micro-propagated plants) or nonsterile (from in vivo grown plants) tobacco leaves (see Note 14). Culture media as well as hormonal concentrations should be adjusted if different plant species are used.

1. Prepare Shoot Medium: Murashige and Skoog basal medium, 3% sucrose, pH 5.5, 1 mg/L 3-indolbutyrric acid (IBA), 1 mg/L 6-benzylaminopurine (BAP), 0.8% agar.
2. Prepare Root Medium: Murashige and Skoog basal medium, 3% sucrose, pH 5.5, 0.8% agar.
3. Transfer sterile leaf explants on Shoot medium in a Petri dish 100 × 20 mm.
4. Place the Petri dish containing the explants in the culture room (24°C, 16 h light/8 h dark) for 3–5 weeks, until shoots form on the dedifferentiated calli (Fig. 3).
5. When two to four leaves have formed, transfer the shoots into Phytatray™ disposable plant cell culture vessels containing

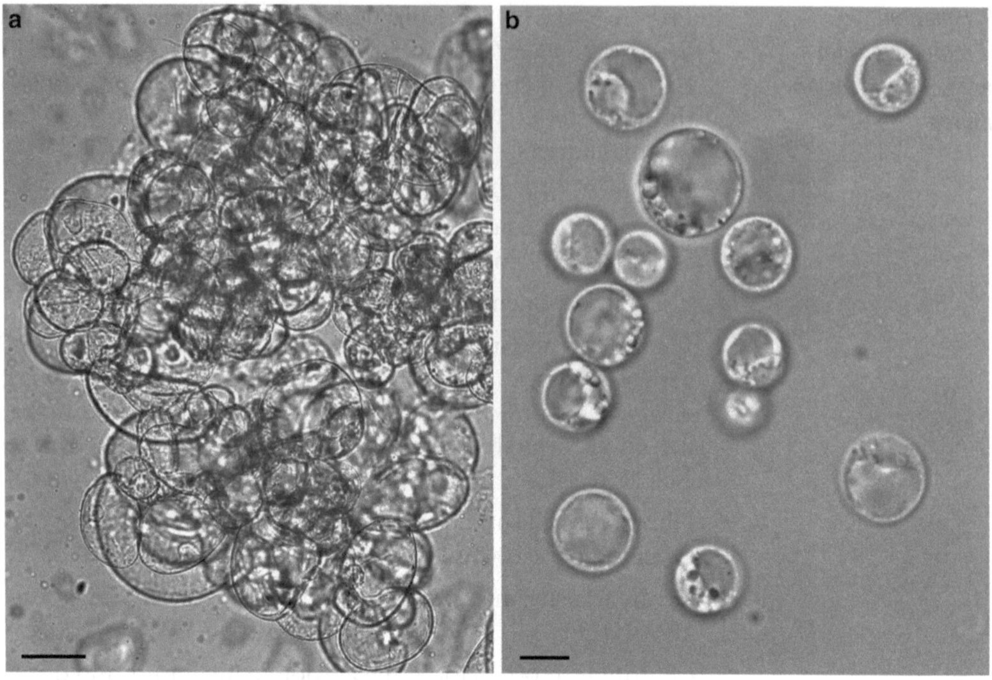

Fig. 2. Isolation of protoplasts from *Daucus carota* L. cell suspension cultures. (**a**) Carrot suspension-cultured cells in mid-exponential phase. Bar, 50 μm. (**b**) Freshly isolated protoplasts. Bar, 10 μm.

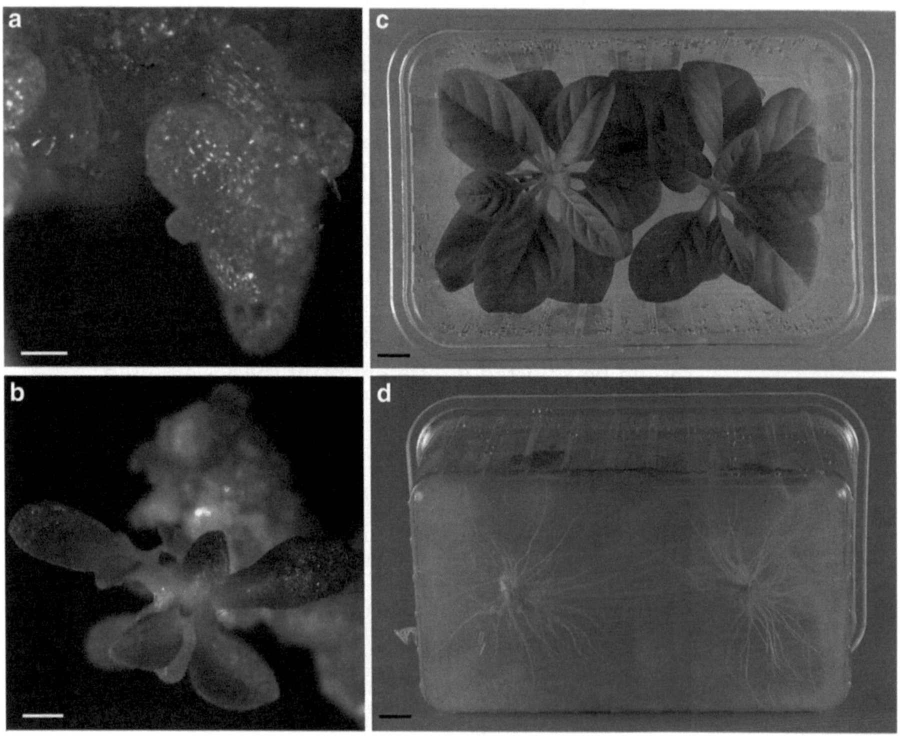

Fig. 3. Plant regeneration via organogenesis in *Nicotiana tabacum* L. (**a**, **b**) Early steps of shoot formation from the callus surface. Bar, 100 μm. (**c**) Shoots of regenerated plants (top view). Bar, 1 cm. (**d**) roots of regenerated plants (bottom view). Bar, 1 cm.

100 mL Root Medium and incubate them in the culture room (24°C, 16 h light/8 h dark) for 2 weeks (Fig. 3).

6. Maintain plants via micropropagation by transferring apical meristems or nodes bearing lateral buds in new Phytatray™ vessels containing 100-mL of the same medium.

3.11. Plant Regeneration via Somatic Embryogenesis

Herewith the protocol for somatic embryogenesis in carrot is described. Conditions should be adjusted if different plant species are used, also on the basis of available literature data.

1. Plate sterile explants from hypocotyls on Gamborg B5 medium, 2% sucrose, pH 5.5, 0.5 mg/L 2,4-D, 0.8% agar to obtain embryogenic calli (Subheading 3.5).
2. After three to four callus subcultures, transfer calli into liquid culture medium (Gamborg B5 medium, 2% sucrose, pH 5.5, 0.5 mg/L 2,4-D) to obtain embryogenic cell suspension culture (Subheading 3.6).
3. Every week subculture an aliquot of cell suspension into 50-mL fresh liquid culture medium to maintain the embryogenic culture (Subheading 3.8).
4. To induce the switch to embryo formation, sieve the cell suspension (6–7 day-old) through a 120-μm nylon mesh mounted on a 100-mL beaker (see Note 11) and wash extensively with liquid culture medium without hormones (Gamborg B5 medium with 2% sucrose, pH 5.5).
5. Sieve the cell suspension, obtained after the 120-μm mesh filtration, through a 50-μm nylon mesh mounted on a 200-mL beaker (see Note 11), wash extensively with the same medium and collect the cells on top of the nylon mesh in a few mLs.
6. Dilute the resuspended cells and count them under a light microscope using a Nageotte chamber (see Note 15).
7. Transfer cells (3,000/mL) in 50 mL fresh liquid culture medium without hormones in a 250-mL Erlenmeyer flask and incubate in the culture room (24°C, 16 h light/8 h dark) on an orbital shaker at 80 rpm.
8. Check the culture after 1 week: by this time, embryos at the globular stage will have developed (Fig. 4).
9. After 10 days, in the suspension culture heart-shaped embryos can be recovered and after 2 weeks embryos at the torpedo stage are obtained (Fig. 4).
10. Select individual embryos and place them on agar (0.8% agar) Gamborg B5 medium with 2% sucrose, pH 5.5, until entire seedlings develop.

Additional detailed information on specific plant cell culture procedures can be found in Loyola-Vargas and Vázquez-Flota (20).

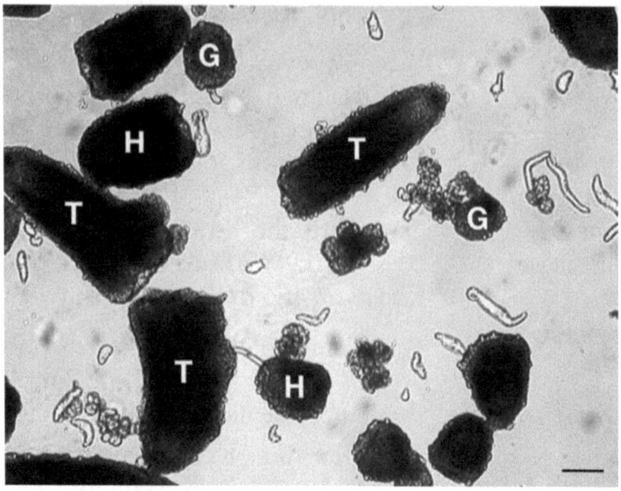

Fig. 4. Different stages of somatic embryogenesis in *D. carota* L. "G," globular embryo; "H," heart-shaped embryo; "T," torpedo. Bar, 100 μm.

4. Notes

1. Use sterile disposable plasticware (centrifuge tubes, Petri dishes, vessels, pipettes, tips). Glassware (Erlenmeyer flasks, beakers) has to be sterilized in the oven at 180°C for 4 h. Cover the Erlenmeyer flasks and beakers with thick aluminum foil before sterilization. Bottles, cell culture media and solutions are sterilized in an autoclave at 121°C for 20 min. Use sterilization tape to label all glassware: after heating/autoclaving, the stripes present on the tape will turn black. The bench under the laminar airflow cabinet needs to be carefully cleaned with denatured alcohol before and after use. Avoid to speak and to pass over open sterile labware with hands and arms while working in the laminar airflow cabinet. Forceps and scalpels can be sterilized by immersing them in ethanol and subsequently heating them over the Bunsen burner until red hot: allow them to cool down before use. To remove the aluminum foil cover of the Erlenmeyer flasks, carefully release it with your hands and place it upside down on the bench in the laminar flow cabinet. Briefly flame the neck of flasks and bottles on the Bunsen burner anytime you either open or close them.
2. Incubation times with ethanol and sodium hypochlorite must be adjusted depending on the plant species used. Carefully choose the best conditions.
3. The time interval needed can vary based on the plant species and the extent to which seeds are contaminated. Usually 3–4 h is enough to reasonably clean the seeds. An overnight

incubation is also acceptable, although some loss of viability of the seeds may occur.

4. If not used immediately after autoclaving, Solution B will solidify. Before use, put the bottle containing Solution B in the microwave for a few min at medium power with a loose cap. The solution is ready to use when the bottle has cooled down enough to be hand-warm. Do not allow Solution B to cool down too much, otherwise the mixture of Solution A + Solution B will not solidify well and homogeneously when poured in the Petri dish.
5. Maintenance of callus cultures is achieved by subdividing calli and transferring them onto fresh agar medium at 4-week intervals.
6. To check the best in vitro growing conditions for a given plant species, maintain cell suspension cultures in diverse culture media, containing different concentrations of sucrose and types/concentrations/ratios of plant hormones. After three to four subcultures, evaluate the growth pattern of the cell suspension culture, as well as the dimension of cell clusters, and select the most suitable culture medium for your specific needs.
7. Make sure that the inoculum (represented by a portion of the callus) is big enough to ensure a reasonable cell density in the initiating suspension culture, otherwise cell proliferation may be impaired.
8. If the cell suspension culture is made of small clumps of cells, use pipettes with a wide tip, otherwise cells will not pass through.
9. Cell suspension cultures are maintained in sterile Erlenmeyer flasks containing one-fifth volume of culture medium to ensure a proper aeration of the cell culture. For example, put 20 mL culture medium into 100-mL Erlenmeyer flasks, 50 mL culture medium into 250-mL flasks, 100 mL culture medium into 500-mL flasks. The inoculum size (1 mL PCV in 20 mL fresh medium) has to be adjusted according to the volume of fresh culture medium to inoculate.
10. The incubation time with the Enzyme Solution may vary according to the plant species considered. Wrap the tube with aluminum foil, because cell wall degrading enzymes are light-sensitive.
11. Place a square of nylon mesh on top of the beaker and secure it to the rims by using a rubber band, so that the nylon mesh forms a concave surface. Wrap the beaker in aluminum foil and sterilize it by autoclaving.
12. At all stages of protoplast isolation, treat protoplasts gently: set the centrifuge with brake off, do not drop them into tubes from any height, do not vigorously resuspend them, use a sterile scalpel blade to remove the end of blue (1 mL) and yellow

(200 μl) tips to give a wider bore. Moreover, always resuspend protoplasts in isoosmotic buffers.

13. For cell counting with the Bürker chamber, make an appropriate dilution of the protoplast preparation (e.g., 10 μl in 100 μl isoosmotic buffer, but it can vary among different protoplast preparations). Staining with Trypan blue can be used to check protoplast viability, by adding an equal volume of Trypan Blue solution (0.4%) to the protoplast suspension before cell counting: nonviable cells stain blue. Aspire sufficient volume (20 μl + 20 μl) to fill both sides of the Bürker chamber. Count the protoplasts contained in the four diagonals (each diagonal is made of 12 squares of 4×10^{-6} cm^3 volume). Calculate the initial protoplast concentration per mL using the following formula:

$$c = \frac{\sum_{i=1}^{48} n_i}{48 \cdot 4 \cdot 10^{-6}} \cdot \text{df}$$

where c = number of protoplasts per mL; n_i = protoplasts in each square (48 squares in total); df = dilution factor. Protoplast viability is usually around 80–90%.

14. If explants from nonsterile plants are used, a sterilization procedure must be carried out (Subheading 3.2). As a general rule, milder sterilization conditions than those used to surface sterilize seeds have to be used. To avoid necrosis of explants (especially on the explant border), carefully check incubation time with 75% ethanol and sodium hypochlorite.

15. For cell counting with the Nageotte chamber, make an appropriate dilution of the suspension culture (for example 1:20 or 1:40, but it can vary among different cell preparations). Aspire sufficient volume (50 μl + 50 μl) to fill both sides of the Nageotte chamber. Count the cells contained in 40 rectangles (of 1.25 mm^3 volume each). Calculate the initial cell concentration per mL using the following formula:

$$c = \sum_{i=1}^{40} n_i \cdot 20 \cdot \text{df}$$

where c = number of cells per mL; n_i = cells in each rectangle (40 rectangles in total); df = dilution factor.

Acknowledgements

This work was supported by Programmi di Ricerca Scientifica di Rilevante Interesse Nazionale (2008WKPAWW to LN) and Ricerca Scientifica quota EX-60% (60A06-2248/11 to LN and 60A06-2448/11 to BB).

References

1. Hall RD (2000) Plant cell culture initiation. Practical tips. Mol Biotechnol 16:161–173
2. Mustafa NR, De Winter W, Van Iren F, Verpoorte R (2011) Initiation, growth and cryopreservation of plant cell suspension cultures. Nature Protoc 6:715–742
3. Murashige T, Skoog F (1962) A revised medium for rapid growth and bioassays with tobacco tissue cultures. Physiol Plant 15:473–497
4. Gamborg OL, Miller RA, Ojima K (1968) Nutrient requirements of suspension cultures of soybean root cells. Exp Cell Res 50:151–158
5. Lecourieux D, Lamotte O, Bourque S, Wendehenne D, Mazars C, Ranjeva R, Pugin A (2005) Proteinaceous and oligosaccharidic elicitors induce different calcium signatures in the nucleus of tobacco cells. Cell Calcium 38:527–538
6. Moscatiello R, Mariani P, Sanders D, Maathuis FJM (2006) Transcriptional analysis of calcium-dependent and calcium-independent signalling pathways induced by oligogalacturonides. J Exp Bot 57:2847–2865
7. Navazio L, Moscatiello R, Genre A, Novero M, Baldan B, Bonfante P, Mariani P (2007) A diffusible signal from arbuscular mycorrhizal fungi elicits a transient cytosolic calcium elevation in host plant cells. Plant Physiol 144:673–681
8. Navazio L, Baldan B, Moscatiello R, Zuppini A, Woo SL, Mariani P, Lorito M (2007) Calcium-mediated perception and defense responses activated in plant cells by metabolite mixtures secreted by the biocontrol fungus *Trichoderma atroviride*. BMC Plant Biol 7:41
9. Caplin SM, Steward FC (1948) Effect of coconut milk on the growth of the explants from carrot root. Science 108:655–657
10. Miller CO, Skoog F, Von Saltza MH, Strong FM (1955) Kinetin, a cell division factor from deoxyribonucleic acid. J Am Chem Soc 77:1392–1393
11. Skoog F, Miller CO (1957) Chemical regulation of growth and organ formation in plant tissue cultures in vitro. Symp Soc Exp Biol 11:118–130
12. Duclercq J, Sangwan-Norreel B, Catterou M, Sangwan RS (2011) De novo shoot organogenesis: from art to science. Trends Plant Sci 16:597–606
13. Zimmerman JL (1993) Somatic embryogenesis: a model for early development in higher plants. Plant Cell 5:1411–1423
14. Feher A, Pasternak T, Dudits D (2003) Transition of somatic plant cells to an embryogenic state. Plant Cell Tissue Organ Cult 74:201–228
15. Wilson SA, Roberts SC (2012) Recent advances towards development and commercialization of plant cell culture processes for the synthesis of biomolecules. Plant Biotechnol J, 10:249–268
16. Rodolfi L, Chini ZG, Bassi N, Padovani G, Biondi N, Bonini G, Tredici MR (2009) Microalgae for oil: strain selection, induction of lipid synthesis and outdoor mass cultivation in a low-cost photobioreactor. Biotechnol Bioeng 102:100–112
17. Lombari P, Ercolano E, El AH, Chiurazzi M (2003) A new transformation-regeneration procedure in the model legume *Lotus japonicus*: root explants as a source of large numbers of cells susceptible to *Agrobacterium*-mediated transformation. Plant Cell Rep 21:771–777
18. Guzzo F, Baldan B, Mariani P, Lo SF, Terzi M (1994) Studies on the origin of totipotent cells in explants of *Daucus carota* L. J Exp Bot 45:1427–1432
19. The Arabidopsis Information Resource (TAIR; http://www.arabidopsis.org/)
20. Loyola-Vargas VM, Vázquez-Flota F (2006) Plant cell culture protocols, 2nd edn. Methods in molecular biology, vol 318. Humana, Totowa, NJ

Chapter 6

Soil Analysis Using Visible and Near Infrared Spectroscopy

Johanna Wetterlind, Bo Stenberg, and Raphael A. Viscarra Rossel

Abstract

Visible-near infrared diffuse reflectance (vis-NIR) spectroscopy is a fast, nondestructive technique well suited for analyses of some of the essential constituents of the soil. These constituents, mainly clay minerals, organic matter and soil water strongly affect conditions for plant growth and influence plant nutrition. Here we describe the process by which vis-NIR spectroscopy can be used to collect soil spectra in the laboratory. Because it is an indirect technique, the succeeding model calibrations and validations that are necessary to obtain reliable predictions about the soil properties of interest are also described in the chapter.

Key words: Diffuse reflectance spectroscopy, vis-NIR, Clay, Organic matter, Calibration, Validation

1. Introduction

Interest among researchers in the use of visible-near infrared (vis-NIR) diffuse reflectance spectroscopy in soil science has increased over the past two decades (1) because there are many advantages with using the technique. It is nondestructive, requires a minimum of sample preparation and does not involve any (hazardous) chemicals. The measurements only take a few seconds and several soil properties can be estimated from a single scan. Moreover, the technique allows for flexible measurement configurations and in situ as well as laboratory-based measurements (2).

Reflectance spectra in the visible (400–780 nm) and near infrared (780–2,500 nm) region are the result of interactions between the radiating energy and the bonds in molecules of soil constituents. In the visible region, the high energy of the radiation causes transitions of electrons between molecular orbits with different energy levels (3). With lower radiation energy, corresponding to longer wavelengths, the absorption of energy occurs due to vibrations in molecular bonds. Absorption in the NIR region is due to

Frans J.M. Maathuis (ed.), *Plant Mineral Nutrients: Methods and Protocols*, Methods in Molecular Biology, vol. 953,
DOI 10.1007/978-1-62703-152-3_6, © Springer Science+Business Media, LLC 2013

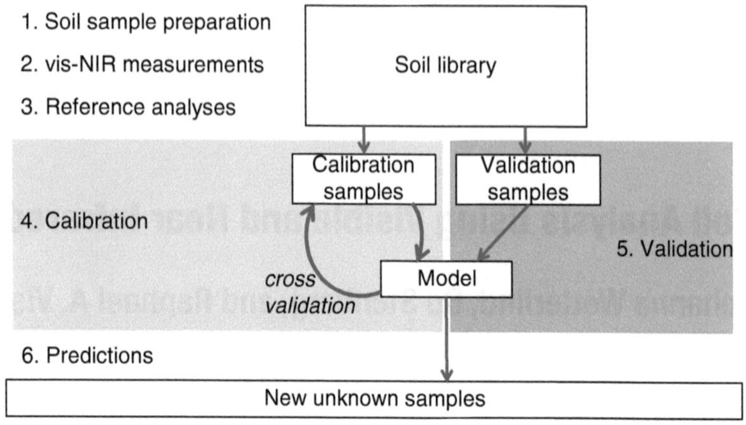

Fig. 1. Procedure for developing vis-NIR prediction models.

overtones and combinations of fundamental vibrations in the mid-infrared region. The energy quanta absorbed are bond specific, but are also affected by the chemical matrix and environmental factors such as type of functional group, neighboring molecules and hydrogen bonds (3). This allows for identification of a range of molecules which may contain the same type of bonds. The same molecule can give rise to several overtones and combination bands over the NIR region with decreasing intensity and increasing overtone order. Because of this, the NIR region is characterized by few, broad, overlapping absorption features. The diffuse reflectance of the soil is also influenced by soil physical properties related to particle size and surface structure, as well as water films on the soil surface.

Due to the lack of specificity and the fact that a spectrum consists of many highly correlated neighboring wavelengths, multivariate calibration techniques are commonly used to correlate the spectra with the targeted soil properties (4). To create calibration models, calibration samples analyzed using conventional laboratory methods are also needed. When the calibration models are established, new samples only need to be analyzed using vis-NIR and the calibration models are used to predict the targeted soil properties. A schematic of the main steps in the procedure from soil sample preparation to prediction is presented in Fig. 1.

The vis-NIR region contains useful information on the organic and inorganic content in the soil and both clay minerals and soil organic matter (SOM), both of which are essential constituents of the soil, have well-recognized absorptions features in this region. Water has strong influences on the spectra with some dominating specific absorption bands near 1,400 and 1,900 nm along with weaker bands in other parts of the spectra (5). In addition, with a water film on soil particles, scattering is more forward directed and moist soils appear darker than dry ones (6). The mineral part of the

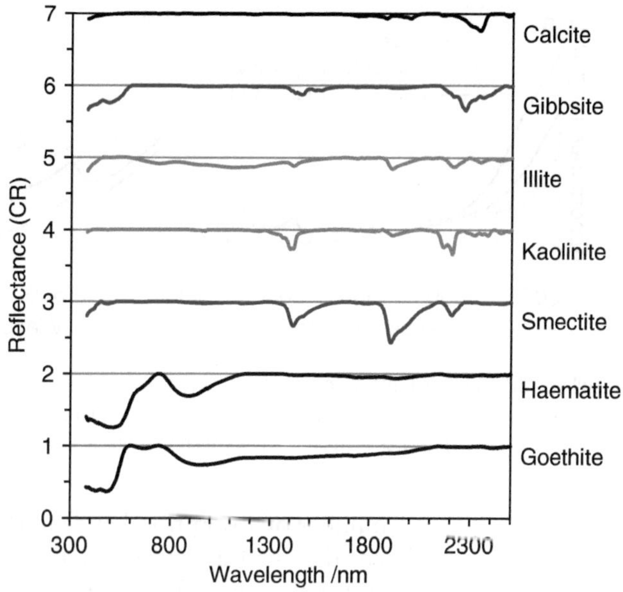

Fig. 2. Continuum-removed spectra of common soil minerals offset by 1 unit, from Stenberg et al. (1).

soil generally accounts for half the soil volume (7), and has pronounced features in the vis-NIR spectrum, both in terms of surface properties affecting the degree of scattering and by absorption. Absorptions in the visible region are primarily associated with iron-containing minerals such as haematite and goethite which show strong absorption bands between 400 and 660 nm (8) (Fig. 2). Both minerals also show absorption bands near 900 nm (880 and 930 for haematite and goethite respectively) but have almost no absorption features at longer wavelengths (9). The absorption by clay minerals in the NIR region is mostly related to absorbed and structural water and Mg-, Al-, and Fe–OH bonds in the mineral crystal lattice. Absorption bands in the 2,200–2,500 nm region, due to combination vibrations involving O-H stretch and metal-OH bend, are evident for several minerals with slightly different peaks depending on metal and mineral (Fig. 2). Absorption bands at 1,400 nm and 1,900 nm are also apparent in a number of minerals to varying degree and are related to absorbed and structurally bound water. Smectite, for example, has a very pronounced absorption band near 1,900 nm due to combination vibrations of water bound in the interlayer lattices as hydrated cations and water adsorbed on particle surfaces (Fig. 2) (10). Carbonates also have several absorption bands in the NIR region, the strongest being near 2,300 nm (9).

Soil organic matter has clear but broad absorption bands in the visible region (Fig. 3a) dominated by chromophores and darkness of the organic matter (11). In the NIR region the absorption is related to stretching and bending of NH, CH, and CO groups (11).

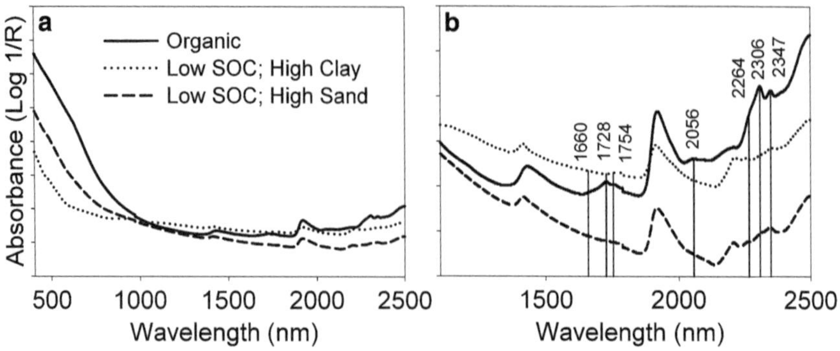

Fig. 3. (**a**) Soil vis-NIR 400–2,500 nm spectra and (**b**) the region 1,100–2,500 nm showing the spectra of three soils: an organic agricultural soil with 40 % SOC and two with 1 % SOC of which one has 87 % sand and 4 % clay, and the other 12 % sand and 44 % clay. Drop lines in (**b**) indicate wavelengths typical of organic matter (from ref. 1).

A number of absorption bands in the 1,100–2,400 nm region has been identified as important for SOM and total N calibrations, however they are often weak and may not be readily apparent to the naked eye. Viscarra Rossel and Behrens (12) present a summary of important fundamental absorptions and their occurrence in the vis-NIR and some examples of the effect of SOM on NIR spectra for different soils are presented in Fig. 3. Stenberg (13) identified some of these as especially important.

Both clay and SOM content are important for the soil structure and thereby soil aeration and water-holding capacity which are factors strongly affecting plant growth. They influence plant nutrition through their possibility to adsorb cations and take part in the transformation of nutrients to plant available forms through weathering and through direct and indirect influences on the microbial soil fauna. However, plant nutrient salts are by themselves not expected to absorb in the vis-NIR region and correlations found to vis-NIR are often weak (1). Nevertheless, good correlations can sometimes be achieved (e.g., (14–17)), possibly through local co-variations with properties that do absorb in the vis-NIR.

In this chapter, we describe the process by which vis-NIR spectroscopy is used to collect soil spectra in the laboratory (Fig. 1). Apart from the actual measurement, which is rather simple and fast, we also discuss quality assessment and normalizations of the spectra. Further, we discuss the steps involved in the creation and assessment of a calibration model (Fig. 1), which requires most of the work and efforts to secure relevant and reliable predictions of the soil properties of interest. As mentioned in the beginning, vis-NIR spectroscopy allows for in situ measurements, and research and development are continuing to facilitate measurements in the field, thus minimizing the need for taking soil samples (18–21). Although this chapter describes the procedures for laboratory measurements, many of the aspects will also have relevance for outdoor measurements.

2. Materials

2.1. Instruments

A variety of vis-NIR spectrophotometers from several manufactures exists today, and these provide a number of different solutions for light dispersion, detectors, and sample presentation configurations. What instrument to choose is largely dependent on the application and basically there is a trade-off between price and performance.

- Resolution and noise
  For scientific purposes, an instrument with high resolution, 10 nm or better, is favorable; however, there is a direct trade-off between resolution and noise (see Note 1).
- Spectral range
  Similarly, a wavelength range covering both visible (400–780 nm) and the entire NIR region (780–2,500 nm) is recommended for scientific purposes, to make sure that as much of the important wavelength bands as possible is included (see Subheading 1).
  If, however the instrument is to be used for a very specific purpose, the need for full vis-NIR spectra may not be necessary.
- Flexibility
  - Possibilities for measurements outside the laboratory
    Different requirements regarding robustness and handling apply depending on whether the instrument is only intended to be used in the laboratory, or if it might be used for outdoor measurements.
  - Special requirements regarding sample presentations
    Will the instrument be used for several very different sample types?
    For outdoor or online measurements flexibility regarding sample presentation is often a priority and a fiber optic for spectra collection is preferred. For these situations a post-dispersive instrument is advisable as shielding of ambient stray light is not crucial (see Note 2).

2.2. Standardized References

Most instruments include all necessary accessories, but depending on type they can be specific to suit that particular instrument or they can be more general. In any case, make sure that you have the following:

- A standardized white reference, for example Spectralon®, which is inert, stable, and has a very high diffuse reflectance in the visible and near infrared: Some instruments have a built in white reference, others need an external one.

- Dark current correction/dark reference is made without the emitting light source, representing 0 reflectance to correct for background electronic noise from the spectrometer. This is often done by using a shutter, but for some instrument-sample configurations an external dark reference might be needed (see Note 3).

3. Methods

3.1. Soil Sample Preparations

- Use air or oven dried soil (see Note 4).
- Grind soil to <2 mm particle size (see Note 5).

3.2. Measurement

Each instrument often has its own sample presentation setup and compatible sample containers. That is, the general recommendation is to follow the instructions for the specific instrument. However, presented below are some general aspects to consider.

- Sample presentation and handling
  - Soils are heterogeneous—therefore, it is very important to measure a representative part of the soil sample and a configuration that allows for a large part of the sample to be scanned is favorable. If the sampled area is very small the use of replicate spectral sampling is recommended.
  - Make sure that the sample is thoroughly mixed in the sample container.Do not shake the sample to get an even surface because this will stratify the sample, with the smaller particles moving down towards the bottom of the container. If an even surface is required, instead use a tool to carefully flatten the sample surface.
  - Pack the containers the same way for all samples. Try to use the same volume of soil and to use the same amount of pressure (22).
  - If the same container is to be used for several samples it is important to clean it between samples, however, avoid using water or alcohol/organic solvents.
  - If the measurement window has direct contact with the soil, it is important to also clean this between samples. Again, avoid using water or alcohol/organic solvents but wipe it clean using a dry dust free tissue.
- White and dark reference

  Depending on instrument, this may be done automatically, however for some instruments it needs to be done manually. This is a crucial step, and to ensure good quality spectra this needs to be done thoroughly and regularly.

- The white and dark reference should be taken every 10 min. In many instruments the dark reference is taken automatically when a white one is taken.
- If a configuration where an external white (and dark) reference is used it is important that the configuration is the same as that used for the sample measurements.
- If it is possible to monitor the spectrum of the white reference, this should represent 100 % reflection at all wavelengths across the 400–25,000 nm range.

- If the measurements are done in such a way that other light sources than that related to the measurement might influence the results, minimize or standardize all other light sources during measurements—e.g., fluorescent light and ambient light from windows.

3.3. Pre-treatment of the Spectra

- Average spectra of repeated vis-NIR scans on the same soil sample to avoid using false replicates in further analyses. This will also increase the signal-to-noise ratio.
- Transform the measured reflectance to apparent absorption through log(1/reflectance) to enhance the linearity between the measured absorbance and the concentration of the chemical of interest (see Note 6).
- To enhance the more chemically related peaks and reduce effects such as baseline shifts and overall curvature, it is often recommended to employ some additional pre-processing transformation of the spectra. Numerous techniques exist and most spectroscopy-dedicated software contains a collection to choose from (e.g., (23)). Unfortunately it is not possible to recommend one single or combination of treatments that would work best for all soil data sets (1) and the benefit of using any is project specific. Because of this it is recommended to test a few different transformations on a representative calibration set (see Notes 7 and 8).

3.4. Reference Soil Analyses

- The calibration statistics can never be better than the quality of the chemical reference analyses. That is, errors related to the traditional chemical analysis to which the spectra are correlated will be included in the calibration model.
- Make a statistical analysis of the soil data before using it for calibration.
  - Check for abnormalities, possible analysis errors and/or outliers. This should be done for both the reference and the spectral data. However, outlier detection and the removing of outliers should be done very restrictively.
  - To plot the data distributions is a good way of examining the data.

3.5. Calibration and Validation

There are many different algorithms that can be used to calibrate soil vis-NIR spectra to predict soil properties. They include multiple linear regression (MLR), principal component regression (PCR) and Partial least squares regression (PLS) as well as data mining techniques like artificial neural networks (ANN), multivariate adaptive regression splines (MARS) and boosted regression trees (12). They all have merits and disadvantages and we will not make specific recommendations on which technique to choose, but the linear ones are more straight forward and most commonly used. At the same time, the use of data mining is increasing, especially for large diverse data sets where data mining is believed to perform slightly better than linear analyses (12). Rather, we will give some general recommendations on what to think about when choosing calibration samples and how to validate your model.

3.5.1. Calibration Set

- The calibration set should cover all possible and relevant variation that is present in the data for which it is meant to be used. That is, if the model is intended to be used for soil samples all over a country, the calibration samples should, as far as possible, include all existing soil types. On the other hand, if the model is intended to be used at a local scale, the calibration samples should capture that variation, and there is no point including soil types that are not present which might actually worsen prediction results.
- The number of calibration samples needed depends on the variation to be covered. As a general rule, the more calibration samples the more robust the calibration model. One hundred to 200 calibration samples might be at the lower limit for a model intended to cover a large geographical area with many and diverse soil types (24). As few as 25 calibration samples can result in good predictions at the farm or field scale, though it is at the very limit (25). The number of calibration samples needed also depends on the target variable. Calibrations for directly measured soil properties with a strong influence on the vis-NIR spectra such as clay could result in good prediction models with rather few calibration samples, whereas indirectly measured soil properties, such as plant nutrients that rely on co-variations with properties that absorb in the vis-NIR, often needs more calibration samples (26).
- When developing a soil spectral library from scratch, soil sampling strategy is very important to capture the variation needed for a good calibration set, independent of scale (26–28).
- Depending on the project you might have a (large) number of vis-NIR scanned samples from where a calibration sample set should be selected. Selecting the calibration samples based on their vis-NIR spectra is a good way to capture as much as possible of the variation in the larger data set. Ideally, the

distribution of the calibration set should be a flat even distribution which can be achieved using, e.g., the Kennard and Stones uniform mapping algorithm (29).

3.5.2. Validation

- To make a true estimate of the prediction performance of a calibration a truly independent validation set is necessary. This means that this set should be sampled and analyzed independently from the calibration set and has no influence on the calibration procedure. Ideally, the validation samples should not even have been sampled at the same time as the calibration samples or at least not with the same strategy.
- If no separate validation set has been collected during the soil sampling and the validation samples are to be selected from the total number of samples, we recommend the following:
  - For field and farm scale analyses:
    If the data consist of a number of soil profiles with several horizons it is recommended that all the samples in a profile follow each other, either in the calibration or in the validation set, since the different horizons in the same profile cannot be considered as independent.
  - For regional or global scale analyses:
    Avoid sample sets with geographical sample clusters, within which dependence between samples can be assumed (e.g., multiple samples from individual field trials or farm soil mapping samples clustered at individual farms or fields). If clustered data are to be used it is recommended that all samples from one cluster follow each other either in the calibration or in the validation set to avoid overoptimistic predictions (e.g., Brown et al. (30)).
- During the calibration an internal validation (before validating the model using an independent validation set) is applied. Depending on calibration method this serves to find the optimum number of components (e.g., PCR, PLSR), to find the best wavelengths to use, or for adjusting and tuning of the calibration model. For this, cross validation or bootstrapping is often recommended. In cross validation, one or more samples are systematically kept out of the calibration and are predicted using that calibration. This is done for all samples, which means that all samples will have been used for both calibration and for validation. The same considerations as applied for the selection of an independent validation set should be applied to the cross validation (see above).
- There is no hard rule regarding what ratio is preferable between the number of calibration and validation samples. However, 2/3 calibration samples and 1/3 validation samples is rather common and can be suggested as a bench mark.

3.5.3. Model Assessment

- There are several numerical measurements describing the performance of the predictions. We recommend the use of the root mean squared error (RMSE), bias (or mean error) and standard deviation of the error distribution (SDE) to account for accuracy and imprecision of the predictions, and the ratio of performance to deviation (RPD) for assessments across units:
  - $$\mathrm{RMSE} = \sqrt{\frac{\sum_{i=1}^{N}(\hat{y}_i - y_i)^2}{N}}$$
  - $$\mathrm{Bias}\ (\text{mean error}) = \sqrt{\frac{\sum_{i=1}^{N}(\hat{y}_i - y_i)}{N}}$$
  - $$\mathrm{SDE} = \sqrt{\frac{\sum_{i=1}^{N}(\hat{y}_i - y_i - \mathrm{Bias})^2}{N-1}}$$
  - $$\mathrm{RPD} = \frac{\text{standard deviation}}{\mathrm{RMSE}}$$

  where $\hat{y}$ is the predicted value and y is the measured value for sample i, with N number of samples.
- Relate the result to what the model it is intended for. Are the estimated prediction errors good enough for the intended implementation?
- When assessing the model, plotting the results (predicted versus measured) is often very helpful.
  - Are there signs of nonlinearity? (This might be improved by using nonlinear data mining calibration techniques).

3.5.4. Software

There are several commercial software packages specially developed for analysis and calibrations of spectral data providing easy to use functions for spectral analyses and calibrations. Many manufacturers provide their instruments with specific software or have close collaborations with commercial software developers to facilitate exporting and importing of measured spectra and sometimes allowing for real-time predictions. Alternatively, most of the analyses and calibration tools can also be found in commercial, shareware and freeware software (23) and in open source environments such as the R-project (31).

4. Notes

1. The signal-to-noise ratio can be improved by a longer integration time (the time interval during which energy is collected), a stronger light source or a higher degree of spectra averaging (number of readings averaged to one collected spectrum). To what degree these are configurable by the user is instrument and sample presentation device dependent.
2. "Post dispersive" means that the sample is subjected to a white light source and dispersed into the separate registered bands after reflectance. That means that only a small part of the energy in the stray light, that corresponding to each registered band, influences the spectrum.
3. Vis-NIR calibrations are largely instrument specific, and even moving or sharing calibrations among instruments of the same type can lead to reduced prediction capacity (22, 32). Pimstein et al. (32) propose the use of a common standard in combination with a common protocol to facilitate sharing and moving calibrations between instruments. In their study, washed and bleached sand (90 % quarts) performed well as such a standard.
4. The two steps in the soil sample pre-treatment are similar to common standard pre-treatments for many chemical and physical analyses, which has practical advantages. It is possible to make good calibrations also on moist soils. Actually there are examples where calibrations on field moist samples have been beneficial (33), etc. and standardized remoistening has led to substantial improved performance of clay and SOC calibrations (13). However, typically calibrations on dry soil in the lab perform better compared to those on field moist soil. This is mainly due to the higher degree of standardization and the fact that broad water bands near 1,400 and 1,900 nm tend to override adjacent absorption bands.
5. The crushing and sieving of soils removes stones and larger plant residues and also forms a basis for representative subsampling. Further grinding and sieving will lead to a more constant particle size, which will have an effect on spectra (16, 34). However, there are few and contradictory reports in the literature comparing the prediction performance based on differences in grinding or sieving procedures (e.g., (33, 35, 36)). Grinding of soil particles increases the overall reflectance and the effect is especially large for clay as aggregates are crushed. However, this effect can be more or less eliminated by a pretreatment step of the spectra discussed in Subheading 3.3.
6. This is recommended because soil spectra also are affected by structural properties of the sample, causing nonlinear light scattering. This means that some of the light that is not measured as reflectance is not directly related to absorbance

but is scattered. Other equations that can be used are for example the Kubelka-Munk transformation $(1 - \text{reflectance})^2/(2 \times \text{refelctance})$ and the Dahm equation (37).

7. Commonly used transformations include the 1st and 2nd derivative combined with smoothing (38) multiplicative scatter correction (39) and standard normal variate combined with D-trending (40). Other transformations are baseline corrections and different types of normalizations. A method that has gained in interest is wavelet transform (41) which, among other things, provides means to divide the variation in the spectra into different scales.
8. It is always possible that information is removed together with what is expected to be noise and because of this care should be taken not to overdo this step.

   This step cannot compensate for having very noisy raw spectra, which is why it is important to maximize the signal-to-noise ratio during measurements (see Subheading 3.2)

References

1. Stenberg B, Rossel RAV, Mouazen AM, Wetterlind J (2010) Visible and near infrared spectroscopy in soil science. In: Sparks DL (ed) Advances in agronomy, vol 107. pp 163–215
2. Viscarra Rossel RA, Walvoort DJJ, McBratney AB, Janik LJ, Skjemstad JO (2006) Visible, near infrared, mid infrared or combined diffuse reflectance spectroscopy for simultaneous assessment of various soil properties. Geoderma 131:59–75
3. Miller CE (2001) Chemical principles of near-infrared technology. In: Williams P, Norris K (eds) Near-infrared technology in the agricultural and food industries, 2nd edn. American Association of Cerial Chemists Inc, Minesota, pp 19–37
4. Martens H, Naes T (1989) Multivariate calibration. Wiley, Chichester, UK
5. Liu WD, Baret F, Gu XF, Tong QX, Zheng LF, Zhang B (2002) Relating soil surface moisture to reflectance. Remote Sens Environ 81:238–246
6. Twomey SA, Bohren CF, Mergenthaler JL (1986) Reflectance and albedo differences between wet and dry surfaces. Appl Opt 25:431–437
7. Schulze DG (2002) An introduction to soil mineralogy. In: Dixon JB, Schulze DG (eds) Soil mineralogy with environmental applications. Soil Science Society of America Inc., Madison, WI, pp 1–35
8. Sherman DM, Waite TD (1985) Electronic spectra of Fe3+ oxides and oxiyhydroxides in the near infrared to ultraviolet. Am Mineral 70:1262–1269
9. Clark RN, King TVV, Klejwa M, Swayze GA, Vergo N (1990) High spectral resolution reflectance spectroscopy of minerals. J Geophys Res Solid Earth Planets 95:12653–12680
10. Bishop JL (1994) Infrared spectroscopic analyses on the nature of water in montmorillonite. Clays Clay Miner 42:702–716
11. Ben-Dor E, Irons JR, Epema GF (1999) Soil refelctance. In: Rencz AN (ed) Remote sensing for the earth sciences: manual of remote sensing, 3rd edn. Wiley, New York, pp 111–188
12. Viscarra Rossel RA, Behrens T (2010) Using data mining to model and interpret soil diffuse reflectance spectra. Geoderma 158:46–54
13. Stenberg B (2010) Effects of soil sample pretreatments and standardised rewetting as interacted with sand classes on Vis-NIR predictions of clay and soil organic carbon. Geoderma 158:15–22
14. van Groenigen JW, Mutters CS, Horwath WR, van Kessel C (2003) NIR and DRIFT-MIR spectrometry of soils for predicting soil and crop parameters in a flooded field. Plant Soil 250:155–165
15. Malley DF, Yesmin L, Wray D, Edwards S (1999) Application of near-infrared spectroscopy in analysis of soil mineral nutrients. Comm Soil Sci Plant Anal 30:999–1012
16. Dunn BW, Beecher HG, Batten GD, Ciavarella S (2002) The potential of near-infrared

reflectance spectroscopy for soil analysis – a case study from the Riverine Plain of south-eastern Australia. Aust J Exp Agric 42:607–614
17. Mouazen AM, De Baerdemaeker J, Ramon H (2006) Effect of wavelength range on the measurement accuracy of some selected soil constituents using visual-near infrared spectroscopy. J Near Infrared Spectrosc 14:189–199
18. Christy CD (2008) Real-time measurement of soil attributes using on-the-go near infrared reflectance spectroscopy. Comput Electron Agr 61:10–19
19. Mouazen AM, De Baerdemaeker J, Ramon H (2005) Towards development of on-line soil moisture content sensor using a fibre-type NIR spectrophotometer. Soil Till Res 80:171–183
20. Sudduth KA, Hummel JW (1993) Soil organic matter, CEC, and moisture sensing with a portable NIR spectrophotometer. Trans ASAE 36:1571–1582
21. Viscarra Rossel RA, Cattle SR, Ortega A, Fouad Y (2009) In situ measurements of soil colour, mineral composition and clay content by vis-NIR spectroscopy. Geoderma 150:253–266
22. Williams PC, Norris K (2001) Variables affecting near-infrared spectroscopic analysis. In: Williams P, Norris K (eds) Near-infrared technology in the agricultural and food industries, 2nd edn. American Association of Cerial Chemists Inc., Minesota, pp 171–185
23. Viscarra Rossel RA (2008) ParLeS: software for chemometric analysis of spectroscopic data. Chemometr Intell Lab Syst 90:72–83
24. Shepherd KD, Walsh MG (2002) Development of reflectance spectral libraries for characterization of soil properties. Soil Sci Soc Am J 66:988–998
25. Wetterlind J, Stenberg B, Söderström M (2007) Farm-soil mapping using NIR-technique for increased sample point density. In: Stafford JV (ed) 6th European conference on precision agriculture. Wageningen Academic publishers, Skiathos, Greece, pp 265–270
26. Wetterlind J, Stenberg B, Soderstrom M (2010) Increased sample point density in farm soil mapping by local calibration of visible and near infrared prediction models. Geoderma 156:152–160
27. Viscarra Rossel RA, Jeon YS, Odeh IOA, McBratney AB (2008) Using a legacy soil sample to develop a mid-IR spectral library. Aust J Soil Res 46:1–16
28. de Gruijter J, Brus D, Bierkens M, Knotters M (2006) Sampling for natural resource monitoring.
29. Kennard RW, Stone LA (1969) Computer aided design of experiments. Technometrics 11:137–148
30. Brown DJ, Bricklemyer RS, Miller PR (2005) Validation requirements for diffuse reflectance soil characterization models with a case study of VNIR soil C prediction in Montana. Geoderma 129:251–267
31. R Development Core Team (2010) R: a language and environment for statistical computing, 2.11.1 edn. R Foundation for Statistical Computing, Vienna, Austria
32. Pimstein A, Notesco G, Ben-Dor E (2011) Performance of three identical spectrometers in retrieving soil reflectance under laboratory conditions. Soil Sci Soc Am J 75:746–759
33. Fystro G (2002) The prediction of C and N content and their potential mineralisation in heterogeneous soil samples using Vis-NIR spectroscopy and comparative methods. Plant Soil 246:139–149
34. Ben-Dor E, Banin A (1995) Near-infrared analysis as a rapid method to simultaneously evaluate several soil properties. Soil Sci Soc Am J 59:364–372
35. Russell CA (2003) Sample preparation and prediction of soil organic matter properties by near infra-red reflectance spectroscopy. Commun Soil Sci Plant Anal 34:1557–1572
36. Barthes BG, Brunet D, Ferrer H, Chotte J-L, Feller C (2006) Determination of total carbon and nitrogen content in a range of tropical soils using near infrared spectroscopy: influence of replication and sample grinding and drying. J Near Infrared Spectros 14:341–348
37. Dahm DJ, Dahm KD (2007) Interpreting diffuse reflectance and transmittance: a theoretical introduction to absorption spectroscopy of scattering materials. NIR Publications, Chichester, UK
38. Savitzky A, Golay M (1964) Smoothing and differentiation of data by simplified least squares procedures. Anal Chem 36:1627–1639
39. Geladi P, Macdougall D, Martens H (1985) Linearization and scatter-correction for near-infrared reflectance spectra of meat. Appl Spectrosc 39:491–500
40. Barnes RJ, Dhanoa MS, Lister SJ (1989) Standard normal variate transformation and de-trending of near-infrared diffuse reflectance spectra. Appl Spectrosc 43:772–777
41. Viscarra Rossel RA, Lark RM (2009) Improved analysis and modelling of soil diffuse reflectance spectra using wavelets. Eur J Soil Sci 60: 453–464

Chapter 7

Mineral Composition Analysis: Measuring Anion Uptake and Anion Concentrations in Plant Tissues

Malcolm J. Hawkesford, Saroj Parmar, and Peter Buchner

Abstract

This chapter describes two basic complementary methods relevant to at least three major macronutrients in plants: NO_3^-, SO_4^{2-}, and phosphate. The first method is the simultaneous determination of tissue content of the oxyanions, NO_3^-, SO_4^{2-}, and phosphate by HPLC, and the second is the determination of tissue uptake (transport) capacity for these same oxyanions. NO_3^-, phosphate, and SO_4^{2-}, as well as other anions including chloride, malate, and nitrite are extracted from milligram quantities of plant tissue and are separated and quantified in a single chromatographic (HPLC) run. Information on uptake (flux) of these same anions through the roots may be obtained using isotopically labeled elements, enabling transport capacity of roots and subsequent translocation to shoot tissues to be determined.

Key words: Nitrate, Sulfate, Phosphate, Anion content, Anion influx, Roots, HPLC

1. Introduction

Measuring the tissue contents of inorganic and organic anions may be achieved by a variety of techniques. Classical single analysis of NO_3^-, phosphate, or SO_4^{2-} may be achieved through colorimetric methods or for SO_4^{2-} through turbidometry by barium precipitation (1). Extremely high throughput can be achieved, for example, by using continuous flow analyzers (for example, Skalar BV, Breda, The Netherlands). The approach described here is a highly sensitive analysis which has the added convenience of the simultaneous measurement of multiple anions, and utilizes HPLC, and by extension, capillary electrophoresis (CE) methods. This method has been applied for the determination of sulfur nutritional status of crops by analysis of SO_4^{2-}:malate ratios, with the advantage that for ratios absolute calibration is not required and additionally the sample quantity is irrelevant (2).

Frans J.M. Maathuis (ed.), *Plant Mineral Nutrients: Methods and Protocols*, Methods in Molecular Biology, vol. 953,
DOI 10.1007/978-1-62703-152-3_7, © Springer Science+Business Media, LLC 2013

Influx measurements of anions, for which there are useful labeled isotopes, allow the influx capacity of roots to be determined. This is readily achieved for SO_4^{2-} and phosphate using radio isotopes. For NO_3^-, it is possible to use ^{13}N; however, the requirement for a cyclotron and the extremely short half-life of this isotope makes this an inconvenient method to use; however, the stable ^{15}N isotope in conjunction with mass spectrometry is usually sufficient although it has a lower sensitivity. Use of these two N isotopes for uptake studies have been compared (3). The stable isotope of sulfur (^{34}S) may also be considered but this is generally a less-sensitive, slower, and more costly procedure compared to using ^{35}S. To assay transporter capacity of the plasma membranes of root cells involved in anion uptake into the roots, generally a short exposure to isotopically labeled solutions is employed followed by analysis of the total amount taken up.

2. Materials

2.1. HPLC Equipment

Many companies, including Dionex (Dionex, Sunnyvale, CA, USA), provide a range of instrumentation. The absolute minimum is an isocratic system with an anion suppressor and conductivity detector and an anion exchange column such as an IonPac AS9-SC column along with an appropriate guard column, e.g., AS9G. A two-pump gradient system is convenient but not essential: advantages of the dual pump setup are that an eluent concentrate may be used and cleaning operations are easier. The anion suppressor is required to improve sensitivity when using a conducting mobile phase. Recent developments are the so-called reagent-free HPLC and CE systems which may also be convenient. A suitable setup is the ICS-2100 Integrated Reagent-Free IC System (Dionex).

2.2. HPLC Reagents

All solutions are prepared using ultrapure water (purify deionized water to attain a resistivity above 18 MΩ cm at 25 °C and filter using 0.2 μm filter) and analytical grade reagents. Solutions should be bubbled (often termed sparging) with helium for 15 min. Prepare and store all reagents at room temperature.

1. The $Na_2CO_3/NaHCO_3$ eluent is used for the measurement of inorganic anions (chloride, NO_3^-, phosphate and SO_4^{2-}) on the AS9SC column. Preparation of a 10× strength eluent of 18 mM Na_2CO_3 and 17 mM $NaHCO_3$ is convenient.
2. Weigh out 1.908 g of Na_2CO_3 (18 mM) and dissolve by gentle shaking.
3. Once the Na_2CO_3 is dissolved, weigh out 1.428 g of $NaHCO_3$ (17 mM) and dissolve in the same solution by gentle shaking.
4. The volume is made up to exactly 1 L.

Table 1
Concentrations of anions present in standards

Anion	Std 1	Std 2	Std 3	Std 4	Std 5	Std 5 (μM)
Fluoride (ppm)	0.08	0.3	0.75	1.5	3.0	158
Chloride (ppm)	0.15	0.6	1.5	3.0	6.0	169
Bromide (ppm)	0.63	2.5	6.25	12.5	25	313
Nitrate (ppm)	0.75	3.0	7.5	15	30	484
Phosphate (ppm)	1.5	6.0	15	30	60	660
Sulfate (ppm)	0.75	3.0	7.5	15	30	312
Malate[a] (μM)	12.5	50	125	250	500	500

[a]Optional

2.3. HPLC Standards

For measuring inorganic anions, five external standards are prepared containing F^-, Cl^-, Br^-, NO_3^-, phosphate, and SO_4^{2-}. Premade mixes are suitable (Mixed standard concentrated solution PO4J/4552/05, Fisher Scientific, Loughborough, UK). Malate or other inorganic acids need to be added separately to the stock solution. For final concentrations of individual anions in the dilution series, see Table 1. The serial dilution is made up as follows:

1. To make standard 5: Add 25 ml mixed anion solution (Fisher J/4552/05) to a 250 ml volumetric and fill up with 225 ml filtered, deionized water. If required, add 0.0168 g of L-malate to give a concentration of 500 μM.
2. To make standard 4: Add 50 ml of standard 5 to a 100 ml volumetric and fill up with 50 ml filtered, deionized water.
3. To make standard 3: Add 25 ml of standard 5 to a 100 ml volumetric and fill up with 75 ml deionized, filtered water.
4. To make standard 2: Add 25 ml of standard 5 to a 250 ml volumetric and fill up with 225 ml deionized, filtered water.
5. To make standard 1: Add 25 ml of standard 2 to a 100 ml volumetric and fill up with 75 ml deionized, filtered water.

2.4. Ion Uptake

2.4.1. Setup

Uptake solutions consist of the same nutrient as used for growing the plants with a defined concentration of the nutrient(s) of interest spiked with a suitable radio or stable isotope. The concentration of the ion of interest will depend upon the object of the experiment and can be above the presumed affinity constant (K_m) for the transport if total influx capacity is required and should cover a range of concentrations if measurements of the K_m are the object of the experiment. For most plant seedling, suitable concentrations are 2 mM NO_3^-, 0.15 mM SO_4^{2-}, and 0.1 mM phosphate in the "uptake solution". Optionally the pH may be controlled with the use of a suitable buffer, for example 2.5 mM Mes/Tris (e.g., at pH 5.5). Assays should be run at controlled temperatures (20–25 °C). The uptake media should be gentle aerated to ensure oxygenation and mixing of the solution. A suitable setup will utilize plastic capillary tubing attached to a rubber tube from an air pump (aquarium type) utilizing hypodermic syringe needles with plastic syringe attachment flanges removed. Volumes of uptake solution without isotope label are also required.

2.4.2. Labeled Isotopes

Add carrier-free $^{35}SO_4^{2-}$, ^{32}P- or ^{33}P-phosphate (as Na_2SO_4, disodium phosphate or orthophosphoric acid) (PerkinElmer NEN Radiochemicals, UK) to final levels of radioactivity of around 12 and 7.5 GBq per mol, respectively. For example, this requires about 50 μCi (1.8 MBq) and 20 μCi (0.75 MBq) of carrier-free $^{35}SO_4^{2-}$ or ^{32}P-phosphate per liter of final uptake solution, respectively. The absolute amounts are not critical as the uptake is calibrated relative to the isotope level in the uptake solution determined experimentally. For ^{15}N labeling, $^{15}N\text{-}NO_3^-$ should be added to a final enrichment of about 25% utilizing 60% enriched $K^{15}NO_3$, Sigma Aldrich. Subsequent analysis is by mass spectrometry.

2.4.3. A Suitable Scintillation Fluid

This should be able to mix with a large aqueous volume. A typical reagent is Packard Elmer Ultima Gold (Waltham, MA, USA).

3. Methods

3.1. Sample Preparation for HPLC

1. Weigh accurately 10–100 mg ground dry (freeze-dried, air-dried, or oven-dried) plant material into a 2 ml microtube.
2. Add 1–2 ml deionized H_2O and heat at 80 °C for a minimum of 30 min (2–4 h is preferable) with frequent mixing or shaking in a heating block.
3. Allow to cool and then centrifuge (microfuge) at 13,000 × *g* for 20 min.
4. Transfer supernatant to a fresh tube and freeze at −20 °C overnight.
5. Allow to thaw and centrifuge at around 13,000 × *g* (microfuge) for 20 min (this leaves a precipitate, most likely of carbohydrates).
6. Filter the supernatant using a 0.2 μm membrane syringe filter. A 4 mm diameter disposable filter is usually suitable.
7. The supernatant is ready for injection onto the HPLC—use 0.1 ml.

3.2. Chromatography Conditions

1. The 10× strength eluent is diluted to 1× strength, either manually or by utilizing a twin-pump HPLC system. General operating conditions will be around 2 ml/min; however, column manufacturers will provide specific guidelines.
2. A sample of standards should be run every ten samples for quality control of the chromatography.
3. A typical standard and typical profiles for shoots and roots of wheat seedlings are shown in Fig. 1.
4. To obtain the calibration curve (Fig. 2), each of the standard solutions (1–5), see Table 1, is run. The data integrator of the HPLC will give peak volumes and these are plotted against the known anion content. A regression curve provides a formula to work out the ion concentration measured in the plant samples.

3.3. Ion Influx Studies

Hydroponically grown seedlings (for wheat plants typically 2 weeks old, for Arabidopsis, any age or size that can be physically manipulated) are grown in an appropriate liquid growth medium, in some cases with manipulated supply (e.g., deficiencies) of individual nutrients to investigate impacts on transporter expression and activity.

1. Plants are gently fixed (using plastic tape) to a rack (readily self-fabricated from a discarded thin plastic bottle) that can be hung from the side of a beaker (see Fig. 3). Plants are attached near the lower part of the stem leaving the roots unimpeded. Several plants can be attached to a single rack; however, this may mean that they are not subsequently separated easily. Alternatively, a flotation device may be used which is placed directly upon the uptake solution. About 50 ml of uptake solution are required per root of around 100–200 mg fresh weight. Small Arabidopsis seedlings will require much smaller volumes.
2. The rack is placed in the uptake solution so that the roots are just submersed and uptake is started.
3. A 10–20 min incubation is sufficient for initial uptake. Longer periods of time may be used if assimilation and/or translocation to the shoot are to be assessed.
4. At the end of the incubation period the rack is used to remove the root(s) and the roots are transferred to a label-free medium for 30 s, with gentle agitation (by hand or by aeration).
5. This is repeated once more in fresh label-free medium.
6. The roots are then gently blotted dry with absorbent paper towel and excised from the shoot with scissors. The fresh weight is recorded.
7. For ^{32}P- and ^{35}S-radioisotope-labeled roots: roots are placed in a graduated pyrex glass tube with 10 ml 0.1 M HCl and heated at 100 °C for 30 min. After cooling the volume is made up to

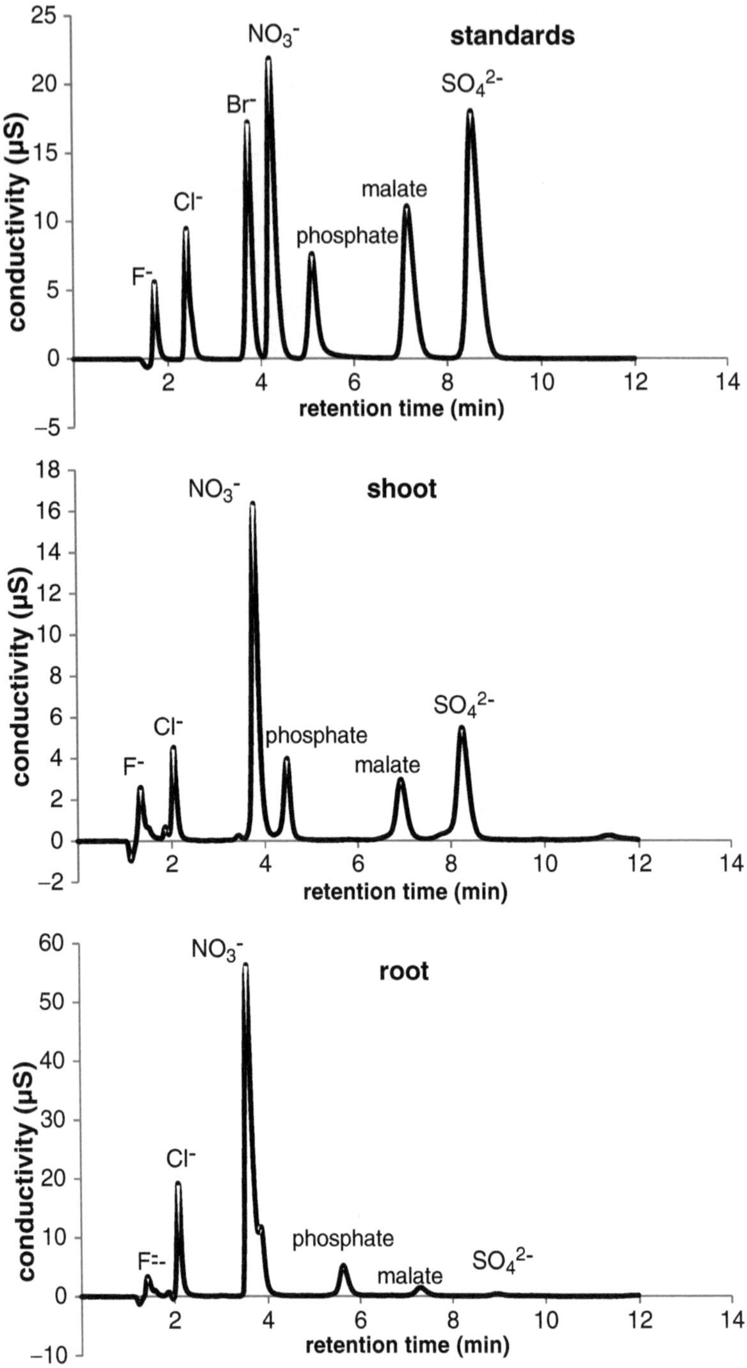

Fig. 1. Typical chromatograms showing separation of a standard mixture and wheat shoot and root extracts. Some small variations in retention times for individual peaks may occur and manual checking of the standards is essential.

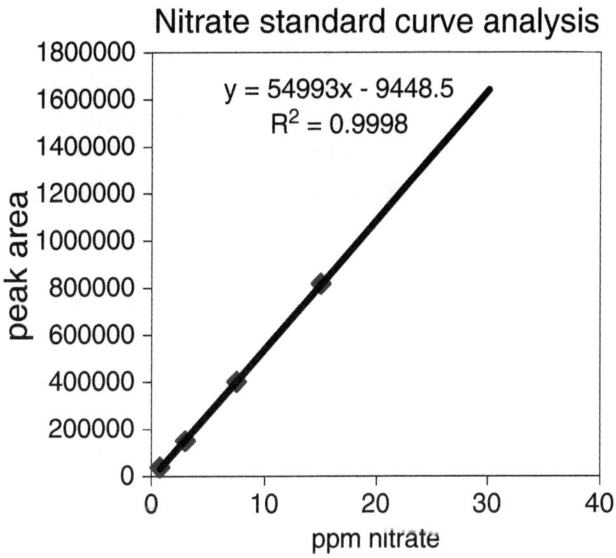

Fig. 2. For each anion a standard curve must be created relating peak volumes to the known anion concentrations of the standard mixture dilution series.

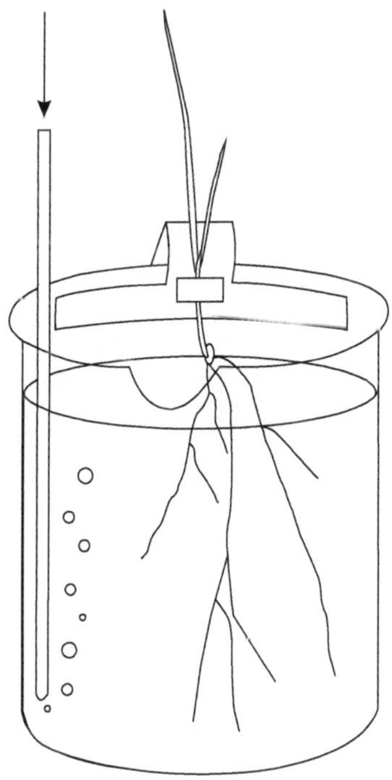

Fig. 3. A simple setup for measuring ion influx into roots. The seedling is attached to a plastic holder using plastic adhesive tape. The top of the root is aligned with the level of the bathing solution. Multiple plants may be attached to the holder. Mixing is achieved by gentle aeration. At the end of the uptake period the holder plus seedling is transferred to sequential wash solutions.

the original 10 ml. For ^{15}N, the roots are dried, reweighed and then ground or milled. A weighed aliquot is analyzed by mass spectroscopy which will give absolute N-content and ^{15}N enrichment of the sample. Uptake may be expressed on a fresh weight or dry weight basis.

8. An aliquot (1 ml) is removed and placed in a scintillation vial, to which scintillation fluid is added and the radioactivity is determined.
9. An aliquot (for example 100 μl) of the uptake medium is sampled and counted. This is vital for the influx calculation and has the additional benefit of obviating any need to account for isotope half-life in the calculation of uptake.
10. Uptake rates are usually referred to as uptake capacities and will vary greatly depending upon nutritional status. The measuring conditions may well influence the transporter expression/activity. Rates are usually expressed per g of fresh weight per hour.
11. Influx calculation examples:
    100 μl of 0.1 mM phosphate solution contains 10 nmol of phosphate and this may be used to determine the moles of P taken up by the root in the 10 min period. If the counts of the 100 μl uptake solution were 5,000 dpm, then this is 500 dpm per nmol. If a single 100 mg root contains radioactivity of 500 dpm, this equates to 1 nmol, or 10 nmol per gper 10 min or 60 nmol/g/h.

    100 μl of 0.15 mM SO_4^{2-} solution contains 15 nmol of SO_4^{2-} and this can be used to determine the moles of sulfate taken up by the root in the 10 min period. If the external solution counts were 15,000 dpm, it represents 1,000 dpm per nmol. If a single 100 mg root contains 5,000 dpm, it equates to 5 nmol per 10 min, or 50 nmol per gper 10 min, or 300 nmol/g/h.

4. Notes

HPLC

1. Samples can be stored frozen at −20 °C for several months.
2. Samples may need diluting with He-sparged (bubbled), filtered water, if peaks are off-scale. In cases where there is a particularly large difference in the content of two anions, two separate runs may be required.
3. If the samples are dark in color, the colored contaminants need to be removed using products such as Amberlite adsorbent resins

(XAD) or polyvinylpyrrolidone (PVP) (Sigma Aldrich, St Louis, MO, USA or Fisher Scientific, Loughborough, UK).

4. If there are pressure fluctuations on the HPLC, check for leaks or blocked tubing.
5. Check column performance every ten samples by running a standard.
6. Column performance degradation (poorly separated peaks) indicates a dirty column and at the first sign of this, columns need to be cleaned.
7. Clean the guard column first. For this, make up 500 ml of 200 mM Na_2CO_3/75 mM $NaHCO_3$. Disconnect the guard column from the separating column. Pump cleaning solution through the columns for 30–60 min. Pump application eluent for a further 60 min. Reconnect the guard column to the separating column and equilibrate until you get a steady baseline.
8. If problems still persist, organic contaminants may need to be removed and both the guard and analytical columns may need cleaning. Prepare two bottles with 500 ml each of the following eluents:

   Bottle 1: 100% acetonitrile

   Bottle 2: 1 M NaCl, using high-quality (18 MΩ) filtered water, adjusted to pH 2 using HCl. Disconnect the analytical column from the anion suppressor and reverse the order of guard and analytical column, i.e., connect the guard column to the outlet line of the analytical column. This is to avoid any contaminants accumulated on the guard column washing onto the analytical column. Set the pump flow rate at 2.0 ml/min. Rinse the column with deionized water for 10 min. Set the gradient pump to mix 80% acetonitrile and 20% 1 M NaCl. Isocratically pump the cleaning solution through the columns for at least 60 min. Rinse the column for 10 min with deionized water. Disconnect the guard column and reconnect in front of the analytical column; connect the outlet from the analytical column to the anion suppressor. Equilibrate the column with application eluent (this can take a long time and is best left running overnight).
9. Decreases in retention time and bunched peaks can be overcome by reducing the flow rate. Do not run at less than 1.0 ml/min. If necessary, clean or replace the guard and/or analytical columns.
10. Typical NO_3^- content is 100–400 μmol/g DW for roots and 200–800 μmol/g DW for shoots.
11. Typical phosphate content (highly dependent on nutrition conditions) is 5–300 μmol/g DW for roots and shoots.

12. Typical SO_4^{2-} content (highly dependent on nutrition conditions and species) of roots is 10–20 μmol/g DW for cereals and 100 μmol/g DW for Brassica and for shoots 100–400 μmol/g DW.
13. Example references: (2, 4–8).

Influx Measurements

1. Use of alternative tissues: other than seedling roots, leaf discs, cultured roots or excised root pieces and cell cultures may all be assayed similarly. Cell cultures require a filtration step, for example using a GF/C filter with suitable porosity (9).
2. Analysis of shoot materials allows for an estimation of translocation to the shoot. Often this will require longer uptake periods.
3. Although not critical, the volume of the uptake solution needs to be sufficient to bath the roots, allow for adequate mixing, and avoid depletion of the anion of interest during the time of incubation.
4. Typical nitrate uptake capacity: 7.5 μmol/g FW of root/h.

   Typical phosphate uptake capacity: these are highly dependent on nutritional conditions (higher if P-starved). Rates for cereals are 200–1,500 nmol/g FW of root/h and >500 nmol/g FW of root/h for Brassica species.
5. Typical SO_4^{2-} uptake rate: these are highly dependent on species and nutritional conditions (higher if S-starved). Rates for cereals and Brassica species are 50–1,000 nmol/g FW of root/h.
6. Examples references: (3, 8, 10, 11).

Acknowledgement

This research is funded by the Biotechnology and Biological Sciences Research Council (BBSRC) of the UK.

References

1. Tabataba MA, Bremner JM (1970) Simple turbidimetric method of determining total sulfur in plant materials. Agron J 62:805–810
2. Blake-Kalff MMA, Hawkesford MJ, Zhao FJ, McGrath SP (2000) Diagnosing sulfur deficiency in field-grown oilseed rape (*Brassica napus* L.) and wheat (*Triticum aestivum* L.). Plant Soil 225:95–107
3. Clarkson DT, Gojon A, Saker LR, Wiersema PK, Purves JV, Tillard P, Arnold GM, Paans AJM, Vaalburg W, Stulen I (1996) Nitrate and ammonium influxes in soybean (Glycine max) roots: direct comparison of N-13 and N-15 tracing. Plant Cell Environ 19:859–868
4. Smith FW, Hawkesford MJ, Ealing PM, Clarkson DT, Vanden Berg PJ, Belcher AR, Warrilow AGS (1997) Regulation of expression of a cDNA from barley roots encoding a high affinity sulphate transporter. Plant J 12: 875–884

5. Blake-Kalff MMA, Harrison KR, Hawkesford MJ, Zhao FJ, McGrath SP (1998) Distribution of sulfur within oilseed rape leaves in response to sulfur deficiency during vegetative growth. Plant Physiol 118:1337–1344
6. Buchner P, Stuiver CEE, Westerman S, Wirtz M, Hell R, Hawkesford MJ, De Kok LJ (2004) Regulation of sulfate uptake and expression of sulfate transporter genes in *Brassica oleracea* as affected by atmospheric H_2S and pedospheric sulfate nutrition. Plant Physiol 136:3396–3408
7. Brown DA, Clark LJ, Howarth JR, Parmar S, Hawkesford MJ (2006) Mechanical impedance and nutrient acquisition in rice. Plant Soil 280:65–76
8. Koralewska A, Buchner P, Stuiver CEE, Posthumus FS, Kopriva S, Hawkesford MJ, De Kok LJ (2009) Expression and activity of sulfate transporters and APS reductase in curly kale in response to sulfate deprivation and re-supply. J Plant Physiol 166:168–179
9. Clarkson DT, Diogo E, Amancio S (1999) Uptake and assimilation of sulphate by sulphur deficient Zea mays cells: the role of O-acetyl-L-serine in the interaction between nitrogen and sulphur assimilatory pathways. Plant Physiol Biochem 37:283–290
10. Lee RB (1982) Selectivity and kinetics of ion uptake by barley plants following nutrient deficiency. Ann Bot 50:429–449
11. Hawkesford MJ, Belcher AR (1991) Differential protein synthesis in response to sulfate and phosphate deprivation – identification of possible components of plasma-membrane transport-systems in cultured tomato roots. Planta 185: 323–329

Chapter 8

Multielement Plant Tissue Analysis Using ICP Spectrometry

T.H. Hansen, T.C. de Bang, K.H. Laursen, P. Pedas, S. Husted, and J.K. Schjoerring

Abstract

Plant tissue analysis is a valuable tool for evaluating the nutritional status and quality of crops and is widely used for scientific and commercial purposes. The majority of plant analyzes are now performed by techniques based on ICP spectrometry such as inductively coupled plasma—optical emission spectroscopy (ICP-OES) or ICP—mass spectrometry (ICP-MS). These techniques enable fast and accurate measurements of multielement profiles when combined with appropriate methods for sample preparation and digestion. This chapter presents state-of-the-art methods for digestion of plant tissues and subsequent analysis of their multielement composition by ICP spectrometry. Details on upcoming techniques, expected to gain importance within the field of multielement plant tissue analysis over the coming years, are also provided. Finally, attention is given to laser ablation ICP-MS (LA-ICP-MS) for multielement bioimaging of plant tissues. The presentation of the methods covers instructions on all steps from sampling and sample preparation to data interpretation.

Key words: Element profiling, ICP-MS, ICP-OES, Laser ablation, Microwave digestion, Multielement analysis, Plant analysis

1. Introduction

Plants are able to complete a full life cycle based on photochemical energy, CO_2, H_2O and 14 mineral elements, the latter classified as the essential plant nutrients. As much as 95 % of plant dry matter is composed of hydrogen (H), carbon (C), and oxygen (O). The remaining 5 % consist of the 6 macronutrients (N, Mg, P, S, K, Ca) and the 8 micronutrients (B, Cl, Mn, Fe, Ni, Cu, Zn, Mo) which are absorbed from the soil solution (1).

T.H. Hansen and T.C. de Bang contributed equally to the chapter.

Frans J.M. Maathuis (ed.), *Plant Mineral Nutrients: Methods and Protocols*, Methods in Molecular Biology, vol. 953, DOI 10.1007/978-1-62703-152-3_8, © Springer Science+Business Media, LLC 2013

1.1. Impact of Multielement Analysis

A balanced supply of the essential plant nutrients in agricultural production systems is a prerequisite for optimal plant growth, high harvest yields and acceptable quality of the harvested products. Multielement plant tissue analysis is increasingly being used as a tool to diagnose the nutritional status of plants. The measured concentration profiles are evaluated by comparison with species-specific reference threshold values for each plant nutrient (2). In most cases an insufficient supply can subsequently be corrected by fertilization. The element composition of plants is of relevance not only to plant producers but also to plant consumers because vegetal food is a major source of inorganic elements that are essential or beneficial for human nutrition and health. Virtually any element present in the soil is to a smaller or greater extent taken up by plants. This does include not only the essential nutrients but also toxic elements such as cadmium, arsenic, and mercury. Multielement analysis is therefore an important tool for evaluation of the nutritional quality of plant-based foods (3, 4).

Element profiling of plant tissue is also an important research tool which has been used to obtain knowledge about the biological controls of plant nutrient acquisition and the functional roles of mineral elements in plant growth and metabolism. Element profiling has for example been used to identify major genes and gene networks of critical importance to plant mineral nutrition and trace element accumulation (5, 6), to obtain increased understanding of the physiological functions of individual genes (7, 8), to evaluate nutrient use efficiency (9), to verify the authenticity of organic plant products (10) and to reveal the distribution of elements within specific plant organs (11, 12). In combination with liquid chromatography, multielement analysis can be used to study the binding forms of inorganic elements which are important in order to get better knowledge about their biological functions and bioavailability (13, 14). The latter aspect is particularly important in relation to solving some of the very severe global problems caused by deficiency of mineral elements, particularly Fe, Zn, Se, and Mg, in human nutrition. The major causes of these deficiencies are low intake due to diets consisting mostly of a few staple foods with low nutrient density, e.g., cereal grain. Possibilities for enriching food staples through plant breeding or genetic engineering are currently investigated in a large number of ongoing research initiatives and international programmes, e.g., the HarvestPlus Challenge Programme (15). Concentrations of mineral elements can vary severalfold among plant genotypes (16), and multielement analysis is an important tool to reveal these differences in breeding programmes focusing on biofortification.

1.2. Techniques for Multielement Analysis

Multielement plant analysis is dominated by techniques based on inductively coupled plasma (ICP) spectrometry such as ICP-mass spectrometry (ICP-MS) and ICP-optical emission spectroscopy

(ICP-OES). These techniques have now almost eliminated the use of single element analysis with techniques such as atomic absorption spectroscopy (AAS). The analytical advantage of ICP spectrometry is founded on the capacity to detect fingerprints consisting of multiple elements in a wide concentration range spanning from below 0.1 to over 50,000 μg/g, i.e., from 0.00001 to 5 % of the plant dry matter (4). Especially ICP-MS has detection limits in the sub-ppt area and a dynamic range of up to 9 orders of magnitude. However, the analytical performance of ICP-spectrometry is highly dependent on appropriate laboratory routines for sample preparation, digestion and analysis. In addition, special attention must be given to the validation of the analytical results as emphasized below.

A diagram of the analytical flow in two different types of multielement plant tissue analyzes is shown in Fig. 1. The left-hand side of the diagram illustrates the flow of total concentration measurements which require preceding digestion of the sample to be analyzed (digestion-dependent analysis). The right-hand side of the diagram illustrates in situ bioimaging of element distribution. All steps of Fig. 1 are detailed in the following sections.

In the digestion-dependent pathway the sample is initially dried and homogenized followed by complete sample digestion

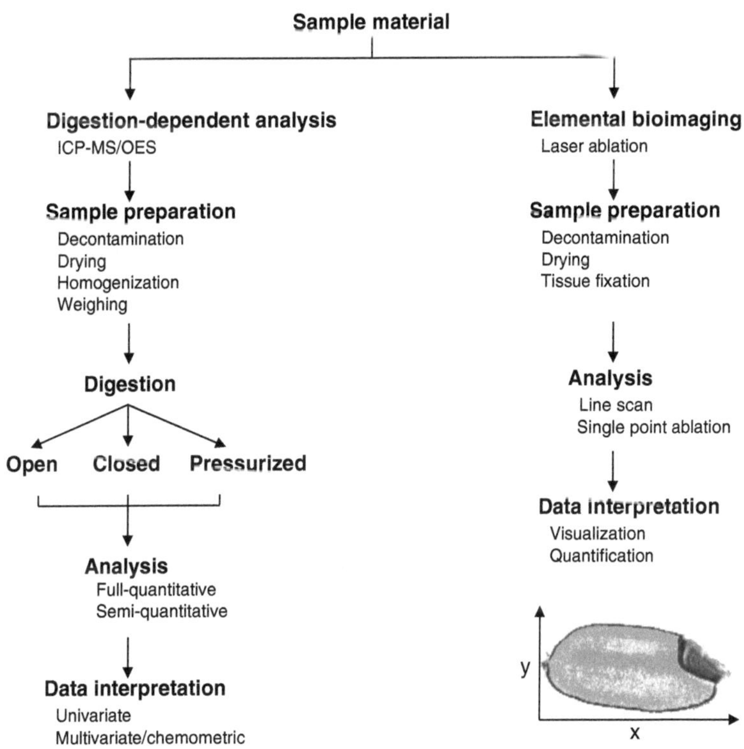

Fig. 1. A schematic flow diagram illustrating the methodological and analytical steps of multielement plant tissue analysis by ICP spectrometry. *Bottom right*: The relative distribution of phosphorus in a rice grain measured by LA-ICP-MS.

using a strong oxidising agent and elevated temperatures which can be reached by various energy sources as elaborated later. In general, 150–250 mg of homogenized material is sufficient for representative sampling. However, the amount of sample material available may in some cases be limited to only a few mg. This limitation has spurred the development of microscaled methods allowing digestion of samples below 20 mg (11). Special care must be taken during drying, homogenization and weighing of small sample amounts to achieve precise analysis. Irrespective of the quantity of material to be digested it is crucial that suitable certified references materials (CRMs), together with adequate numbers of blank samples, are included in order to evaluate the quality of the obtained results. After digestion and dilution, samples, CRMs and blanks are typically analyzed along with external calibration standards enabling full-quantitative measurements. ICP-MS, but not ICP-OES, also offers the possibility of conducting semiquantitative measurements of multielement fingerprints which are useful for fast sample comparisons (17). The final step of the analysis consists of data processing by suitable univariate or multivariate (chemometric) statistical tools (Fig. 1).

Element bioimaging of plant tissue is possible using laser ablation ICP spectrometry (LA-ICP). This technique has enabled the distribution of elements within a tissue to be studied without preceding sample digestion. The LA-ICP technique utilizes a high-energy laser beam with a diameter ranging from 5 to 500 μm. When fired directly at the solid surface of the sample, the highly concentrated energy from the laser creates an aerosol of the material, which by use of a carrier gas is transported directly from the ablation chamber to the plasma torch of the ICP spectrometer. Laser ablation can be carried out in different ways but the most frequent application is the scan mode which can provide a continuous stream of sample material to the ICP at speeds of up to 1 mm/s. Laser ablation is highly matrix dependent and data quantification remains a challenge for plant samples.

The first step of a typical LA-ICP bioimaging analysis involves sampling and decontamination of the material (Fig. 1). Next, the sample is fixed on a microscope glass slide and lyophilized. Samples of leaf material are now ready for analysis, whereas grain samples typically have to be sliced in order to ensure a smooth sample surface. The dried samples are positioned in the ablation chamber, where the sampling occurs, while the multielement detection is carried out in the ICP part of the platform. Instrumental drift can be monitored by use of CRMs, while compressed plant material with known concentrations of elements can be used for quantification purposes. The large amount of data generated in LA bioimaging studies is typically organized in spreadsheets and can be sorted and visualized using specialized macros or dedicated software packages.

2. Materials

Essentially all types of plant tissue can be analyzed by ICP spectrometry; e.g., samples of roots, leaves, grain, xylem and phloem sap as well as isolated cell compartments such as chloroplasts. The only limitation is the amount of sample material available or, more specifically, the quantity of elements contained therein. The acquired quantity depends on several factors, including digestion method, type of sample introduction (laser ablation or liquid solution) and the sensitivity of the ICP instrument. Common sample requirements are outlined in the method section.

2.1. Sample Preparation

2.1.1. Digestion-Dependent Analysis

1. Milli-Q water (Milli-Q Plus, Millipore Corporation, Bedford, MS, US).
2. Tween-20 solution (1 g/L) in Milli-Q water for surface decontamination.
3. Freeze-drier.
4. A grinding mill equipped with a titanium rotor to avoid trace element contamination.
5. Acid bath consisting of 5–10 % HNO_3.
6. Clean consumables (plastic tubes, flasks, bottles, etc.).

2.1.2. Laser Ablation

1. Milli-Q water.
2. Tween-20 solution (1 g/L) in Milli-Q water for surface decontamination.
3. Freeze-drier.
4. Vibrating blade vibrotome (e.g., a Leica VT1000 S, Leica Microsystems, Wetzlar, Germany).
5. Microscope glass slides.
6. Double adhesive tape.
7. Contact glue.

2.2. Sample Digestion (Macro- and Microscaled)

1. For open digestion, a heating device (e.g., the ModBlock™ system, CPI International, Amsterdam, Netherlands) and clean sample vials that can withstand high temperatures.
2. For closed digestion, a microwave oven (e.g., the Multiwave 3000, Anton Paar GmbH, Graz, Austria).
   (a) For microscaled digestion (1–20 mg/dry matter sample), a 64MG5 rotor (Anton Paar GmbH, Graz, Austria) with a capacity of 64 samples is used. This rotor accommodates 5 mL digestion bombs, e.g., 5 mL glass digestion vials equipped with lip seals and screw caps capable of tolerating pressures up to 20 bar.

(b) For macroscaled digestion (typically 100–300 mg/dry matter sample), a 16HF100 rotor (Anton Paar GmbH, Graz, Austria) accommodating 16 digestion bombs can be used. The bombs can be 100 mL Teflon liners inserted into ceramic vessels and closed with vessel jackets (max 70 bar, 240 °C) or similar. Other systems with different specifications are available from other manufacturers. The same is the case for rotors suitable for sample amounts in the range 20–100 mg (e.g., the 48 position rotor from Anton Paar GmbH, Graz, Austria).

3. For pressurized digestion, a combined microwave and pressurizing system (e.g., the UltraWave system, Milestone Srl, Sorisole, Italy).
4. Certified reference material (CRM), representing the matrix and elements of interest, e.g., the NIST 1515 (apple leaves) or NIST 1567a (wheat flour) from the National Institute of Standards and Technology, Gaithersburg, MD, USA (see Note 1).
5. 30 % H_2O_2 (with suitable element purity) and 15 % H_2O_2 prepared from 30 % H_2O_2 and Milli-Q water.
6. 69 % HNO_3 (with suitable element purity, see Note 2).
7. Milli-Q water.
8. 70 mL HD polyethylene vials (Capitol Vial, Fulton Ville, NY, USA).
9. All liquids used should be stored in plastic flasks and bottles to avoid contamination.

2.3. ICP Analysis

2.3.1. ICP-MS Analysis

1. An ICP-MS equipped with an octopole reaction cell, such as the Agilent 7500ce (Agilent Technologies, Manchester, UK).
2. Tuning solution containing 10 μg/L Li, Y, Ce, Tl and Co (Agilent technologies, Manchester, UK).
3. The ICP-MS should be equipped with an auto-sampler for automatic direct injection.
4. Inorganic standards for ICP-MS calibration (e.g., the non-equimolar standards P/N 4400-ICP-MSCS, P/N4400-132565A and P/N4400-132565B, CPI International, Amsterdam, Netherlands).
5. A perfluroalkoxy (PFA) micro-flow nebulizer for nebulization of liquid samples.
6. Wash solution; 1.75 % HNO_3 and 0.2 % HF, prepared from 69 % HNO_3, 40 % HF and Milli-Q water.
7. Auto-sampler needle wash; 2 % HNO_3, prepared from 69 % HNO_3 and Milli-Q water.
8. Extra wash solution; 3.5 % HNO_3, prepared from 69 % HNO_3 and Milli-Q water.

9. An internal standard solution of an element not present in the plant sample (e.g., Er at 50 µg/L) which can be spiked to all samples and standards or introduced via a T-piece.

2.3.2. ICP-OES Analysis

1. An ICP-OES preferably capable of dual viewing such as the Optima 5300 DV (PerkinElmer, Waltham, MA, US).
2. Tuning solutions containing Mn as a minimum in 0.1 or 1 mg/L concentrations.
3. The ICP-OES should be equipped with an auto-sampler for automatic direct injection.
4. Inorganic standards for ICP-MS calibration (e.g., nonequimolar standards P/N 4400-ICP-MSCS, P/N4400-132565A and P/N4400-132565B, CPI International, Amsterdam, Netherlands).
5. A Meinhard nebulizer and a cyclonic spray chamber for optimum sensitivity. Numerous other combinations of nebulizers and spray chamber are possible.
6. Wash solution; 3.5 % HNO_3 prepared from 69 % HNO_3 and Milli-Q water.
7. Internal standard solution (e.g., Y at 1 mg/L) which can be spiked to all samples and standards or introduced via a T-piece.

2.4. LA-ICP-MS Analysis

1. An ICP-MS suitable for LA hyphenation, such as an Agilent 7500ce (Agilent technologies, Manchester, UK).
2. Tuning solution containing 10 µg/L Li, Y, Ce, Tl and Co (Agilent technologies, Manchester, UK).
3. A laser ablation (LA) system as for instance the NWR-193 (New Wave Research, Electro Scientific Industries, Fremont, CA, US).
4. Certified reference material, such as the SRM 612 Trace elements in glass (National Institute for Standards and technology, Gaithersburg, US).
5. Flexible tygon® tubes for LA and ICP-MS connection.
6. A Y-piece coupling the ICP-MS carrier gas to the ablation transport stream.

2.5. Software for Data Analysis

1. ICP software (manufacturer specific).
2. Data processing software such as MatLab R2011a (Mathworks, US) or Microsoft Excel (Microsoft, US).
3. Visualization software such as SigmaPlot 11.0 (Systat Software Inc., US), Microsoft Excel (Microsoft, US) or IgorPro 6.2 (WaveMetrics Inc, US).

3. Methods

3.1. Sample Preparation (Drying, Milling, Etc.)

Plant tissue samples can originate from a wide variety of cultivation methods, e.g., growth chambers, green houses or field trials. Representative sampling has to be performed and potential sample surface contaminants eliminated in order to prevent biases. Airborne dust deposits for instance are very pronounced in field trials and surface accumulation of soil particles or applied fertilizers can lead to overestimations of the element concentrations.

3.1.1. ICP-MS and ICP-OES Measurements

1. Rinse the plant material once in a Tween-20 solution followed by three times in Milli-Q water to remove dust-borne contamination (see Note 3).
2. Lyophilize the samples (this is preferred compared to ovendrying to avoid surface contamination).
3. Homogenize the samples using a grinding mill to allow representative sub-sampling.
4. Prior to sample weighing, dry the samples and reference materials for 2 h at 60 °C. The final sample must be completely dry and well homogenized.
5. Normally, 150–300 mg is needed for representative sampling depending on the particle size.

3.1.2. LA-ICP-MS

Laser ablation enables the direct sampling of solid material and consequently minimal sample preparation is required. The ablation efficiency of an element is highly dependent on the analyzed matrix and thus absolute quantification per se, is an analytical challenge for nonhomogenized material and the procedure is therefore not presented here. The method described both involves grain and leaf analysis (see Note 4).

Leaf analysis

1. Rinse the leaf once in Tween-20 solution and two times in Milli-Q water in order to remove surface contaminants.
2. Mount the leaf on a glass slide using double adhesive tape.
3. Dry the fixed leaf in the freeze-drier for a minimum of 12 h.

Grain analysis

1. Wash the grain one time in the Tween-20 solution and two times in Milli-Q water.
2. Dry the grain in a freeze-drier until complete dryness.
3. Fix the sample on a glass slide in the proper position using contact glue.
4. Cut slides of the grain using a vibrating blade vibrotome until the surface of interest appears (see Note 5).

3.1.3. Cleaning Procedure of Consumables

Clean consumables (glass and plastic ware) are mandatory both in relation to precision and accuracy of the analysis. In general, plastic is produced with negligible amounts of element contaminants, and especially polytetrafluoroethylene (PTFE) is preferred due to its high melting point enabling digestion at high temperatures. However, because of the production costs, most PTFE is reused and therefore has to be thoroughly cleaned between individual runs. Furthermore, quartz glass of high purity can be purchased and is preferred for some elements, e.g., Se. A standard cleaning procedure for PFTE and glassware is given below:

1. Place the containers in a 5–10 % HNO_3 acid bath (10 % for PFTE and 5 % for glassware) for at least 2 h. Glassware can become fragile using higher percentages of acid.
2. Wash the containers three times with Milli-Q water.
3. Let the containers air dry in a noncontaminated environment (e.g., a sterile bench).
4. Acid baths can be reused several times (see Note 6).

3.2. Sample Digestion (Macro- and Microscaled)

Plant samples consist of an organic matrix that may cause analytical bias and block the sample introduction system if not fully decomposed. The most commonly applied methods for destruction of organic matter are based on high-temperature oxidation (dry ashing) or wet digestion in vessels using strong acids, both with or without oxidising solvents. In the following, only methods for wet digestion are described (see Note 7).

3.2.1. Open Digestion

The boiling point of the digestion reagent in open digestion systems is reached at atmospheric pressure. This decreases the digestion efficiency and limits the method to easily digestible matrices such as leaf material. However, the method is associated with relatively low costs and is advantageous in this perspective. The method described below is based on a multi-position open digestion system (see Note 8). Do also treat blanks and CRMs as described below.

1. Weigh 200–250 mg sample in 70 mL heat-resistant tubes.
2. Add 5 mL 35 % HNO_3. Mix the slurry and place the samples in the heating device over night (optional) with the lid closed.
3. Set the heating device's temperature to give an effective temperature of 90–95 °C inside the tubes. Do not shake the samples at this stage.
4. Cover the tubes with HD-PE watch glasses and heat the samples to 95 °C. Reflux 10–15 min without boiling. Leave the watch glasses on until the digestion is completed (step 11).
5. Remove the samples from the heating device and let them cool down.

6. Add 2.5 mL 69 % HNO_3 and place the samples in the heating device once more. Heat the samples and let them reflux for 30 min.
7. Remove the samples and let them cool down.
8. Add 1.5 mL 30 % H_2O_2 and put the samples in the heating device again. Cover the samples and heat up to 95 °C. This will start the peroxide reaction.
9. When the reaction is over, repeat the procedure starting from step 8, but now only with 1 mL H_2O_2. Continue with 1 mL aliquots until no reaction is observed (use no more than 5 mL, normally 3 mL is enough).
10. Reflux the samples until the volume is reduced to 5 mL (approximately 60 min).
11. Remove the samples from the heating device to cool down and wash any sample droplets from the watch glasses using Milli-Q water into the respective tubes. Dilute samples to 50 mL with Milli-Q water.
12. The samples now approximately contain 7 % HNO_3.
13. Directly before ICP analysis dilute samples 1:1 with Milli-Q water, yielding a final HNO_3 concentration of 3.5 %.

3.2.2. Closed Digestion

Systems utilizing closed vessels have several advantages compared to open digestion systems, including significantly reduced risks of contamination and loss of volatile analytes. In addition, the boiling point of the acid is elevated when the internal pressure increases upon heating, leading to a much faster and more thorough matrix digestion (see Note 9).

Macroscaled Digestion (100–300 mg Plant Material)

For macroscaled digestion, a rotor with a capacity for 16 samples, designed for digestion of samples with dry matter masses between 100 and 300 mg, is employed. It is recommended to include at least one CRM and one true blank for each duty cycle; however, a minimum of 7 blanks and 7 CRMs should be reached for the subsequent ICP analysis.

1. Around 250 mg of plant material is transferred to the microwave oven tube. Record the exact weights.
2. Add 5 mL 69 % HNO_3 and 5 mL 15 % H_2O_2.
3. Leave the samples overnight to pre-digest (optional).
4. Assemble the tubes with vessel jackets and screw caps and tighten the decomposition vessel by hand (special tools are required to widen the screw caps and tighten the vessels, respectively).
5. Place all 16 pressure vessels in the rotor body and assemble the top and the bottom of the rotor.

6. Place the rotor in the microwave oven.
7. Make a program that ramps the energy to 1,400 W over a period of 10 min. Keep the energy level for 18 min with fan speed 1 and an additional 20 min with fan speed 2. Hereafter let the system cool down to 50 °C using maximum fan speed.
8. For very reactive samples it is recommended to use an internal temperature and pressure sensor to avoid sample loss.
9. When the program has finished, place the microwave rotor in a fume hood.
10. Open the vessels slowly with the vessel tool, while the top of the rotor is still in place.
11. Remove the rotor top (in the fume hood).
12. Remove the screw cap and rinse the inner side of the seal and transfer all liquid quantitatively with Milli-Q water to HD polyethylene vials and add Milli-Q water to 50 mL. The HNO_3 concentration is approximately 7 %.
13. Directly before analysis by ICP-MS, samples are diluted 1:1 with Milli-Q water, giving a final HNO_3 concentration of 3.5 %.

Microscaled Digestion (5–20 mg of Plant Material)

In special cases, limited plant material is available and macroscaled microwave digestion is not applicable. In these cases an alternative method based on a 64 position rotor can be employed. This rotor has been optimized for the digestion of dry matter in the mass range of 2–20 mg (see Note 10). It is recommended to include at least three CRMs and three true blanks in each duty cycle, in order to monitor possible fluctuations between individual duty cycles. However, the total number of blanks and CRMs for the subsequent ICP analysis must reach 7.

1. Transfer 5–20 mg of dried plant material to the digestion vials. Record the weight of the sample.
2. Add 250 μL 69 % HNO_3 and 125 μL of 30 % H_2O_2 and leave overnight (optional)
3. Close the vials with seals and screw caps. Record the total weight of the vial (including sample, H_2O_2, HNO_3, seal and screw cap).
4. Microwave the digestion bombs for 100 min using the following program: 10 min ramping to a maximum temperature of 140 °C, keep this temperature for 80 min and afterwards let the digested samples cool down for 10 min. The maximum radiation energy input should match the number of samples to be digested (400 W for 32 samples and 600 W for 64 samples).
5. When the rotor has cooled down to room temperature, transfer samples to a –20 °C freezer for 30 min.

6. Take the samples out of the freezer and quickly release the pressure of all samples, either by using the tip of a thin hypodermic needle or by gently untying the cap.
7. Record the weight of the vials (including sample, H_2O_2, HNO_3, seal and screw cap) to quantify any potential sample loss. If the sample loss is greater than 3–5 %, adjustments of the final analysis results are necessary.
8. Dilute the samples using Milli-Q water to a final concentration of 3.5 % HNO_3 based on the remaining sample weight in a 5 mL plastic tube.

3.2.3. Pressurized Microwave Digestion

In a pressurized system, the boiling point of the acid is increased before sample heating by exposing samples to an external pressure exerted by an inert gas, e.g., nitrogen. This facilitates sample digestion without boiling and, hence, several samples can be digested in the same bomb without the risk of cross contamination. A further advantage is that different sample masses and digestion media can be included in the same run. Include blanks and CRMs as described above.

1. Add a few drops of Milli-Q water to the bottom of each digestion vial before weighing of the sample. This will ensure that samples are completely wet prior to digestion.
2. Plant material is weighed directly into the digestion vials in the mass range of 5–100 mg (22 sample position holder) or 50–300 mg (15 sample position holder).
3. Add digestion medium to the vials (see Note 11).
4. Turn on the chiller, set the cooling temperature to 10 °C, and let it run throughout the complete digestion procedure.
5. Turn on the pressurized digestion unit.
6. Add 120 mL Milli-Q water and 5 mL 30 % H_2O_2 to the main PFTE vessel and place it in the base of the pressurized digestion unit.
7. Cap all digestion bombs.
8. Position the rack into the reaction chamber and close the chamber.
9. Tighten the clamps in order to create a reaction bomb.
10. Increase the pressure in the chamber by adding 40 bars of nitrogen. This will increase the boiling temperature of the digestion medium.
11. Run a program using the maximum microwave power (1,500 W) until a temperature of 230 °C is reached which will take around 15 min. Keep this temperature for 10 min and cool to 80 °C before releasing the pressure automatically.
12. Loosen the clamps and open the main chamber.

13. Dilute samples to a final concentration of 3.5 % HNO_3 using Milli-Q water. This can either be done directly in the vials or by sample transfer. When transferring the sample to another tube, rinse the vial at least three times.

3.3. ICP Analysis

After digestion, the liquid samples are ready for multielement analysis which can be performed by either ICP-MS or ICP-OES. The advantage of ICP-MS is the multielement capacity in combination with superior precision and sensitivity. In addition, the sample consumption can be as low as 100 μl/min if using a micro-nebulizer, allowing the analysis of multiple elements even in small sample volumes. However, the matrix tolerance and long-term stability of the ICP-OES is usually superior to the ICP-MS, but this comes at the expense of a lower sensitivity for several important trace elements in plants. These include the heavy metals Cd, Pb, and Hg, as well as the essential plant micronutrients Ni and Mo.

3.3.1. ICP-MS Analysis

In the ICP-MS from Agilent, a built-in sample injector is available that includes 89 samples (5 mL) and 3 large samples (100 mL). One of the large samples contains a washing solution and a washing cycle is typically included for every ten samples to minimize memory effects in the system. Between each sample, a short rinsing of the sample uptake system is performed using a 2 % HNO_3 solution.

Full-quantitative ICP-MS analysis

1. Tune the machine as described by the manufacturer.
2. Build a method including the elements of interest, selecting interference free masses if possible. Use a collision/reaction cell if available to eliminate element-specific interferences (e.g., $^{40}Ar^{16}O^+$ on $^{56}Fe^+$). Do optionally include several analysis modes for the same element as day to day variability in instrument performance and differences in matrix can influence which mode gives the best results.
3. Perform an external calibration comprising at least 10 standards including the elements of interest and covering a linear range of a minimum of 3 orders of magnitude.
4. After analyzing the calibration standards, include two extra wash cycles (3.5 % HNO_3) to ensure that the system is fully decontaminated. After each acquisition the auto-sampler needle is washed for 10 s in 2 % HNO_3. Then follows a 30–60 s wash in the HNO_3/HF mixture depending on the length of the tubing from auto-sampler to nebulizer.
5. Include a representative sample (possibly a CRM matching the sample matrix) for every 10 samples in order to evaluate and correct possible drift throughout the run series (drift control sample).

6. Thereafter analyze at least seven independently digested CRM samples. Use the CRMs for validation and do only accept elements that are determined with accuracy above 90 %.
7. Run two extra wash cycles followed by the seven blanks. Blanks are used for estimating the limit of detection (LOD) and the limit of quantification (LOQ), represented by 3σ (three times the standard deviation) and 10σ, respectively.
8. Now include your samples in the sequence.
9. When the analysis is complete, check the calibration curves (exclude outliers) and recalculate.
10. Export both counts per second (cps) and dry weight concentration (μg/g) data.

Semiquantitative ICP-MS analysis

1. Perform the analysis as described for full-quantitative ICP-MS analysis but exclude the external calibration step. Instead, include a single multielement standard, preferably nonequimolar, in a concentration matching the expected mid range of all elements in the samples.
2. When analysis is completed, adjust the element-specific response factors according to the ICP manufacturer's recommendations and calculate the approximate element concentrations.
3. Export both counts per second (cps) and dry weight concentration (μg/g) data.

3.3.2. ICP-OES Analysis

Using the ICP-OES from PerkinElmer an external auto-sampler is needed. A Cetac auto-sampler can be configured in multiple ways suitable for this purpose. For multielement analysis a 5–10 mL volume per samples is optimal as ICP-OES analysis often requires more sample material than an ICP-MS. As for the ICP-MS analysis, large vials of wash and CRM are desirable for the continuous rinsing of the sample introduction system and drift monitoring, respectively.

1. Tune the machine as described by the manufacturer. As an absolute minimum, check the sensitivity in a 0.1 and 1 mg/L Mn solution in axial and radial viewing. Do optionally also tune the instrument according to the sensitivity of high-importance elements.
2. Build a method including the elements of interest. Select wavelengths according to the ICP manufacturer's recommendations. When selecting lines, attention must be given to sensitivity and precision requirements as well as possible spectral interferences. For unknown samples it is recommended to include several wavelengths for each element in both axial and radial mode if available.
3. Perform an external calibration of the elements of interest, covering a linear range of at least 3 orders of magnitude (8–10 calibration points).

4. After analyzing the calibration standards, include two extra wash cycles (3.5 % HNO_3) to ensure that the system is fully decontaminated.
5. Include a representative sample (possibly a CRM matching the sample matrix) for every 10 samples in order to evaluate and correct possible drift throughout the run series (drift control sample).
6. Thereafter analyze the seven replicate CRM samples equivalent to the ICP-MS analysis. Use the CRMs for accuracy validation and do only accept elements that are determined with accuracy above 90 %.
7. Run two extra wash cycles followed by the seven blanks. These are used for estimating the limit of detection (LOD) and the limit of quantification (LOQ), represented by 3σ (three times the standard deviation) and 10σ, respectively.
8. Now include your samples in the sequence.
9. When the analysis is complete, check the calibration curves (exclude outliers) and recalculate.
10. Export dry weight concentration (μg/g) data.

3.3.3. Data Extraction and Analysis

During data extraction and analysis several pitfalls exist. However, if the analytical sequence is correctly constructed by insertion of sample weight, dilutions, etc., all modern ICP software types enables automatic calculation of sample concentrations. In addition, automatic drift control, LOD calculations, etc., are also possible. However, it is still recommended that the final data are carefully evaluated and that test calculations are performed to eliminate errors.

1. Check the stability according to the internal standard and correct element concentrations accordingly.
2. Calculate the RSD on the drift control samples. If higher than 5 % adjust element specific drift accordingly.
3. Check that accuracy of the minimum 7 CRMs is > 90 % of the certified values and that the standard deviation is acceptable. Calculate the LOD and LOQ from the standard deviation between the 7 true blanks and compare with a typical sample. Ensure that all samples for all individual elements are above LOD and/or LOQ depending on the required quality of the final data. Please notice that sample results close to the LOD might not be reproducible.

3.4. LA-ICP-MS Analysis

Samples have to be fixed in the ablation chamber during analysis as they might otherwise move during chamber evacuation or stage movement. Scotch tape is a simple way to fix the sample position. At all times during analysis the laser has to be kept in focus at the sample surface in order to ensure that the same amount of material is ablated. The focus of the laser is controlled by the LA software. Using the LA scan mode during analysis, the LA unit provides a continuous flow of sample material to the ICP-MS which is why time-resolved analysis is the acquisition method of choice. If sequences are started by means of an external trigger signal, it is crucial to align the time spent on each scan to the ICP-MS acquisition time of the sequence. In relation to bioimaging, the total dwell time spent for element detection has to be congruent with the spot size and the ablation speed during analysis, in order to ensure proper resolution of the image (see Note 12).

1. Tune the ICP-MS according to the manufacturer's guidelines.
2. Shut down the ICP-MS and decouple the spray chamber from the ICP-MS.
3. Adjust any ICP-MS software settings to carry out LA analysis if necessary.
4. Connect the LA unit to the ICP-MS via the transport tube and add argon as a make-up gas to the transport stream via a y-peace. The argon flow can for example be controlled by the carrier gas flow controller on the ICP-MS and is typically in the range of 0.70–1.00 L/min.
5. Place the sample horizontally in the ablation chamber.
6. Turn on the LA system, initialize laser warm up and exchange gas if needed (only of relevance if an excimer laser is used).
7. Evacuate the chamber of atmospheric air using the chosen transport gas (typically helium or argon) at a flow rate of 500–1,000 mL/min. Employ the guidelines given by the manufacturer (see Note 13).
8. Decrease the transport flow to 100 mL/min and turn on the ICP-MS.
9. Increase the transport gas flow gradually until the desired flow is obtained (500–1,000 mL/min). Especially if helium is used as transport gas, great care must be taken not to shut down the ICP-MS plasma (see Note 14).
10. Let the plasma stabilize for 30 min.
11. Ablate the SRM 612 using the following settings; energy 1.0 GW/cm^2, spot 100 μm, speed 20 μm/s and a shot repetition rate of 20 Hz.
12. Measure the 248/232 and 232/238 m/z ratios. Adjust the carrier gas flow and sample depth until the values are below 1 % and 1, respectively. Typical settings are a sample depth of 5.5 mm and a carrier gas flow of 0.75 L/min.
13. Create the desired ablation lines in the LA software and make sure to keep the laser focal point in focus. Typical parameters

for leaf analysis are spot size: 25–50 μm, speed: 25–50 μm/s, shot repetition rate: 10–20 Hz and laser energy: 0.20 GW cm^2. To ensure proper comparisons between day to day measurements include five standard samples such as compressed pellets of CRM at the beginning and end of each sample set.

14. Generate a method in the ICP-MS software using time resolved analysis and include the elements of interest together with an internal standard evenly distributed in the sample matrix, such as ^{13}C. The data acquisition time should be aligned with the ablation time for each line.
15. Set up a sequence in the ICP-MS software with the same number of sequences as ablation lines. Utilizing a trigger cable enables automated start of each sequence at the same time as the ablation begins.
16. Start the ablation.

3.4.1. Data Extraction and Analysis: LA-ICP-MS

Depending on the type of analysis the data can be interpreted in different ways. Simple chromatograms are sufficient in some cases while an image is desired in others.

1. Assemble the data for each element in separate spreadsheets. This can either be done manually or by utilizing a macro created directly in Excel or MatLab.
2. Import the assembled data files to the chosen visualization software program for data treatment.

4. Notes

1. Reference material with certified values of multielement concentrations are commercially available, however not all matrices are available and in such cases in-house reference materials with known element compositions are valid alternatives. The ideal reference material reflects the matrix of the material analyzed as well as the concentration ranges of the elements of interest. Several certification institutes supply certified reference material, e.g., NIST (US Department of Commerce, National Institute of Standards and Technology, Gaithersburgh, MD, USA), NACIS-CN (China National Analysis Center for Iron and Steel, Beijing, PRC) and JCR-IRMM (Joint Research Centre, Institute for Reference Materials and Measurements, Geel, Belgium). References can be purchased either directly from these institutes or from suppliers such as LGC Standards, Middlesex, UK. When analyzing the reference material the accuracy should not deviate more than +/− 10 % from the certified values.

2. Depending on the desired detection limits of the individual elements in each analysis, the purity of HNO_3 is decided. For most applications analytical grade HNO_3 is of sufficient purity and background signals can be directly subtracted in the data analysis process. However, a higher acid purity is required when conducting trace element analysis. High background levels of particular elements will also incite the use of ultra pure HNO_3, which we have observed in the case of sulfur. Ultra pure HNO_3 is commercially available, however in-house distillation is possible by means of sub-boiling systems such as the Spex CertiPrep (Thomas Scientific, US) or DuoPUR (Milestone Inc, Italy, EU) systems.
3. Root material from hydroponic experiments needs to be desorbed for cell wall-bound nutrients. This can be done by washing the roots one time in a Tween-20 solution (1 g/L), followed by two washes in a desorption solution (0.2 mM $CaCl_2$ and 12.5 μM H_3BO_3) and a final rinse in Milli-Q water. If Ca and B are among the analytes, exchange the desorption solution with 5 mM EDTA and 5 mM MES-Tris (pH = 6.0).
4. Basically any solid material can be analyzed by LA-ICP and in cases where more delicate samples are to be analyzed, the application of other sample preparation methods is necessary. For example, transverse sections of leaf or root tissue need to be embedded in a supporting material such as an epoxy resin, followed by sectioning into thin slides. However, the slides need to be fixed prior to analysis in the same manner as a leaf sample.
5. Sectioning by use of a vibrating blade vibrotome provides a smooth sample surface suitable for LA analysis. It is our experience that the vibrotome causes negligible contamination.
6. Instead of cleaning consumables in acid baths, utilization of the low boiling point of HNO_3 for acid refluxing is an alternative procedure (e.g., the TraceClean system from Milestone Srl, Italy). This has the advantage that any contamination remains in the acid solution and does not affect the steam-based cleaning of the consumables. Do optionally follow build up of impurities in acid baths by weekly semiquantitative ICP-MS analyzes.
7. Several oxidising agents are suitable for sample digestion including $HClO_3$, H_2SO_4, HCl, and HNO_3, the latter being the typical choice in ICP analyzes as HNO_3 offers a minimum of analytical biases, ultra pure quality and a great capacity for element dissolving. Since the complete digestion of plant material in HNO_3 requires a temperature of 300 °C, the addition of H_2O_2 to the digestion medium is advantageous as complete digestion will take place at lower temperatures. The typical ratio of HNO_3 to H_2O_2 is 2:1. Furthermore, the addition of

HF is a prerequisite for proper digestion of plant material containing high levels of Si, which is the case for cereal straw particularly from rice (*Oryza sativa* L.). The amount of HF needed depends on the amount of Si in the sample. For straw material the typical ratio of HNO_3 to HF is 2:1. When HF is used in the digestion procedure, an equimolar amount of H_3BO_3 should be added after digestion to complex any HF excess.

8. Because of the relatively low temperatures used in the open digestion procedure, there is a risk that even easily digestible matrices are not fully decomposed. Even clear solutions can contain carbon residues and small particles which will lead to analytical bias. This can be circumvented by increasing the digestion time. However, due to the inherent acid evaporation in open digestion systems several digestion cycles are needed. Be aware that especially S and Fe are often analytically underestimated due to insufficient digestion. For these reasons it is recommended that samples from the open digestion are centrifuged or filtered prior to dilution and analysis.
9. Analyzes of grain material are often biased by underestimation of Fe and S when compared to the certified reference values. By increasing the digestion time it is possible to overcome this problem due to a more efficient digestion. An alternative strategy is to increase the temperature, but this requires digestion bombs capable of withstanding both higher temperature and pressure which is a challenge for most closed digestion systems.
10. An alternative microscaled digestion method is carried out by introducing a small closed vial to a macroscaled digestion bomb containing solvent for heat transfer. This method is termed the "vial in vial" procedure (18).
11. The amount of digestion media needed in a pressurized system depends on the amount of sample to digest and also on the ability to create a sufficient pressure. As a minimum requirement, enough digestion media to soak the sample and to keep elements in solution after digestion should be added. For a sample of 250 mg no more than 5 mL 69 % HNO_3 and 2.5 mL 30 % H_2O_2 should be added and for a <20 mg sample 250 μL 69 % HNO_3 and 125 μL of 30 % H_2O_2 is used.
12. Image resolution is dependent on several factors controlled both by the LA unit and the ICP. The resolution in the y-dimension (Fig. 1) depends on the spot size of the laser beam and the distance between the line scans, while the resolution in the x-dimension is reflected by the scan speed, repetition rate as well as the spot size. In addition, the resolution depends on the time spent on each scan cycle. In a quadrupole ICP-MS the total scan time is determined by the number of elements analyzed and the individual element scan times are based on the

concentrations of each element in the sample. Please refer to Lear et al. (19) for details on optimization of speed, spot, and total scan time in relation to image resolution.

13. The purge and evacuation time intervals can be changed in the LA software, allowing customization of the chamber evacuation cycle. This is beneficial if the ablation cell is equipped with an adjustable floor, enabling a reduction of the effective ablation chamber volume.
14. In general the flow needed for optimal signal intensities should be optimized for each LA set-up, as it largely depends on the effective ablation chamber volume and sample type. A gentle way of introducing helium into the argon plasma is by constant ramping which can be controlled via the LA software.

Acknowledgements

This work was financially supported by the EU-FP6 projects META-PHOR (FOOD-CT-2006-03622) and PHIME (FOOD-CT-2006-016253), the Danish Strategic Research Council (NUTRIEFFICIENT; grant no. 10-093498), the ICROFS research program OrgTrace (3304-FOJO-05-45-01), and the CORE Organic II EU project AuthenticFood.

References

1. Marschner H (1995) Mineral nutrition of higher plants, 2nd edn. Academic, London
2. Reuter DJ, Robinson JB (eds) (1997) Plant analysis – an interpretation manual, 2nd edn. CSIRO Publishing, Collingwood, Australia
3. Persson DP et al (2006) Multi-elemental speciation analysis of barley genotypes differing in tolerance to cadmium toxicity using SEC-ICP-MS and ESI-TOF-MS. J Anal Atom Spectrom 21:996–1005
4. Husted S et al (2011) Review: the role of atomic spectrometry in plant science. J Anal Atom Spectrom 26:52–79
5. Atwell S et al (2010) Genome-wide association study of 107 phenotypes in Arabidopsis thaliana inbred lines. Nature 465:627–631
6. Buescher E et al (2010) Natural genetic variation in selected populations of Arabidopsis thaliana is associated with ionomic differences. PLos One 5:e11081
7. Pedas P et al (2008) Manganese efficiency in barley: identification and characterization of the metal ion transporter HvIRT1. Plant Physiol 148:455–466
8. Lee S et al (2011) Bio-available zinc in rice seeds is increased by activation tagging of nicotianamine synthase. Plant Biotechnol J 9: 865–873
9. Hebbern CA et al (2005) Genotypic differences in manganese efficiency: field experiments with winter barley (*Hordeum vulgare L.*). Plant Soil 272:233–244
10. Laursen KH et al (2011) Multielemental fingerprinting as a tool for authentication of organic wheat, barley, faba bean, and potato. J Agr Food Chem 59:4385–4396
11. Hansen TH et al (2009) Micro-scaled high-throughput digestion of plant tissue samples for multi-elemental analysis. Plant Methods 5:12
12. Lombi E et al (2011) Megapixel imaging of (micro)nutrients in mature barley grains. J Exp Bot 62:273–282
13. Lee S et al (2009) Iron fortification of rice seeds through activation of the nicotianamine synthase gene. Proc Natl Acad Sci USA 106:22014–22019
14. Persson DP et al (2009) Simultaneous iron, zinc, sulfur and phosphorus speciation analysis

of barley grain tissues using SEC-ICP-MS and IP-ICP-MS. Metallomics 1:418–426

15. Bouis HE et al (2011) Biofortification: a new tool to reduce micronutrient malnutrition. Food Nutr Bull 32:S31–S40
16. Gomez-Becerra HF et al (2010) Genetic variation and environmental stability of grain mineral nutrient concentrations in Triticum dicoccoides under five environments. Euphytica 171:39–52
17. Laursen KH et al (2009) Multi-elemental fingerprinting of plant tissue by semi-quantitative ICP-MS and chemometrics. J Anal Atom Spectrom 24:1198–1207
18. Jakobsen MK et al (2005) Pollen development and fertilization in Arabidopsis is dependent on the male gametogenesis impaired anthers gene encoding a type VP-type ATPase. Gene Dev 19:2757–2769
19. Lear J et al (2012) Improving acquisition times of elemental bio-imaging for quadrupole-based LA-ICP-MS. J Anal Atom Spectrom 27: 159–164

Chapter 9

Mapping Element Distributions in Plant Tissues Using Synchrotron X-ray Fluorescence Techniques

Erica Donner, Martin D. de Jonge, Peter M. Kopittke, and Enzo Lombi

Abstract

Synchrotron-based X-ray fluorescence (XRF) is allowing substantial advances in several disciplines of plant science by allowing the in situ examination of elements within plant tissues. Continual improvements in detector speed, sensitivity, and resolution are increasing the diversity of questions that can be addressed using this technique, including the in situ analysis of elements (such as nutrients or toxicants) within fresh and hydrated tissues. Here, we describe the general principles for designing and conducting experiments for the examination of elemental distributions in plant material using micro-XRF.

Key words: Synchrotron, X-ray fluorescence, Element mapping, Tomography, Metals, Nutrients, Trace elements, Plants

1. Introduction

Synchrotron X-ray fluorescence (XRF) techniques offer powerful approaches for probing and mapping the in situ distribution of a wide range of elements in plant tissues and organs. This chapter focuses on two of these methods, namely micro-XRF (μ-XRF) element mapping and X-ray fluorescence micro-computed tomography. The first of these methods provides a means of obtaining quantitative 2D maps of elemental concentration, whilst the second can provide 2D virtual cross sections and 3D volumetric distributions of internal elemental distributions. Both techniques have been demonstrated to be feasible using plant tissues (e.g., (1, 2)). As the analysis is completed in situ, requiring a minimum of sample preparation, the potential for elemental redistribution artifacts is very low. Nevertheless, the analytical challenges of imaging low element concentrations in the native state are considerable, and as such these techniques require highly specialized facilities.

Frans J.M. Maathuis (ed.), *Plant Mineral Nutrients: Methods and Protocols*, Methods in Molecular Biology, vol. 953,
DOI 10.1007/978-1-62703-152-3_9, © Springer Science+Business Media, LLC 2013

Micro-XRF at ultimate resolution and sensitivity can be conducted only using a synchrotron light source and a specialized μ-XRF beamline.

There are over 50 synchrotron light sources around the world (http://lightsources.org) and researchers can access these facilities through peer-reviewed beamtime application processes. This chapter provides guidance on how to apply for beamtime and how to prepare for and conduct a synchrotron μ-XRF experiment. While there are some general guidelines that can be applied across most facilities, one should always consult with the local beamline staff, as local approaches do differ. Although it will not be covered in this chapter, it should also be noted that many XRF beamlines have the capacity to collect both spatial distribution data and micro-focussed X-ray absorption near edge spectroscopy data (i.e., μ-XANES spectra). Micro-XANES spectra can be used to infer the likely speciation of a target element at a particular point within the sample (e.g., (3, 4)). Therefore many of these beamlines not only provide element maps which can be used to investigate element distributions and colocation, they can also provide speciation information at the microscale. Nanoscale mapping (resolution below 100 nm) is also available at some beamlines.

Typically, each XRF beamline is unique, being designed and constructed by specialized beamline scientists (an example is given in Fig. 1). As such, some beamlines are more suited to biological sample analysis than others. In recent years there have been several significant developments in fluorescence detector technology and these advances have been highly beneficial in relation to plant science (6–8). State-of-the-art detectors (9) offer vastly increased sensitivity, thereby enabling low-concentration element mapping (e.g., (10)) and rapid scanning of large specimens. They also offer much faster data acquisition times, allowing the radiation exposure of the sample to be significantly reduced compared with older systems. At leading beamlines with fast fluorescence detection capabilities, data acquisition is now fast enough to allow tomography of freshly hydrated tissues such as roots without causing radiation damage (2). In a parallel development, the speed of detector electronics has also increased in recent years (11); however, without matching advances in detector sensitivity, these systems are unlikely to realize their full potential.

These developments are increasingly being harnessed to support novel plant research projects. Recent studies employing the use of synchrotron μ-XRF techniques have provided new insights into the mechanisms of micronutrient uptake and transport (e.g., (8)), mechanisms of metal toxicity and tolerance (4, 10), and spatial gene expression and transporter gene function (12).

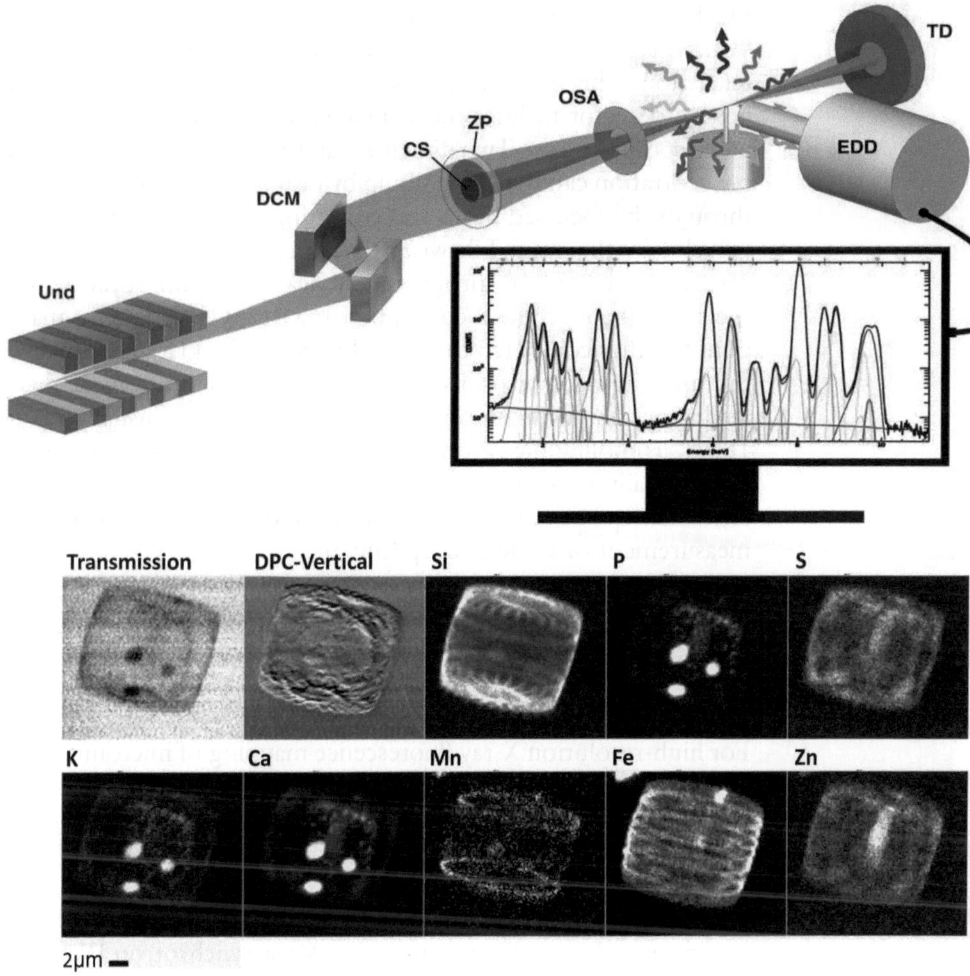

Fig. 1. Schematic representation of an X-ray microprobe with data reduction. An X-ray beam can be produced using an undulator (Und), monochromated using a double-crystal monochromator (DCM) and focussed into a small spot onto a specimen by use of a central stop (CS), zone-plate (ZP), and order-sorting aperture (OSA). At each scan position a transmission detector (TD) gauges the absorption and might be used to acquire phase contrast information. An energy dispersive detector (EDD) is used to determine the energy of characteristic X-rays emitted by the specimen, as well as Rayleigh and Compton scattered X-rays. X-ray fluorescence spectra obtained at each pixel in the scan are fitted in silico to determine elemental content: the XRF peak positions are a direct measure of the photon energies (and thus of which elements were illuminated), and the peak area is a measure of the elemental quantity. By interrogating the spectra at each pixel, elemental spatial distributions can be mapped, as shown here for the diatom *Cyclotella meneghiniania*. The first two panels show the absorption and the vertical component of the differential phase contrast; the remaining eight panels present elemental distributions (scale bar: 2 μm). Figure from de Jonge and Vogt (5).

1.1. Overview: Scanning X-Ray Fluorescence Microscopy

By placing a specimen into the focus of an X-ray beam, a narrow column through the specimen is illuminated with X-rays. Typically, over 90 % of the focussed light passes straight through the specimen, and the entire column is thereby evenly illuminated. The X-rays ionize the atoms in the column, liberating an electron from the K-, L-, or M-shell depending on the incident X-ray energy. "Characteristic" X-ray fluorescence is emitted in the process of relaxation. This fluorescence can be detected using an energy dispersive detector to infer the atomic species present in the illuminated specimen area. Two dimensional maps of projected elemental concentration can then be built up by raster-scanning the specimen through the focussed beam and collecting signal at each point (or "pixel") in the scan. Likewise, tomographic images (sinograms) can be measured by progressively scanning and rotating the specimen through the focussed beam. XRF-μCT is directly analogous to other computed tomography modalities (13), and involves the measurement of projected elemental concentrations at many angles over as wide an angular range as possible (usually up to 180°). Tomographic reconstruction requires a computation step, and typically uses one of the established methods. In both cases elemental quantification is usually achieved by comparison to a measurement of a reference specimen.

2. Materials

For high-resolution X-ray fluorescence mapping of micronutrients, macronutrients and contaminants in plant tissues, access to a synchrotron light source and suitable X-ray fluorescence microscopy beamline is required.

2.1. Beamline Selection

1. Web links to synchrotron light sources around the world are given at http://lightsources.org. Most synchrotron Web sites contain summary information about the beamlines available at that light source, together with information on their capabilities and characteristics.
2. Generally speaking, "soft" X-ray beamlines are not suitable for X-ray fluorescence studies. "Tender" X-ray beamlines produce X-rays in the range of 2–4 keV, and may be optimal for high-resolution studies of P, S, Cl, and K. "Hard" X-ray beamlines operate at energies between 8 keV and 20 keV and are used to measure a range of elements from Si to Sr (K-shell ionization) and heavier (L and M-shell ionization).
3. All beamlines have an optimal X-ray energy range. This energy range directly determines the elements that can be mapped at a given beamline (see Notes 1–4). For example, a beamline with an upper energy limit of 13.5 keV will be able to excite elements lighter than Br in the K series, lighter than Bi in the L series, and without limit in the M-series (see ref. 14 for a

listing of X-ray emission energies). Note that tender X-ray beamlines may require analysis under vacuum conditions, but hard X-ray beamlines usually do not, unless this is specifically required for the specimen analysis environment (e.g., cryo-cooling).

4. Some beamlines may have a number of energy dispersive detectors available for use. These may differ primarily in terms of their sensitivity (efficiency of conversion of characteristic emission into signal), speed (minimum time between consecutive measurements), energy range (there will be roll off at both low and high energy), and energy resolution (the intrinsic width of features in the resulting spectrum). Each of these factors will play a part in your investigation, so it is important to be clear about your experimental priorities: for example, whether you require ultimate trace-element sensitivity, or whether you put a higher priority on measuring a larger sample area. The type of detector also plays a role in determining which elements can be detected and the sensitivity of detection (i.e., minimum element concentration required for reliable mapping). To map "dilute" samples, such as trace elements in plant tissues, highly sensitive detectors with high energy resolution are preferred.
5. Energy dispersive detectors have readout overheads ranging from 10 ns to around 1 s. When these overheads are incurred on a per-pixel basis, the overhead can have a significant impact on the experimental plan. A fast detector does not necessarily mean that useful measurements can be obtained quickly, because unless the detector is also extremely sensitive insufficient signal may be measured when mapping at speed. However, highly sensitive detectors do facilitate faster mapping, as good quality data can be collected relatively quickly. Therefore high sensitivity also equates to shorter required dwell time per pixel. Detector speed and efficiency should be considered as a very positive feature when working with samples that are highly susceptible to radiation damage or dehydration (e.g., fresh hydrated tissues).
6. When choosing a beamline, it is also important to check the lateral resolution of the microprobe. This must be able to match the size of the smallest feature you need to resolve. For example, if the smallest feature to be investigated is 200 nm in diameter, there will be no point using a beamline with a spot size of 5×5 μm. The spatial resolution is determined by the beamline optics (see refs. 15, 16 for reviews on X-ray microoptics) and their illumination. Microprobes can be designed using a variety of focussing optics, but typical microprobe resolutions range between 50 nm and 20 μm.

7. Clearly, there are several factors that must be taken into account when choosing a beamline. New synchrotron users should contact the local beamline scientist for advice on the above issues, and should seriously consider collaborating with an experienced user for their first experiment.

2.2. Consumables for Sample Mounting

The two most significant considerations for the choice of specimen mount are (1) elemental purity and (2) thinness (see Notes 5–16).

1. The specimen mount should contain none of the elements of interest, as these will adversely affect elemental sensitivity. Furthermore, presence of other elements with spectral features within the X-ray energy region of interest will complicate the interpretation of the fluorescence spectrum and will ultimately reduce elemental sensitivity.
2. For hard X-ray μ-XRF of most plant tissue specimens, over 90 % of the incident beam will be transmitted directly through the specimen (see Note 16). If the transmitted X-rays interact strongly with the specimen mount they will produce unwanted background fluorescence which will adversely affect the measurement sensitivity.
3. Several substrates are suitable for μ-XRF. Silicon-nitride windows are both thin (less than 1 μm) and extremely pure. Ultralene film can be used to make large mounting windows and can also be used to sandwich a specimen to prevent sample drying or oxidation (see Note 7). Kapton film can be used where a more robust support is required, and can be obtained in a wide variety of thicknesses.
4. If a thick substrate is necessary (for example for specimen preparation such as thin sections), consider using quartz microscope slides as these will not introduce elemental contamination (see Notes 9 and 10).
5. Glues and resins should be pure where possible. Note that methyl methacrylates are generally suitable (particular brands may vary) and readily available (see Note 11).
6. The support/substrate should not prevent the sample from being brought close to the detector as this is required to optimize the measurement efficiency.

3. Methods

3.1. Pre-beamtime Preparation

1. Identify a suitable beamline at an appropriate synchrotron research facility (see Notes 17 and 18).
2. Estimate the amount of beamtime required to collect the necessary maps (see Notes 19–25). Be sure to include detector

and beamline overheads in your estimate (see Note 26). This is influenced by the dimensions of the sample to be analyzed, the spatial resolution required, the sensitivity/speed of the detector, and the concentration of the element(s) of interest in the tissue (see Note 27). It is usually wise to ask the beamline staff for typical scan times for your specimen type and sensitivity requirement. While it is sometimes possible to acquire a data-set in a single day, most experiments aim for a multi-dimensional data-set with higher statistical power, and so 2 or 3 days of beamtime is a more typical beamtime request.

3. Determine how the samples will be prepared and mounted on the scanning stage (see Notes 5–15). This is best discussed with a beamline scientist in advance.
4. Apply for beamtime. Proposal requirements differ depending on the particular synchrotron facility, but generally require: background information about the proposed experiment; its scientific relevance, purpose, and expected outcomes; information about the research team; and the required beamline set-up.
5. Merit proposals will be peer reviewed and beamtime will be granted on the basis of scientific merit and feasibility. Successful merit proposals are commonly granted beamtime free of charge, and depending on the facility, travel funding may also be provided. At some facilities, it is also possible to purchase beamtime.
6. Prepare your samples and plan their transport to the synchrotron (see Notes 5–11). Be aware of international quarantine laws and clearance times. You should also be aware of any occupational health and safety requirements that may be relevant your experiment (e.g., restrictions on chemical use and handling) and ensure that appropriate protocols are in place and observed. Also be prepared to take your specimens away with you at the end of the experiment as the synchrotron may not have the appropriate disposal facilities (see Note 28).

3.2. At the Beamline: 2D Mapping Protocol

1. The beamline scientist(s) will prepare the beamline for your experiment. The time required for set-up will depend on the complexity of the beamline and the degree of similarity with the previous experimental set-up. Several hours of preparation time is not unusual and may be necessary to ensure the optimization of the beamline for your experiment.
2. In conjunction with the beamline scientists, select the resolution you wish to map with, and set the size of the X-ray beam accordingly (see Notes 27, 29, 30).
3. Set the energy of the incident X-ray beam to an appropriate excitation level for the elements you wish to map. The energy of the incident beam should be set above the absorption energy of all the elements of interest.

4. If data collection is limited to Regions of Interest (ROI) rather than full fluorescence spectra at the beamline where you are working, you must select ROIs for all the elements you are interested in, as only data from these ROIs will be captured. If full spectra are collected then emission lines for all the fluorescing elements can be extracted post analysis and ROIs do not need to be selected prior to scanning (see Notes 31, 32).
5. Scan a reference sample to ensure that the beamline is operating correctly and that the quantification is adequately calibrated (if quantification is relevant at that beamline and for your experiment). Each beamline has its own protocols for reference specimen measurement, but typically you should aim to measure this around once a day or two-to-three times during an experiment if this is not too onerous.
6. It can be extremely useful to have access to test specimens with which to test the performance of the system. If ultimate sensitivity is your goal, bring along a specimen with an order of magnitude more of the element of interest than is likely to be found in your sample; this can be used to establish a positive signal and to tune sensitivity.
7. Mount the sample on the sample stage. Sample presentation is a major issue to consider. For example, Kopittke et al. (10) used two layers of Ultralene film to contain excised plant roots in a humid environment in order to help maintain root cell turgor and prevent dehydration during analysis. In some cases analysis of leaves still attached to a growing plant may be possible (e.g., (17)).
8. Identify the relevant coordinates of the sample area you wish to scan (see Note 33). Where possible, you should have a very good idea of the regions you want to measure before you go to the synchrotron. If possible, bring a zoom-in sequence of images to assist with finding the required part of the specimen as the visualization tools at the beamline may not be very sophisticated.
9. Select an appropriate per-pixel dwell (also referred to as scan speed). This will depend on the concentration and distribution of the elements of interest in the sample as well as some key beamline characteristics (e.g., photon flux, detector efficiency).
10. Calculate the length of time required to map the target area at the selected speed and assess whether this is realistic within the confines of the experiment. Also assess whether the sample is likely to move, either due to growth or dehydration in that time.
11. Check the fluorescence signal from an area of the sample that you suspect to be representative of the element concentration in the rest of the sample. If possible check this on a test specimen

or at a location on your specimen that you do not plan to map as part of your experiment. This is recommended in order to avoid any problems with beam damage induced artifacts.

12. If the count rate is too low/high, adjust the beamline controls to obtain an appropriate count rate. This should be performed in conjunction with the beamline scientist, as each beamline is unique. This may require opening/closing the slits to increase/decrease the photon flux hitting your sample.
13. Launch the scan.
14. If the beamline offers real-time data display check your images during scanning to ensure that you are scanning the correct area and that the image being collected is useful (e.g., is it of sufficient resolution and are the elements of interest being detected). If a mistake is noted, abort the scan and re-select the scan area coordinates.
15. After scanning, remove your sample from the stage and inspect the sample for X-ray beam damage (e.g., by observing the specimen under an optical or electron microscope) (see Notes 33–36).
16. If possible, analyze your data while the next scan is being acquired! At-beamline data analysis improves experimental outcomes significantly, and allows real-time improvements to experimental conditions and research directions.

3.3. At the Beamline: Microtomography Protocol

See steps 1–6 in Subheading 3.2 above.

7. Mount the sample on the rotation stage. The sample should usually be mounted with its long axis parallel to the rotation axis and should be well-aligned to the rotation center to minimize overscanning requirements.
8. Again, sample presentation is a major issue to consider. For example, Lombi et al. (2) used polyimide (Kapton®) capillaries (available in a wide range of diameters) to contain excised plant roots in a humid environment and circumvent dehydration during analysis. A small amount of water was sucked into the end of the capillary but was not in contact with the plant root, and the ends of the capillary were sealed with wax to retain the moisture.
9. Select the relevant scan parameters, such as the coordinates of the line or area you wish to scan, and the number of angles you wish to scan. For a certain desired resolution of reconstruction, the number of angles should be at least the width of the measurement divided by the desired resolution. Typically, this means that the number of angles should equal the number of pixels across your scan.

See steps 9–14 in Subheading 3.2 above.

15. After scanning, consider performing further measurements to assess beam damage (see Notes 34–38). Always take the 180° complementary image for your tomographic series, as differences between this and the 0° projection are a strong indicator of damage, drift, motion, or even growth! If possible, also take a two-dimensional scan that encompasses the tomographic scan region as this may reveal any changes to the elemental distribution in the scanned region due to beam damage. Then remove your sample from the stage and further inspect the sample for X-ray beam damage (e.g., by observing the specimen under an optical or electron microscope).

See step 16 (Subheading 3.2) above.

3.4. Data Analysis

1. Preprocess your data at the beamline. Micro-XRF experiments can result in a very large amount of data being collected, particularly if full fluorescence spectra are collected for all images.
2. Download and save the data (see Note 39).
3. Use an appropriate software package to analyze the data and produce appropriate images. Image analysis is typically carried out using a software package such as GeoPIXE, MAPS, or AXIL. Many beamlines have software developed specifically for that beamline although the data collected there may also be compatible with other data analysis packages.
4. Software features that are typically used in element mapping analysis include element association functions, element profiles, Pearson scatter plots, Principal Component Analysis, and three-color (elemental) overlays.
5. Data may be presented as scatterplots, single color maps, 3-color maps, movies, etc.
6. Tomography data are reconstructed to give a virtual cross section or 3D volume using any one of a number of algorithms. Corrections for self-absorption may be required depending on sample size and elements of interest. The beamline staff will usually be an excellent resource for reconstructing your data.

4. Notes

1. The incident beam energy needs to be high enough to photoionize the elements of interest. For lighter elements this may involve K-shell photoionization; for heavier elements, L-shell or M-shell photoionization may be appropriate. On some occasions the energy selected may be deliberately low enough so as not to stimulate an element that may otherwise dominate

the spectrum. The incident energy chosen should be sufficiently high enough to ensure that the Compton tail on the emission spectrum does not compromise sensitivity. For X-ray ionization and emission energies, consult one of the following reference guides (14, 18, 19).

2. As an example, if you wish to map Zn, Cu and Fe, the relevant K-shell absorption edges are 9.66 keV, 8.98 keV and 7.11 keV respectively. In this case, an incident beam energy of approximately 10 keV should be suitable for data collection as it is above the absorption energy of all elements of interest.
3. When the K-shell absorption energies are not in the range of the beamline you are using, you may consider L-shell absorption. For example, imagine you are working at a beamline where the tunable energy range varies from 6 to 20 keV and you wish to analyze for Cu and Pb. In this case the beamline will be able to excite K-shell emission for Cu, but not for Pb. However, the Pb L-1 edge is at 15.8 keV and this could be measured instead.
4. Note that the lower energy of the beamline tuneable energy range does not restrict investigation of elements with ionization potentials below this energy. The low energy sensitivity is determined only by that of the energy dispersive detector.

5. Well-planned sample preparation is vital to the success of μ-XRF imaging studies. It is particularly important to ensure that sample preparation methods do not introduce element redistribution artifacts in the tissues that will be analyzed. Sample preparation artifacts scale to the resolution of the measurements. Hence, elemental redistribution at the level of 1 μm is inconsequential for a study at 5 μm resolution, but fatal to a study at 100 nm resolution.
6. In the case of freeze dried samples, redistribution of mobile elements is possible (e.g., (20)), so fresh material analysis may be preferred.
7. The nature of the sample must always be taken into account when designing the sample mounting system. For example, special preparation may be required to ensure that hydrated samples are protected from dehydration during the scanning period. This is important as analysis is normally conducted at room temperature and it can potentially become quite hot in the experimental hutch.
8. Samples can also be prepared as thin sections. In this case the samples are often snap frozen and sectioned using various techniques. High pressure freezing (HPF) is considered to be the best method for preserving the most challenging materials, but is limited to specimens under 200 μm thick. After freezing,

samples could be freeze-dried or cryo-substituted prior to sectioning, or cryo-sectioned and substituted or dried afterwards. A review of sample preparation options for plant material can be found in Appendix 1 of Moore et al. (21).

9. Where no alternatives exist, thin sections may be left mounted on high-quality quartz slides. When this approach is considered the beamline staff should be consulted as the slide may render the analysis difficult or impossible. In any case, ultimate sensitivity will be compromised.
10. Regular glass microscope slides often contain rare-earth and transition metals which may compromise sensitivity.
11. Resin-embedded double-polished thin sections, for samples such as seeds, can be fixed to their preparation slides with methyl methacrylates ("super-glue") during the polishing process, and then detached by soaking them in acetone.
12. In order to ensure that the beam is correctly focussed across the whole sample area, the sample must be at least as uniform and flat as the depth-of-field of the X-ray microprobe. Consult the beamline staff for requirements if you have concerns about the flatness of your specimens.
13. Acquisition of elemental maps for multiple samples within the one scan enables clear and direct assessment of elemental distribution, etc., and may aid later exposition of the results. However, you should also be aware that this will often not be possible.
14. In some cases, the field of view of the X-ray scan can be quite restricted; however, this varies greatly between beamlines. In any case, it is well advised to prepare the samples in such as way that, if the system will allow it and the samples are suitable, they can be mounted closely together on the specimen stage. This will help minimize the time lost between scans, as individual scans can potentially be run as a batch and less time will be required for sample changeover. If you are working with hydrated samples this approach is probably not suitable as it would be preferable to minimize the potential dehydration time of the sample by mounting them immediately prior to analysis instead of using batch scans. Consult the beamline scientist for advice well in advance of your allocated beamtime.
15. It may be possible under some circumstances to use a cryostage during analysis, but in such cases, great care must be taken to ensure that the sample which has been frozen in preparation for analysis remains frozen during transfer to the sample stage and this may be difficult to achieve, particularly when working with thin sections. Transfer should not expose the specimen to air as atmospheric water will result in condensation and ice formation.

16. For thin specimens the fluorescence signal originates from throughout the entire thickness of the specimen; and the resulting map is a projected map of elemental content. Accordingly, specimen thickness variations will appear as a concentration gradient on the elemental map. This should always be considered when preparing thin sections as small differences in sectioning thickness can greatly complicate the data interpretation.

17. Most synchrotron facilities require that you acknowledge your facility access in any publications arising from your experiment; these acknowledgements are counted and used to justify facility funding. This should appear in the acknowledgement section of your publication to make it easy to find.
18. Coauthorship with beamline staff should also be considered as beamline scientists will often provide a level of service that goes beyond a general user facility. They will usually set up the measurement with you, discuss your specimen and science requirements, and may also help analyze your data, or teach you how to do it, or both. They can also help edit your articles for technical and scientific accuracy.
19. When choosing a scanning speed, the radiation resilience of the sample must be assessed. For example, small hydrated roots would be expected to have low resilience to radiation damage, and the scanning speed must therefore be fast enough to beat the onset of beam damage to the sample during scanning as this may result in the collapse of the root and/or redistribution of target elements.
20. Secondly, the amount of beamtime available must also be considered as this can be a major constraint. The aim is to reach a compromise between the speed of scanning and hence the number of samples that can be scanned, and the quality of the data that will be collected. Data quality and statistics will also be dependent on the concentration of the target elements in the sample and the distribution of these elements, either whether they are very diffusely distributed or concentrated in hotspots of particular interest. Samples with higher concentrations of the target elements can typically be mapped more quickly than samples with lower element concentrations.
21. It is often necessary to estimate the likely scan duration of each scan so that sample changeovers can be planned appropriately.
22. As an example of a 2D mapping exercise, consider the following. Mapping an area of 200 μm × 200 μm with a dwell time/pixel transit time of 1 s/pixel at 2-μm resolution will require 100 × 100 s or 2.8 h. Note that this does not include stage motion or detector overheads.

23. As an example of a single slice tomography experiment, consider the following. We desire measurement of the Zn distribution in a specimen of 500 μm diameter at 5 μm resolution. We believe that self absorption effects may be neglected, and so we wish to measure over 180°. Measuring 100 pixels in a traverse across the specimen indicates that data should be collected for 100 angles, i.e., at 1.8° intervals. The traverses will need to overscan the specimen and need to ensure that it remains in the field of view at all angles; and so a 700 μm traverse may be required, depending on the accuracy of the alignment to the rotation axis. For a dwell of 30 ms this scan will take $(700\ \mu m/5\ \mu m) \times (180°/1.8°) \times 0.03 = 420\ s = 7$ min of acquisition, *plus beamline overheads.*
24. When calculating scan time requirements you should also allow extra time for detector and motion-control overheads. This may involve time per pixel or time per line, and the difference is critical for timing estimates.
25. You may also need to allow sufficient time to scan standards, particularly if quantitative analyses are required.

26. Try to have your beamtime well planned to ensure that change-over times between samples are as fast as possible to maximize the available beamtime. For labor-intensive experiments requiring a lot of additional lab work in relation to sample preparation (e.g., hydroponic growth systems with periodic sampling) you may need to organize your research team to work in shifts to ensure that samples can be changed around the clock as your beamtime will be granted for 24 h a day and it is expected that you use it continuously throughout that period.
27. Some beamlines operate in stepping mode where the sample is moved by a fixed distance followed by data collection for a given dwell time, and so on. At other beamlines, data collection occurs ‘on the fly’, and the sample stage moves continuously in one direction. In this case the dwell time is better expressed as pixel transit time. This latter approach greatly reduces overheads.
28. The equipment needed to conduct your experiment will differ depending on the experiment you wish to conduct. Most synchrotron facilities also have chemistry/biology/microbiology preparation labs where samples can be prepared prior to scanning, but they differ considerably in the available consumables and instruments (e.g., centrifuges, microscopes, microtomes).
29. The most common compromise that is made during an experiment is the trade-off between spatial resolution and elemental sensitivity. By increasing the size of the source that is used to illuminate the focussing optics the spatial resolution is degraded

(the spot size increases) but the sensitivity improves (more X-rays are in the focus).

30. In principle, maximum information is obtained by stepping in intervals of half of the beam size. However, due to time limitations, a 'balanced sampling' approach is often adopted, where step size is equal to spot size. Under sampling—where beam size is much smaller than step size—is common for overview scans, but cannot reliably be used to gauge total elemental content except where the specimen feature size is much larger than the step size.
31. Some beamlines use Regions of Interest (ROIs) to define the fluorescence energy ranges for the elements of interest and only collect data from those predefined regions. Other beamlines collect full fluorescence spectra, in which case the fluorescence signal for all elements which fluoresce at the energy you conduct the experiment at can be extracted post facto. Be aware that all elements of interest must be known beforehand for ROI analysis, and that there is no chance to consider other distributions after the measurement.
32. Possible interferences must be considered during data analysis. For instance, the fluorescence peaks of some elements may overlap such that a high signal of one element results in a heightened signal for a different element—e.g., cobalt and iron. Such overlaps may be fatal for ROI analysis, but can usually be fitted in full-spectral approaches.
33. When scanning roots, begin the scan from the tip of the roots to ensure that the root tip will still be within the relevant mapping frame even if the roots continue to grow during the scan.
34. Given the intense nature of synchrotron radiation, it is imperative to check for evidence of beam damage. Several studies have demonstrated that damage from the beam can influence both elemental distribution as well as speciation (where µ-XANES data are being collected). See Scheckel et al. (17) or Leinweber et al. (22) for examples. Therefore, care must be taken to ensure that samples are not damaged during the acquisition process or else the data obtained will be confounded by experimental artifacts.
35. It is recommended that for each sample type, time be devoted during the experiment explicitly for checking for the occurrence of beam damage.
36. Severe physiological damage such as dehydration and cell collapse may be visible to the naked eye, but a microscope should also be used to look for less severe damage as this may still lead to some redistribution of elements during the scanning process.

37. If tomography has been performed, a subsequent 2D map covering the area where the tomography was conducted may also be collected to reveal any inconsistencies in element distribution which may indicate beam damage. See Lombi et al. (2) for an example.
38. If beam damage is considered likely it is also possible to collect the tomogram through a full 360° rotation and then compare the reconstructions from the first and second half of that series (i.e., 0–180° and 180–360°). If beam damage is not an issue then the two reconstructions should be identical. Changes in the tomogram can also be used to detect beam damage.
39. Large amounts of data can be generated during μ-XRF experiments (often many gigabytes, if not terabytes) so it is advisable to take a portable hard drive with you to the beamline to ensure that you can save your data and take them away with you for further analysis.

Acknowledgements

The method described in this chapter is based on work carried out at the X-ray Fluorescence Microscopy (XFM) beamline at the Australian Synchrotron. We would like to acknowledge the XFM beamline scientists David Paterson and Daryl Howard, and the GeoPIXE image analysis software developer Chris Ryan (CSIRO) for their expertise and support. We also thank Dr Simon A. James (Australian Synchrotron) for proof reading and commenting on the manuscript.

References

1. Takahashi M, Nozoye T, Kitajima N, Fukuda N, Hokura A, Terada Y, Nakai I, Ishimaru Y, Kobayashi T, Nakanishi H, Nishizawa NK (2009) In vivo analysis of metal distribution and expression of metal transporters in rice seed during germination process by microarray and X-ray Fluorescence Imaging of Fe, Zn, Mn, and Cu. Plant Soil 325:39–51
2. Lombi E, de Jonge MD, Donner E, Kopittke PM, Howard DL, Kirkham R, Ryan CG, Paterson D (2011) Fast X-ray fluorescence microtomography of hydrated biological samples. PLoS One 6:e20626
3. Sarret G, Harada E, Choi Y-E, Isaure M-P, Geoffroy N, Fakra S, Marcus MA, Birschwilks M, Clemens S, Manceau A (2006) Trichomes of tobacco excrete zinc as zinc-substituted calcium carbonate and other zinc-containing compounds. Plant Physiol 141:1021–1034
4. Terzano R, Al Chami Z, Vekemans B, Janssens K, Miano T, Ruggiero P (2008) Zinc distribution and speciation within rocket plants (*Eruca vesicaria* L. Cavalieri) grown on a polluted soil amended with compost as determined by XRF microtomography and micro-XANES. J Agric Food Chem 56:3222–3231
5. de Jonge MD, Vogt S (2010) Hard X-ray fluorescence tomography – an emerging tool for structural visualization. Curr Opin Struct Biol 20:606–614
6. Lombi E, Susini J (2009) Synchrotron-based techniques for plant and soil science: opportunities, challenges and future perspectives. Plant Soil 320:1–35

7. Punshon T, Guerinot ML, Lanzirotti A (2009) Using synchrotron X-ray fluorescence microprobes in the study of metal homeostasis in plants. Ann Bot 103:665–672
8. Lombi E, de Jonge MD, Donner E, Ryan CG, Paterson D (2011) Trends in hard X-ray fluorescence mapping: environmental applications in the age of fast detectors. Anal Bioanal Chem 400:1637–1644
9. Ryan CG, Siddons DP, Kirkham R, Dunn PA, Kuczewski A, Moorhead G, De Geronimo G, Paterson DJ, de Jonge MD, Hough RM, Lintern MJ, Howard DL, Kappen P, Cleverley J (2010) The New Maia Detector System: methods for high definition trace element imaging of natural material. AIP conference series 1221: 9–17
10. Kopittke PM, Menzies NW, De Jonge MD, McKenna BA, Donner E, Webb RI, Paterson DJ, Howard DL, Ryan CG, Glover CJ, Scheckel KG, Lombi E (2011) In situ distribution and speciation of toxic Cu, Ni and Zn in hydrated roots of cowpea. Plant Physiol 156:663–673
11. Rivers M (2010) 4,000 Spectra or 4,000,000 ROIs per second: EPICS support for high-speed digital X-ray spectroscopy with the XIA xMap", XRM 2010, 10th international conference on X-ray microscopy
12. Kim SA, Punshon T, Lanzirotti A, Li LT, Alonso JM, Ecker JR, Kaplan J, Guerinot ML (2006) Localization of iron in Arabidopsis seed requires the vacuolar membrane transporter VIT1. Science 314:1295–1298
13. Kak AC (1988) Slaney M: principles of computerized tomographic imaging. IEEE Press, New York
14. Thompson A, Attwood D, Gullikson E, Howells M, Kim K-J, Kirz J, Kortright J, Lindau I, Liu Y, Pianetta P, Robinson A, Scofield J, Underwood J, Williams G, Winick H (2009) Centre for X-ray optics and advanced light source X-ray data booklet. Lawrence Berkeley National Laboratory, University of California, Berkeley, CA. Available online http://xdb.lbl.gov
15. Ice GE (1997) Microbeam-forming methods for synchrotron radiation. X-ray Spectrom 26:315–326
16. Snigirev A, Snigireva I (2008) Hard X-ray microoptics. Springer series in optical sciences. In: Erko A, Idir M, Krist T, Michette AG (eds) Modern developments in X-ray and neutron optics. Springer, New York, pp 255–285
17. Scheckel KG, Lombi E, Rock SA, McLaughlin MJ (2004) In vivo synchrotron study of thallium speciation and compartmentation in *Iberis intermedia*. Environ Sci Technol 38: 5095–5100
18. Ravel B, Newville M (2005) ATHENA, ARTEMIS, HEPHAESTUS: data analysis for X-ray absorption spectroscopy using IFEFFIT. J Synchrotron Radiat 12:537–541
19. Webb S (2010) X-ray utilities. App store, Apple. http://itunes.apple.com/us/app/x-ray-utilities/id386510883?mt=8
20. Tylko G, Mesjasz-Przyblowicz J, Przybylowicz WJ (2007) In-vacuum micro-PIXE analysis of biological specimens in frozen-hydrated state. Nucl Instr Meth Phys Res B 260:141–148
21. Moore KL, Lombi E, Zhao FJ, Grovenor CRM (2012) Elemental imaging at the nanoscale: NanoSIMS and complementary techniques for element localisation in plants. Anal Bioanal Chem 402:3263–3273
22. Leinweber P, Kruse J, Walley FL, Gillespie A, Eckhardt KU, Blyth RIR, Regier T (2007) Nitrogen K-edge XANES – an overview of reference compounds used to identify 'unknown' organic nitrogen in environmental samples. J Synchrotron Radiat 14:500–511

Chapter 10

Flux Measurements of Cations Using Radioactive Tracers

D.T. Britto and H.J. Kronzucker

Abstract

Standard procedures for the tracing of ion fluxes into roots of plants are described here, with emphasis on cations, especially potassium (K^+). We focus in particular on the measurement of unidirectional influx by use of radiotracers and provide a brief introduction to compartmental analysis by tracer efflux (CATE).

Key words: Influx, Efflux, Potassium, Cations, Radiotracers

1. Introduction

Plant root transport systems are major portals through which the mineral substances that support life enter terrestrial food webs, usually in ionic form. Not surprisingly, the uptake and distribution of nutrient ions by plants has long been a major area of research in plant physiology, ecology, and biophysics. In particular, major strides have been made by means of radioactive and stable isotopes, with two of the most notable pioneers being G. de Hevesy (1), starting in the 1920s, and later E. Epstein (2), from the 1950s on.

The main advantage of isotope methodology over other approaches is that it enables the measurement of unidirectional fluxes, which are indispensable in the determination of kinetic flux parameters (e.g., K_M, V_{max} in a Michaelis-Menten sense), and therefore in the study of specific transporter capacities, energetics, mechanisms, and regulation. This information cannot be precisely determined by use of net flux measurement. The need to isolate a flux in one direction is particularly important under conditions where the flux in the opposite direction is high, and the turnover of intracellular pools is rapid (3). In addition, because fluxes of one isotope are usually observed against a background of a different isotope of the same element, tracer measurement can be conducted under fairly high nutrient conditions, relative to other methods.

Frans J.M. Maathuis (ed.), *Plant Mineral Nutrients: Methods and Protocols*, Methods in Molecular Biology, vol. 953,
DOI 10.1007/978-1-62703-152-3_10,

Isotopes can also be used in the measurement of net fluxes, but for this purpose a host of other well-established methods can also be applied, including: measurements of depletion in the external medium and/or accumulation in the tissue, use of ion-selective vibrating electrodes (e.g., MIFE, SIET), and the use of ion-selective fluorescent dyes.

In this article, we shall provide a basic procedure for the measurement of ion transport in plant roots, with special emphasis on cation uptake. Of all ions transported into plant roots, potassium (K^+) has probably been studied in the greatest detail historically and is one of the most rapidly absorbed and highly accumulated cations, along with calcium (Ca^{2+}), and (when present) ammonium (NH_4^+) and sodium (Na^+). For these reasons, we shall take the potassium ion as a prime example; by and large, however, the methodology presented here can be extended to uptake measurements for other cations, as well as anions, with few changes to protocol (see Note 1).

2. Materials

1. Plant material, typically intact, hydroponically grown seedlings, or excised roots (see Notes 2–4).
2. Glass or plastic vessels for pre-absorption, labeling, and desorption steps, typically 50–500 ml.
3. Equipment for aeration and/or stirring of solutions, often consisting of a compressed air source, a distribution system such as an aquarium-style manifold, and plastic tubing.
4. Dissecting tools (scalpel, forceps, scissors).
5. Labeling solutions (see Notes 5 and 6).
6. Pipettors and pipette tips (1 ml) for specific activity samples
7. Radiotracer, typically 5–20 millicuries for ^{24}Na and ^{42}K (see Notes 7–9).
8. Safety equipment, including protective clothing, eyewear, and shielding (see Note 7).
9. Radiometric equipment, usually a gamma or scintillation counter, plus a Geiger-Müller counter for safety purposes (see Note 9).
10. Low-speed (clinical-type) centrifuge and centrifuge tubes (see Note 10).
11. Drying oven (see Note 10).
12. Analytical balance.

3. Methods

1. Measure specific activity of uptake solution (see Note 11)
2. Pre-equilibrate plants for 5–10 min, if appropriate (see Note 12)
3. Immerse roots in radioactive solution (see Note 13)
4. Remove plants from radioactive solution after appropriate labeling period, typically 2–10 min (see Notes 14 and 15)
5. Desorb roots of extracellular tracer, typically for 5–10 min (see Note 16)
6. Detach roots from shoots, if appropriate (see Note 17)
7. Weigh plants, subsequent to centrifugation or drying (see Note 17)
8. Count radioactivity in plant samples, usually by gamma- or scintillation-counting (see Note 9)
9. Calculate the flux, using absorbed counts, specific activity of labeling solution, root mass, and labeling time (see Notes 18 and 19)

4. Notes

1. A few interesting differences between cations and anions, with respect to plant roots, might be mentioned here. The uptake of cations can substantially exceed that of anions when ammonium is the main nitrogen source, while the reverse is true when nitrate is the source (4, 5). In either case, due to the metabolism of anions, the accumulation of cations often exceeds that of anions (cations other than NH_4^+ are rarely metabolized), with charge balance being achieved via the synthesis of organic acids and the transport of protons (4, 6). Interestingly, the uptake of both anions and cations is at least partially driven by the inwardly negative charge on the plasma membrane of root cells; in the case of anions, this is typically facilitated by the cotransport of protons (7).
2. Plants should be grown hydroponically, so that culture conditions can be precisely controlled, and roots are directly accessible for examination. It is important to consider the use of plants from a variety of developmental stages, as their nutrient requirements will change with age. It should also be considered that seed reserves play an important role in the nutrition of a young seedling.
3. Excised roots are frequently used in tracer-flux experiments, as they reduce complications arising from transpiration and

translocation to the shoot. However, their use also entails tissue damage at the point of excision, which can be propagated to the remainder of the root segment. Therefore, excised tissue should be aged for several hours prior to measurement, ideally in conjunction with an indicator of recovery (8). It may also be necessary to supplement this heterotrophic tissue with a source of energy (e.g., sucrose). Even when such precautions are taken, however, valuable information is inevitably lost when excised roots are used—e.g., information about the partitioning of a substance between organs of the plant, or the influence of transpiration as a potential driver of the flux. Thus, we recommend that, when possible, intact plants be used for tracer flux measurements.

4. Often, several plants are bundled together at the shoot base and treated as a single replicate. This can improve statistics, and measuring accuracy when seedlings of low root mass are used, or when specific activity is low.
5. For steady-state investigations, labeling solutions should be identical to plant growth solutions, except for the addition of radiotracer. Volumes of solution should be chosen to ensure that depletion of nutrients by plant roots is not significant over the duration of the protocol, especially when working with low concentrations of rapidly accumulated substrate (200 ml of solution is typically sufficient for a 5-min uptake procedure). In addition, all other growth conditions, especially temperature, light, and humidity, should be maintained during experimentation and, therefore, simple bench experimentation, in marginally controlled lab spaces, should be avoided. For non-steady-state conditions, the uptake solutions and/or ambient conditions are often modified, e.g., to include variations in substrate concentrations (as in the development of flux isotherms), or the provision of metabolic inhibitors (see Note 12).
6. Particularly in older studies, the uptake of a labeled substance is often monitored against a background of only that substance, plus a small amount of Ca^{2+} (usually as $CaSO_4$), typically at about 100–200 μM, to maintain membrane integrity and basic membrane function. While this simplified approach may reduce complications due to interactions between the traced substance and other materials in solution, it also may change the nature of the experimental system. For example, removing the K^+ provided during plant growth (or the NH_4^+, if present) will likely result in electrical hyperpolarization of the plasma membrane, thus changing the driving force for ions across the membrane and therefore, quite possibly, their fluxes. In general, we recommend using complete nutrient solutions, unless the hypothesis guiding a particular study requires the removal or addition of solution components (e.g., ref. 9).

7. Often the choice between stable and radioactive isotope is a matter of convenience, and will depend upon the availability of appropriate isotopic material or instrumentation (see Note 9). In general, the processing time for radioactive counting is much shorter than that for measurement of stable isotopes by use of mass spectrometry. On the other hand, radioisotopes of some elements (e.g., N, O) are very short-lived and are only available to researchers working in close proximity to a production facility such as a cyclotron. While less widely available and affording less handling time, however, such tracers are advantageous from the perspective of radioactive waste; for example, a sample of ^{13}N will have more or less completely decayed into stable ^{13}C within a few hours of its production. Still, longer-lived tracers are much more commonly used; these include tracers for K^+ and Na^+ that have slightly longer half-lives, on the order of several hours (see Note 8). In all cases involving radiotracers, appropriate safety measures must be taken. This often involves a combination of shielding types, such as Plexiglas and/or lead, depending on the nature of the isotopic decay.
8. Perhaps surprisingly, the most widely used isotope for K^+ tracing is not potassium at all, but ^{86}Rb, a radioisotope of rubidium, an alkali-metal "analog" of potassium. This is due in large part to its long half-life (18.65 days) relative to isotopes of K (see below). However, there is substantial evidence that Rb is an imperfect substitute for K, for example with respect to its translocation to the shoot (10). ^{40}K, with its extremely long half-life (1.25 billion years), is rarely used as a tracer but rather as an environmental indicator (e.g., ref. 11). The other radioisotopes of K are all too short-lived to be of significant practical use, except for ^{42}K and ^{43}K (half-lives, respectively, of 12.36 and 22.3 h). Of these, ^{42}K is almost exclusively used as a tracer in plant systems, although ^{43}K has been successfully used, for example, in the microautoradiographic tracing of K^+ fluxes in stomatal cells (12). In the case of Na^+ tracing, only two radioisotopes of sodium are sufficiently long-lived to be useful: ^{22}Na and ^{24}Na (half-lives of 2.6 years and 14.96 h, respectively); both have been used extensively in plant systems. It is worth noting that the production of ^{42}K and ^{24}Na typically involves bombardment of nonradioactive ^{39}K and ^{23}Na, respectively. Because these nuclear transformations are generally incomplete, researchers must consider the presence of (sometimes substantial) residual amounts of "cold" isotope in their preparation of experimental solutions. This can be contrasted with the production of ^{13}N, via the proton bombardment of the oxygen atom of water; in such a case, there is no background ^{14}N to contend with.
9. Almost all radiotracing in biological systems involves the measurement of electromagnetic radiation associated with the

decay of beta (plus or minus) -emitting nuclides (i.e., positrons or electrons). Liquid scintillation counters and gamma counters are used to measure most biologically important radioisotopes. Scintillation is the most widely used counting method, despite the need to use a scintillation "cocktail" in which the sample is dissolved or suspended; gamma counting has no such requirement. Regardless of counting device, it is essential that it correct for radioactive decay, particularly when using short-lived tracers. As well, care must be taken to ensure that the geometry of the sample-detector system is uniform from one sample to the next (or corrections be made to account for differences), since the position of the detector relative to the sample can influence the quantity of radiation measured. This phenomenon is not always fully acknowledged by providers of detection instrumentation and usually must be worked out by the individual researcher. Similarly, manufacturer claims about the effectiveness of detector shielding against ambient radiation (e.g., from nearby samples within the counter) are at times exaggerated, and such issues must also be worked out for individual measuring systems.

10. When fluxes are to be expressed as dry weight, centrifugation is not required; conversely, when expressed as fresh weight, oven drying is not required (also see Note 17).

11. Prior to placing plants into an uptake solution, a small sample of solution (typically, 1 ml or less) must be removed to determine its specific radioactivity. Specific activity is usually expressed as cpm/μmol (cpm = counts per minute, which is related, by the counting efficiency of the detecting instrument, to the dpm, or disintegrations per minute), and is often denoted "SA," or S_o (the subscript "o" indicating the outside solution, bathing the roots). Typical values for SA are in the range of $1–2 \times 10^5$ cpm. See Note 18 for details of flux calculation.

12. When plants are to be measured under non-steady-state conditions, their roots are often first exposed to modified solution without radiotracer, for an equilibration period of 5–10 min, prior to their exposure to modified solution containing tracer. If inhibitors or other additives are to be used, sufficient time must be allowed for their effects to take place.

13. When roots of intact plants are immersed in radioactive uptake solution, care must be taken to ensure that the aerial parts of the plant do not come into contact with the solution. This is particularly important when solutions are vigorously stirred or aerated, or with plants having a short stem, such as *Arabidopsis*. A shielding collar is sometimes used to isolate stems and leaves from solution. If they are not isolated from radioactive solution, substantial translocation artifacts may be obtained (sometimes identifiable due to their high variability).

14. The duration of tracer absorption (and desorption; see Note 16 below) by roots is an important aspect to consider when tracing influx at a high substrate concentration, because such "low-affinity" influxes can be extremely high and are often associated with very high efflux rates and rapid turnover times (7). Under such conditions, a failure to consider the simultaneous efflux of a substrate over the course of influx measurement will result in an underestimate of the flux. Such errors can be minimized, however, by reducing the duration of labeling and desorption (3), although this courts the danger of adding increased proportions of background counts from the apoplast. In general, when such high efflux rates are observed, it is particularly important to identify the source(s) of tracer release, by using, for example, compartmental analysis by tracer efflux (CATE; refs. 13, 14; see Note 19). While such a situation can occur when tracing low-affinity fluxes of K^+ (or fluxes of Ca^{2+} in any range, due to its very rapid cytosolic turnover; see ref. (15)), the measurement of sodium influx under salinity conditions may be the most pronounced example of this situation, since external concentrations of Na^+ can be 100 mM or more, greatly exceeding the naturally occurring, and experimentally provided, levels of other ions (16). In the case of K^+, recent evidence (17) suggests that above 1 mM external K^+, the efflux component (which can become very pronounced; see ref. (18)) may be apoplastic in nature, since inhibitors such as Cs^+ and Ba^{2+} (which have powerful effects on K^+ efflux in the high-affinity range) fail to affect this component.

15. In practice, it can be very difficult to distinguish extracellular events from those occurring across cell membranes. This presents a dilemma to the researcher: should one attempt to prevent underestimates of influx across the membrane caused by tracer efflux from the cell during measurement, by opting for very short labeling and desorption times (see Note 16)? Alternatively, should one minimize artifacts associated with extracellular accumulation of tracer, by opting for the opposite? The answer to this depends at least partly on the intention of the study. Because the accumulation of sodium in plant tissues is central to the toxicology of salinity stress, the latter approach (basically a net-flux measurement) can be quite appropriate and satisfactory in many situations. On the other hand, if the study is to focus on a classical enzyme-kinetics evaluation of membrane transport systems (19), or an analysis of the energetics of unidirectional transport processes (20), the requirement for unidirectional flux measurements is stringent, and thus the former approach would be recommended. However, the veracity of the efflux component of the flux as a transmembrane phenomenon must also be determined in such

cases, for the particular experimental system, e.g., by performing CATE analyses (see Note 19).

16. In all cases, care must be taken to clear (desorb) tracer from known apoplastic phases, once labeling is complete. These phases include the surface water-film of roots, and the electrostatically binding "Donnan" phase of cell walls, which consists of fixed extracellular charges. Since these charges are mostly negative, this is a particularly important issue when measuring cation fluxes, especially in the case of the divalent cation Ca^{2+}. Desorption is generally done by immersing roots in a solution identical to the uptake solution, except that it contains no radiotracer; this solution is sometimes chilled to 4 °C to minimize loss of tracer from the symplast. Multiple desorption steps, each in fresh solution, are often used. The length of the desorption period is often 5–10 min, but can often be more precisely determined for extracellular phases by means of CATE (see Note 19).

17. Once desorption is complete, roots are typically detached from shoots (if intact plants are used) for separate counting, to estimate translocation rates. At this stage, roots can be weighed prior to counting if fluxes are to be normalized to root fresh weight. If so, a brief, low-speed centrifugation of root tissue (e.g., in a clinical centrifuge at $5000 \times g$) is required to remove surface and interstitial water. If dry weight is the standard, radioactivity of samples may be counted before or after drying and weighing.

18. The influx or net flux into the plant can be calculated quite straightforwardly using the formula $\phi = \frac{Q^*}{S_o w t_L}$, where ϕ is the flux (e.g., μmol/g h), Q^* (cpm) is the quantity of tracer accumulated in tissue (usually root and shoot combined), S_o is the specific activity of the uptake solution (cpm/μmol), w is the root weight (g), and t_L is the labeling time (h). More sophisticated calculations can also be made that account for simultaneous tracer efflux from root cells during labeling and desorption, based on parameters obtained using CATE analysis (for details, see ref. 3; also see Note 19, and caveats above). The absolute quantification of transport to the shoot is more problematic, because (1) the specific activity of the translocating pool is difficult to estimate, and (2) a lag phase may retard the appearance in the shoot of some labeled ions, particularly K^+ (e.g., ref. 21). Thus, investigations of root-to-shoot transport may require longer labeling times and is sometimes expressed as % of total tracer absorbed that is found in the shoot. One additional issue is the possibility of apoplastic bypass flow of Na^+ to the shoot, well documented in species such as rice (22), which contributes to the noncellular component of influx into the plant (see Note 14).

19. Compartmental Analysis by Tracer Efflux (CATE) has been used extensively in biology and medicine (23) to quantify ion fluxes and metabolic pool sizes, including those of K^+ in plants and algae (16, 24, 25). While its use in plant science has diminished somewhat in recent years, it remains an important methodology in medical science, particularly pharmacokinetics (e.g., ref. 26). This method uses long labeling periods (typically one to several hours) followed by a partial washing out of tracer from roots by means of a timed series of nonradioactive eluates. When done correctly, CATE can provide a more comprehensive view of unidirectional fluxes than the procedure outlined above, as well as compartmentation data on both cellular and whole-plant scales. However, it is more labor-intensive, generally limited to steady-state conditions, and prone to its own set of heuristic problems. While a detailed exposition of CATE is beyond the scope of this chapter (but *see* refs. 13, 14, 27, 28 for rationale and procedures), a few points are worth mentioning. Exponential half-times of tracer exchange in intra- and extra-cellular phases of plant roots, as well as the ratio of unidirectional efflux to influx, can be estimated using CATE. This information is pertinent to the design of "direct influx" protocols, because it facilitates the choice of: (1) labeling time, during which tracer efflux from an absorbing and releasing root can be minimized (hence reducing underestimates of the flux); and (2) desorption time, to maximize the release of tracer from extracellular spaces. For details on this approach, see ref. (3). CATE can also be useful as an independent line of investigation against which direct-influx results may be compared. When discrepancies appear between the two systems of measurement, a comparison can yield useful insights, both biological and methodological. Requirements for the correct interpretation of CATE data, however, are stringent. Phases of tracer release (e.g., surface film, cell wall, cytosol, vacuole) must be correctly identified (13, 14) and should also be sufficiently distinct, kinetically speaking, from one another to be resolved (29). In addition, steady state must be ensured, limiting the use of CATE for some investigations, such as those involving flux isotherms.

References

1. de Hevesy G (1962) Adventures in radioisotope research: the collected papers of George Hevesy. Pergamon, New York
2. Epstein E (1953) Mechanism of ion absorption by roots. Nature 171:83–84
3. Britto DT, Kronzucker HJ (2001) Can unidirectional influx be measured in higher plants? A mathematical approach using parameters from efflux analysis. New Phytol 150:37–47
4. van Beusichem ML, Kirkby EA, Baas R (1988) Influence of nitrate and ammonium nutrition on the uptake, assimilation, and distribution of nutrients in *Ricinus communis*. Plant Physiol 86:914–921

5. Speer M, Brune A, Kaiser WM (1994) Replacement of nitrate by ammonium as the nitrogen source increases the salt sensitivity of pea plants. I. Ion concentrations in roots and leaves. Plant Cell Environ 17:1215–1221
6. Britto DT, Kronzucker HJ (2005) Nitrogen acquisition, PEP carboxylase, and cellular pH homeostasis: new views on old paradigms. Plant Cell Environ 28:1396–1409
7. Britto DT, Kronzucker HJ (2006) Futile cycling at the plasma membrane: a hallmark of low-affinity nutrient transport. Trends Plant Sci 11:529–534
8. Glass ADM (1978) Influence of incision and aging upon K^+ influx on developing barley roots. Plant Physiol 61:481–483
9. Horie T, Costa A, Kim TH et al (2007) Rice OsHKT2;1 transporter mediates large Na+ influx component into K+-starved rice roots for growth. EMBO J 26:3003–3014
10. Britto DT, Kronzucker HJ (2008) Cellular mechanisms of potassium transport in plants. Physiol Plant 133:637–650
11. Brownridge JD (1984) The radial distribution of ^{137}Cs and ^{40}K in tree stems. J Plant Nutr 7:887–896
12. Maier-Maercker U, Jahnke A (1980) Microautoradiography with ^{43}K: a method for the reliable tracing of ion transport in stomata. Z Pflanzenphysiol 100:35–42
13. Kronzucker HJ, Siddiqi MY, Glass ADM (1995) Analysis of $^{13}NH_4^+$ efflux in spruce roots: a test case for phase identification in compartmental analysis. Plant Physiol 109:481–490
14. Britto DT, Kronzucker HJ (2003) Trans-stimulation of $^{13}NH_4^+$ efflux provides evidence for the cytosolic origin of tracer in the compartmental analysis of barley roots. Funct Plant Biol 30:1233–1238
15. Britto DT, Kronzucker HJ (2003) Cytosolic ion exchange dynamics: insights into the mechanisms of component ion fluxes and their measurement. Funct Plant Biol 30:355–363
16. Malagoli P, Britto DT, Schulze LM et al (2008) Futile Na^+ cycling at the root plasma membrane in rice (*Oryza sativa* L.) – kinetics, energetics, and relation to salinity tolerance. J Exp Bot 59:4109–4117
17. Coskun D, Britto DT, Kronzucker HJ (2010) Regulation and mechanism of potassium release from barley roots: an *in planta* ^{42}K analysis. New Phytol 188:1028–1038
18. Szczerba MW, Britto DT, Kronzucker HJ (2006) Rapid, futile K^+ cycling and pool-size dynamics define low-affinity potassium transport in barley. Plant Physiol 141:1494–1507
19. Epstein E (1966) Dual pattern of ion absorption by plant cells and by plants. Nature 212:1324–1327
20. Britto DT, Kronzucker HJ (2009) Ussing's conundrum and the search for transport mechanisms in plants. New Phytol 183:243–246
21. Lazof D, Cheeseman JM (1988) Sodium and potassium compartmentation and transport in the roots of intact lettuce plants. Plant Physiol 88:1279–1284
22. Yeo AR, Yeo ME, Flowers TJ (1987) The contribution of an apoplastic pathway to sodium uptake by rice roots in saline conditions. J Exp Bot 38:1141–1153
23. Jacquez JA (1996) Compartmental analysis in biology and medicine, 3rd edn. U Michigan, Ann Arbor
24. Kronzucker HJ, Szczerba MW, Moazami-Goudarzi M et al (2006) The cytosolic Na^+:K^+ ratio does not explain salinity-induced growth impairment in barley: a dual-tracer study using $^{42}K^+$ and $^{24}Na^+$. Plant Cell Environ 29: 2228–2237
25. Hajibagheri MA, Flowers TJ, Collins JC et al (1988) A comparison of the methods of X-ray microanalysis, compartmental analysis and longitudinal ion profiles to estimate cytoplasmic ion concentrations in two maize varieties. J Exp Bot 39:279–290
26. Ashworth S, Rabiner EA, Gunn RN et al (2010) Evaluation of ^{11}C-GSK189254 as a novel radioligand for the H_3 receptor in humans using PET. J Nucl Med 51:1021–1029
27. Walker NA, Pitman MG (1976) Measurement of fluxes across membranes. In: Lüttge U, Pitman MG (eds) Encyclopedia of plant physiology, vol 2 Part A. Springer, Berlin
28. Siddiqi MY, Glass ADM, Ruth TJ (1991) Studies of the uptake of nitrate in barley. III. Compartmentation of NO_3^-. J Exp Bot 42: 1455–1463
29. Cheeseman JM (1986) Compartmental efflux analysis – an evaluation of the technique and its limitations. Plant Physiol 80:51–64

Chapter 11

Ion Flux Measurements Using the MIFE Technique

Sergey Shabala, Lana Shabala, Jayakumar Bose, Tracey Cuin, and Ian Newman

Abstract

Noninvasive microelectrode ion flux measurements (the MIFE™ technique) allow the concurrent quantification of net fluxes of several ions with high spatial (several μm) and temporal (ca 5 s) resolution. The MIFE technique has become a popular tool for studying the adaptive responses of plant cells and tissues to a large number of abiotic and biotic stresses. This chapter briefly summarizes some key findings on spatial and temporal organization of plant nutrient acquisition obtained by the MIFE technique, as well as the MIFE contribution towards elucidating the mechanisms behind a plant's perception and signaling of major abiotic stresses. The full protocols for microelectrode fabrication, calibration, and use are then given, and two basic routines for mapping root ion flux profiles and studying transient ion flux kinetics are given.

Key words: Ion flux, Membrane transport, Ion-selective microelectrodes, Deficiency, Toxicity, Acidity, Salinity, Oxidative stress

1. Introduction

The MIFE™ technique (acronym MIFE derived from *M*icroelectrode *I*on *F*lux *E*stimation) enables measurements of ion fluxes by using noninvasive microelectrodes. Having been developed at the University of Tasmania, the MIFE technique has become a popular tool in studying adaptive responses of plant cells and tissues to a large number of abiotic and biotic stresses (see below). Similar technologies (known under acronyms SIET or SERIS) have been developed elsewhere (1–4). Overall, at least 20 laboratories around the world already employ this technology, and their number is rapidly growing. The main advantage of this method is that it is noninvasive; it allows researchers to study the kinetics of net ion flux into and out of plant tissues in conditions as close to natural as possible. Another advantage of the MIFE method is its high temporal (ca 5 s) and spatial (several microns) resolution and the possibility

Frans J.M. Maathuis (ed.), *Plant Mineral Nutrients: Methods and Protocols*, Methods in Molecular Biology, vol. 953,
DOI 10.1007/978-1-62703-152-3_11, © Springer Science+Business Media, LLC 2013

to measure fluxes of several ions concurrently (hence, study the stoichiometry of transporters) for an essentially unlimited amount of time (hours or days).

Based on anatomy and growth patterns, plant roots are divided into five distinct longitudinal zones: (1) root cap, (2) meristem, (3) transition zone, (4) elongation zone, and (5) mature zone (5). Each of these zones may differ distinctly in nutrient acquisition and plant signaling in response to environmental stimuli. From this perspective, noninvasive vibrating microelectrodes form the most convenient and, arguably, the *only* technique available that provides the required spatial resolution along the longitudinal axis. It has been demonstrated that H^+, K^+ (6), NH_4^+, NO_3^- (7), Ca^{2+}, and Mg^{2+} (8) fluxes differ between the different root zones. Moreover, the MIFE technique can be used for functional analysis of individual transporters at the molecular level. For example, Guo et al. (8) demonstrated that the AtCNGC10 transporter (*Arabidopsis* cyclic nucleotide gated channel) is involved in transporting Ca^{2+} and Mg^{2+} ions. Similarly, functional analysis of AtHELPS, an *Arabidopsis* DExD/H box RNA helicase, revealed that AtHELPS is a negative regulator of high-affinity K^+ transporters (e.g., AKT1—*Arabidopsis K*$^+$ *transporter1*) that are required during K^+ deprivation (9).

Application of the MIFE technique has also provided insights into the complexity of the temporal organization of plant nutrient acquisition. Ultradian oscillations (period range between several minutes to 1–1.5 h) were shown to be a quintessential part of nutrient acquisition processes (10–13). These oscillations were shown to be causally linked with root circumnutation (11, 13, 14), turgor adjustment (15), and encoding environmental information (16–18).

The MIFE technique has also been successfully applied to study receptor activity at the plasma membrane and elucidate signal transduction pathways in response to various hormonal factors (19–21), light (22, 23), and pathogens (24). Also, it has been used as an early diagnostic tool for plant magnesium deficiency (25)—a nutritional problem without clear visual symptoms.

In acidic soils, aluminum toxicity poses a major threat to plant growth by inhibiting root growth (26). Al^{3+} ion is the most rhizotoxic form (27); its activity peaks at around pH 4.2–4.3 (28). While these two stresses normally occur together, the mechanisms of plant tolerance to each of these individually may be strikingly different. It was shown that H^+ toxicity causes irreversible damage to primary and lateral roots in *Arabidopsis*, with the pattern of damage being different from the one caused by Al^{3+} rhizotoxicity (29). Furthermore, an *Arabidopsis* QTL analysis suggested that Al^{3+} tolerance and H^+ tolerance are controlled by different genes (30).

A comprehensive functional characterization of *Arabidopsis* root responses to low-pH and combined low-pH/Al^{3+} stresses has recently been undertaken using the MIFE technique. It was shown that in the absence of Al^{3+}, low-pH stress induces H^+ influx so causing rhizosphere alkalinization. In contrast, the presence of Al^{3+} inhibited H^+ influx, resulting in less rhizosphere alkalinization (6). Moreover, the Al^{3+}-sensitive *Arabidopsis* mutant *als5* grew well under low-pH stress and poorly under Al^{3+} stress, whereas *als3* was sensitive and *alr104* tolerant to both stresses. An independent MIFE study has reported that acid-soil-tolerant species are able to maintain higher H^+ efflux and NH_4^+ influx compared with their acid-soil-sensitive counterparts (31).

To deal with the Al^{3+} toxicity problem, plants release various organic acids such as malate or citrate (32) to modify the rizhosphere pH and reduce Al^{3+} solubility. The Al^{3+}-induced release of organic anions from plant roots is accompanied by K^+ efflux to maintain electroneutrality (33). The MIFE technique has revealed that this release occurs only at the elongating zone and not in the meristematic zone, providing new temporal and spatial information on the Al^{3+}-activated efflux of K^+ from intact plants (34).

Finally, the MIFE technique has been widely used to elucidate plant nutritional responses and signal transduction pathways to a wide range of abiotic stresses such as salinity, osmotic, and oxidative stresses; chilling; and anoxia. Amongst other things, we have (1) revealed the critical role of inorganic ion uptake in cell osmoregulation (35, 36); (2) elucidated specific details of SOS (37) and MAPK (38) pathways in salt and osmotic stress signaling; (3) shown that a plant's ability to maintain an optimal cytosolic K^+/Na^+ ratio is central to salt tolerance (35, 39–41); (4) revealed the role of specific membrane transporters such as NSCC, GORK, AKT, and SOS1 in adaptation to salinity and osmotic stress (35, 37, 41, 42); (5) explored the ionic mechanisms of amelioration of salinity stress by divalent cations (43, 44), compatible solutes (42, 45) and polyamines (46, 47); (6) studied the genetic aspects of inheritance of salt tolerance in crops (39, 48); (7) demonstrated the involvement of GORK and NSCC channels in oxidative stress responses (49–51); (8) demonstrated an in situ ameliorating role of compatible solutes in oxidative stress (42); (9) revealed spatial and temporal heterogeneity and differential sensitivity of HACC and NSCC channels to H_2O_2 and OH^- (49, 50, 52); (10) demonstrated the involvement of GORK and NSCC channels in salinity- and oxidative-stress-induced PCD (51, 53); (11) identified the central role of posttranslational regulation of H^+-ATPase in plant adaptive responses to salinity (41); (12) revealed the involvement of Ca^{2+} efflux systems in oxidative stress tolerance induced by pathogens (24, 54); (13) revealed mechanisms of NADPH oxidase activation by $(Ca^{2+})_{cyt}$ (55); and (14) elucidated effects of anoxia (56) and secondary metabolites produced in waterlogged soils

(57) on root nutrient acquisition. Taken together, the results obtained have identified several key membrane transport systems that are the cornerstone of plant adaptive responses to numerous environmental stresses.

2. Materials

2.1. MIFE™ System Components

1. MIFE main amplifier/controller and preamplifier.
2. Multi-manipulator providing 3-axis positioning and driven by the stepping-motor (see Note 1).
3. PC with ISA-bus, running Windows 98 or ME, for system control and data acquisition.
4. CIO-DAS08 card for analogue to digital conversion.
5. MIFE CHART and MIFEFLUX software.
6. Handbook DOC files, explaining the hardware and software.

2.2. Other Equipment Required for Flux Studies

1. Microscope with long-working distance objectives (100× or 200×). This can be an inverted microscope or a compound microscope lying on its back.
2. Anti-vibration table.
3. Faraday cage.
4. Vertical electrode puller (see Note 2).
5. Borosilicate glass capillaries, without filling fiber, OD = 1.5 mm (GC150-10; Harvard Apparatus Ltd, Kent, UK).
6. Specific ion-selective cocktails (LIX; see Note 3).
7. Silanizing agent (tributylchlorosilane; Fluka Chemicals 90796).
8. Electrode holders (e.g., E45W-F15PH; Warner Instruments, USA).
9. Nonmetallic syringe needle for filling micropipettes (MF34G-5, 0.1 mm ID; WPI, Sarasota, FL, USA).
10. Electrode filling station consisting of two simple micromanipulators and a stereomicroscope.
11. Small oven, to 250 °C, with gloves and metal electrode racks having metal covers.
12. Fume cabinet.
13. Measuring chambers to immobilize plant roots.

2.3. Solutions and Plant Material

1. Basic measuring solution (see Note 4).
2. Equipment to grow plants in aerated hydroponic solutions under sterile conditions.
3. Temperature-controlled room.

3. Methods

3.1. Pulling Electrode Blanks

1. Insert nonfilamentous borosilicate glass capillaries into a vertical pipette puller.
2. Pull the blanks to <3 μm diameter tips.
3. Store pulled electrodes in a stainless steel or aluminum covered rack in a vertical position.

3.2. Baking and Silanizing the Blanks

1. Place electrode blanks uncovered in a rack with tips upright and base down.
2. Oven-dry the blanks at 220 °C overnight. They remain in the 220 °C oven for the next three steps.
3. Ten to fifteen minutes before silanization, place a steel cover over the electrode blanks.
4. Add two drops (approximately 50 μL) of tributylchlorosilane on the rack under the cover using a yellow-tip pipettor.
5. Ten minutes later, remove the lid and bake electrode blanks for a further 30 min.
6. Switch the oven off and let the blank electrodes cool down.
7. Take electrodes from the cold oven and keep them in a closed container.

3.3. Filling the Electrodes with an Appropriate Liquid Ion Exchanger

1. Prepare a liquid ion exchanger (LIX)-containing tube by dipping a broken-tip glass micro-capillary (tip diameter approximately 50 μm) into the LIX, taking up a column of cocktail of approximately 1 mm.
2. Mount the pulled microelectrode blank horizontally on a three-dimensional micromanipulator. Likewise, also mount the LIX tube on another micromanipulator. Observe both under a dissecting microscope.
3. Break back the blank tip to achieve an external tip diameter of 2–3 μm by placing it against a flat glass surface.
4. Position the blank coaxially with the LIX-containing tube at 100–200 μm distance.
5. Fill the blank with appropriate back-filling solutions (see Note 5) using a syringe with a nonmetallic needle.
6. Front-fill the blank tip with LIX by putting it briefly into contact with the LIX-containing tube to achieve a column length between 100 and 150 μm.
7. Store electrodes with their tips immersed in the proposed measuring solution until use (up to 8–10 h).

3.4. Calibrating Electrodes

1. Mount electrodes in the holder.
2. Fabricate a reference electrode by inserting a chlorided silver wire (galvanized at 1.5 V in 0.25 N HCl for 15 min) into a broken glass micro-capillary (ca. 50 μm tip diameter) filled with 1 M KCl in 2% agar and sealing it with Parafilm. Alternatively, use a commercial reference electrode.
3. Prepare an appropriate set of at least three standards covering the expected range of the ion to be measured, using about equal factors change between their concentrations.
4. Calibrate electrodes using the MIFE CHART calibration routine below (see Note 6).

3.5. Calibration Routine

1. Put electrodes into your first standard solution, close the Faraday cage and start data acquisition.
2. Press the function key F7. In the window that opens, correct the temperature, type the names of all the ions to be used and press <ENTER>.
3. When the measurement for the ion of the first solution is stable and in range, press F7 again and enter the concentration for that ion. After you press <ENTER> the data recorded will be used in the calibration calculation.
4. After about 20 s, a third press of F7 terminates that set, giving you the option of keeping it or rejecting it.
5. Repeat the process of three presses of F7 for the other concentrations and any other ions.
6. When all the required calibration data have been recorded, type ALT-H to stop data acquisition.
7. From the <E> lectrometer menu, choose the <A> verage option and the <C> alibration average. CHART will do the least squares fit to the data for each ion, create the AVC file and display in a window the parameters of the electrodes.

3.6. Conducting Measurements

1. Mount roots in the measuring chamber filled with an appropriate solution.
2. Under the microscope, position microelectrode tips in one plane, separated laterally by 1–2 μm, 20–30 μm away from the root surface.
3. From the DOS prompt or from Windows Explorer, select your working directory.
4. Start the CHART program.
5. Press ALT-S and start data acquisition.
6. Accept the default file name by typing S.
7. When a window opens to let you set the starting time, duration of measurements and other values, accept all the defaults by typing G.

8. Put appropriate description into the one-line window and press <ENTER> to begin the data acquisition. This is recorded with the data and also in the DAY file.
9. Navigate the Chart Display screen, using the "buttons" at the bottom and right of the screen, to view any or all of the data as desired.
10. Stop data acquisition by pressing ALT-H or the <H> option in the <M> ain Menu.

3.7. MIFE Protocols for Mapping Root Ion Flux Profiles

1. Sterilize seeds and grow plants in hydroponics until required age/size (see Note 7).
2. Prepare and calibrate ion-selective microelectrodes (see Note 8).
3. One hour prior to measurement, immobilize a root in a measuring chamber (see Note 9).
4. Transfer the measuring chamber onto microscope stage.
5. Position electrodes 20–30 μm above the root surface, with their tips closely aligned.
6. Start data acquisition and record net ion fluxes for 10–15 min (as described in Subheading 3.6).
7. Ensure the steady state was reached, i.e., that fluxes are not changing with time.
8. Reposition electrodes to another spot and continue measurements (see Note 10).
9. Once the entire root surface is mapped, position electrodes at the first spot and repeat the recording. This will ensure the absence of any temporal variations in nutrient acquisition.
10. Analyze data as described below in Subheading 3.10.

3.8. MIFE Protocols to Study Transient Ion Flux Responses to External Stimuli

1. Repeat steps 1–5 in Subheading 3.7.
2. Start data acquisition and record net ion fluxes for ca. 10 min. Ensure the absence of any transients.
3. With MIFE still running, add appropriate chemicals (see Note 11). Alternatively, apply some other external stimulus (see Note 12).
4. Mix the chemical as fast as practically possible without causing major disturbance to the plant roots (see Note 13). Avoid root exposure to the air.
5. Ensure the electrode position did not change and continue recordings for as long as required (see Note 14).
6. Stop the data acquisition and calculate fluxes (as below).

3.9. Data Analysis: Produce Files of Average Voltage V and ΔV

1. Produce an AVM file by using the <M>anipulator cycle average routine in the <A>veraging option in the <E>lectrometer menu.
2. Specify a "Valid Time" at the end of each stage of manipulator movement for which the actual "Stage Time" is provided (see Note 15). Pressing <ENTER> moves the highlight to the next stage. Pressing UPARROW moves up to the previous one to allow you to correct an error.
3. Specify the root diameter and the starting electrode position in appropriate boxes (<Z> and <U>, respectively).
4. Press <ENTER> to accept the order of the curve ("Kind of<F>it") to fit the data (typing F to cycle through the orders).
5. Quit the CHART program by pressing ALT-Q.

3.10. Data Analysis: Produce the Flux File

1. Run the MIFEFLUX program under DOS in your current directory that contains the AVC and AVM files for the flux calculations.
2. The program first asks you for the AVC file(s) you want, then for the AVM file of your averaged flux data, and finally for the kind of analysis (planar, cylindrical, or spherical). Select "cylindrical" if you are measuring fluxes from the root tissue.
3. After the calculation is done, you can calculate more fluxes using the same calibration files.
4. Once calculations are completed, the resulting FLX file (ASCII text format) can be opened by any spreadsheet and evaluated in a conventional way.

4. Notes

1. Our default configuration includes an SM-17 3-electrode holder, mounted on a hydraulic manipulator MHW-4 (1-axis) which, in turn, is assembled on the top of a MX-2 (3-axis) mechanical micromanipulator (all from Narishige, Tokyo, Japan). A custom-made MIFE stepper motor control unit drives the hydraulic manipulator to measure fluxes. Manipulator details are available at http://www.narishige.co.jp/english/.
2. Any standard puller may be used. Because the electrode tip diameter is controlled at a later stage, the pulling process is much less demanding than for patch clamp, and can be performed in a single step.
3. The following commercial ionophore cocktails (LIX) are used in our laboratory: K^+ (60031); Na^+ (71138); Ca^{2+} (21048);

Mg^{2+} (63048); H^+ (95297); NH_4^+ (09882); Cl^- (24902); all from Fluka (Buchs, Switzerland). More types of LIX can be found at http://www.sigmaaldrich.com/.

4. The specific ionic composition of measuring solutions depends on the specific purpose of the experiment. Nonetheless, a few general principles must be observed: (a) solution ionic composition should be as simple as possible to avoid potential confounding effects of interfering ions or LIX poisons; (b) concentrations of ions whose fluxes are to be measured should be kept as low as practically (physiologically) possible to maximize signal-to-noise ratio of the measured signal; (c) for H^+ fluxes, avoid alkaline pH, keep buffer concentrations as low as possible and choose a pK at least 0.5 unit above the pH.
5. The backfilling solutions must contain some background chloride ions plus the ion of interest. The routinely used backfilling solutions in our laboratory are: H^+—15 mM NaCl + 40 mM KH_2PO_4; K^+—200 mM KCl; Na^+—500 mM NaCl; Ca^{2+}—500 mM $CaCl_2$; Mg^{2+}—500 mM $MgCl_2$; Cl^-—500 mM NaCl; NH_4^+—500 mM NH_4Cl.
6. Good electrodes must have a slope above 50 mV per decade for monovalent ions, and at least 25 mV per decade for divalent ions, and have correlation coefficients 0.999 or better (for the three-point calibration).
7. From a practical point of view, it is easiest to work on roots of 30–80 mm length. Smaller roots may be more difficult to immobilize in the measuring chamber, and longer roots require large chambers and, hence, may be heavy to be moved by the hydraulic manipulator.
8. The standard MIFE configuration allows simultaneous measurement of three different ions.
9. Due to methodological reasons, it is more convenient to measure ion fluxes from horizontally positioned roots. However, measurements in the root apex may be confounded by the root bending due to gravitropism. If a root apex is to be measured for a prolonged (e.g., hours) time, then vertical root immobilization should be used.
10. Assuming fluxes are steady and not changing with time, 1.5–2 min measurements at each position is sufficient. Longer measurements at each spot will significantly increase the overall time required to scan the entire root surface and may be confounded by diurnal or other changes in root nutrient acquisition patterns, over the time. From practical experience, 1–1.5 h is a maximum advisable time to complete the entire root mapping.
11. The type of chemical treatment is entirely dependent on the purpose of the experiment, so cannot be specified here. Several methodological aspects, however, should be kept in mind.

First, ensure that the chemical used is not interfering with any ion-selective LIX. This can be achieved by calibrating electrodes in the presence and absence of the chemical. If either electrode slope or intercept is affected, then two calibration files should be used: one calibration set to calculate fluxes before treatment (e.g., in the absence of interfering chemical), and another set for flux calculations after the treatment (in the presence of chemical in the calibration buffers). Second, it is important to remember that signal-to-noise ratio is inversely proportional to the background concentration of the ion in question. Hence, measurements on net fluxes of ions are often problematic when high concentrations of appropriate salts are used (e.g., measuring net Na^+ uptake under saline conditions, or net H^+ fluxes at very low pH).

12. This may include a range of environmental factors such as light, temperature, and osmotica.
13. The easiest way of doing this is to prepare a solution of equal volume containing a double concentration of the chemical of interest, and then simply mix volumes at 50:50 ratio. This achieves the desired final concentration.
14. In most cases, transient responses take between 20 and 50 min. This may vary significantly though between treatments and should be determined empirically.
15. The default setting is that the first 20% of the recording at each electrode location is discarded to account for electrode movement and stabilization. This may need to be increased when electrode responses are slow. Having the "valid time" set as 50% of manipulator cycle is the safest option.

Acknowledgements

This work was supported by the Australian Research Council and Grain Research and Development Corporation grants to Sergey Shabala.

References

1. Smith PJS, Hammar K, Porterfield DM et al (1999) Self-referencing, non-invasive, ion selective electrode for single cell detection of trans-plasma membrane calcium flux. Microsc Res Tech 46:398–417
2. Kunkel JG, Cordeiro S, Xu Y et al (2006) Use of non-invasive ion-selective microelectrode techniques for the study of plant development. In: Volkov AG (ed) Plant electrophysiology – theory and methods. Springer, Berlin
3. Messerli MA, Robinson KR, Smith PJS (2006) Electrochemical sensor applications to the study of Molecular physiology and analyte flux in plants. In: Volkov AG (ed) Plant electrophysiology – theory and methods. Springer, Berlin

4. Porterfield DM (2007) Measuring metabolism and biophysical flux in the tissue, cellular and sub-cellular domains: Recent developments in self-referencing amperometry for physiological sensing. Biosens Bioelectron 22:1186–1196
5. Verbelen JP, De Cnodder T, Le J, Vissenberg K, Baluška F (2006) The root apex of *Arabidopsis thaliana* consists of four distinct zones of cellular activities: meristematic zone, transition zone, fast elongation zone, and growth terminating zone. Plant Signal Behav 1:296–304
6. Bose J, Babourina O, Shabala S, Rengel Z (2010) Aluminum-dependent dynamics of ion transport in Arabidopsis: specificity of low pH and aluminum responses. Physiol Plant 139:401–412
7. Hawkins B, Boukcim H, Plassard C (2008) A comparison of ammonium, nitrate and proton net fluxes along seedling roots of Douglas fir and lodgepole pine grown and measured with different inorganic nitrogen sources. Plant Cell Environ 31:278–287
8. Guo KM, Babourina O, Christopher DA, Borsic T, Rengel Z (2010) The cyclic nucleotide gated channel AtCNGC10 transports Ca^{2+} and Mg^{2+} in *Arabidopsis*. Physiol Plant 139:303–312
9. Xu RR, Qi SD, Lu LT, Chen CT, Wu CA, Zheng CC (2011) A DExD H box RNA helicase is important for K deprivation responses and tolerance in *Arabidopsis thaliana*. FEBS J 278:2296–2306
10. Shabala SN, Newman IA, Morris J (1997) Oscillations in H^+ and Ca^{2+} ion fluxes around the elongation region of corn roots and effects of external pH. Plant Physiol 113:111–118
11. Shabala SN, Newman IA (1997) Proton and calcium flux oscillations in the elongation region correlate with root nutation. Physiol Plant 100:917–926
12. Shabala S, Knowles A (2002) Rhythmic patterns of nutrient acquisition by wheat roots. Funct Plant Biol 29:595–605
13. Shabala S (2003) Physiological implications of ultradian oscillations in plant roots. Plant Soil 255:217–226
14. Shabala SN, Newman IA (1997) Root nutation modelled by two ion flux-linked growth waves around the root. Physiol Plant 101:770–776
15. Shabala SN, Lew RR (2002) Turgor regulation in osmotically stressed Arabidopsis epidermal root cells. Direct support for the role of inorganic ion uptake as revealed by concurrent flux and cell turgor measurements. Plant Physiol 129:290–299
16. Shabala SN, Newman IA (1998) Osmotic sensitivity of Ca^{2+} and H^+ transporters in corn roots: effect on fluxes and their oscillations in the elongation region. J Membr Biol 161:45–54
17. Tyerman SD, Beilby M, Whittington J, Juswono U, Newman I, Shabala S (2001) Oscillations in proton transport revealed from simultaneous measurements of net current and net proton fluxes from isolated root protoplasts: MIFE meets patch-clamp. Aust J Plant Physiol 28:591–604
18. Shabala S, Demidchik V, Shabala L, Cuin TA, Smith SJ, Miller AJ, Davies JM, Newman IA (2006) Extracellular Ca^{2+} ameliorates NaCl-induced K^+ loss from Arabidopsis root and leaf cells by controlling plasma membrane K^+-permeable channels. Plant Physiol 141: 1653–1665
19. Maryani MM, Shabala SN, Gehring CA (2000) Plant natriuretic peptide immunoreactants modulate plasma-membrane H^+ gradients in *Solanum tuberosum* L. leaf tissue vesicles. Arch Biochem Biophys 376:456–458
20. Babourina O, Shabala S, Newman I (1998) Auxin stimulates Cl^- uptake by oat coleoptiles. Ann Bot 82:331–336
21. Ludidi N, Morse M, Sayed M, Wherrett T, Shabala S, Gehring C (2004) A recombinant plant natriuretic peptide causes rapid and spatially differentiated K^+, Na^+ and H^+ flux changes in Arabidopsis thaliana roots. Plant Cell Physiol 45:1093–1098
22. Babourina OK, Newman IA, Shabala SN (2003) Electrophysiological localization of blue light sensory sites in etiolated dicotyledon seedlings. Plant Cell Environ 26:1505–1514
23. Babourina O, Newman I, Shabala S (2002) Blue light-induced kinetics of H^+ and Ca^{2+} fluxes in etiolated wild-type and phototropin-mutant Arabidopsis seedlings. Proc Natl Acad Sci USA 99:2433–2438
24. Nemchinov LG, Shabala L, Shabala S (2008) Calcium efflux as a component of the hypersensitive response of *Nicotiana benthamiana* to *Pseudomonas syringae*. Plant Cell Physiol 49:40–46
25. Shabala S, Hariadi Y (2005) Effects of magnesium availability on the activity of plasma membrane ion transporters and light-induced responses from broad bean leaf mesophyll. Planta 221:56–65
26. Kochian LV, Hoekenga OA, Piñeros MA (2004) How do crop plants tolerate acid soils? Mechanisms of aluminum tolerance and phosphorous efficiency. Annu Rev Plant Biol 55:459–493

27. Kochian LV (1995) Cellular mechanisms of aluminum toxicity and resistance in plants. Annu Rev Plant Physiol Plant Mol Biol 46:237–260
28. Taylor GJ et al (2000) Direct measurement of aluminum uptake and distribution in single cells of *Chara corallina*. Plant Physiol 123: 987–996
29. Koyama H, Toda T, Hara T (2001) Brief exposure to low-pH stress causes irreversible damage to the growing root in *Arabidopsis thaliana*: pectin-Ca interaction may play an important role in proton rhizotoxicity. J Exp Bot 52:361–368
30. Ikka T et al (2007) Natural variation of *Arabidopsis thaliana* reveals that aluminum resistance and proton resistance are controlled by different genetic factors. Theor Appl Genet 115:709–719
31. Hawkins BJ, Robbins S (2010) pH affects ammonium, nitrate and proton fluxes in the apical region of conifer and soybean roots. Physiol Plant 138:238–247
32. Ryan PR, Delhaize E, Jones DL (2001) Function and mechanism of organic anion exudation from plant roots. Annu Rev Plant Physiol Plant Mol Biol 52:527–560
33. Ryan PR, Delhaize E, Randall PJ (1995) Characterisation of Al-stimulated efflux of malate from the apices of Al-tolerant wheat roots. Planta 196:103–110
34. Wherrett T, Ryan PR, Delhaize E, Shabala S (2005) Effect of aluminium on membrane potential and ion fluxes at the apices of wheat roots. Funct Plant Biol 32:199–208
35. Shabala S, Cuin TA (2008) Potassium transport and plant salt tolerance. Physiol Plant 133:651–669
36. Chen ZH, Cuin TA, Zhou M, Twomey A, Naidu B, Shabala S (2007) Compatible solute accumulation and stress-mitigating effects in barley genotypes contrasting in their salt tolerance. J Exp Bot 58:4245–4255
37. Shabala L, Cuin TA, Newman IA, Shabala S (2005) Salinity-induced ion flux patterns from the excised roots of Arabidopsis *sos* mutants. Planta 222:1041–1050
38. Lew RR, Levina NN, Shabala L, Anderca MI, Shabala SN (2006) Role of a mitogen-activated protein kinase cascade in ion flux-mediated turgor regulation in fungi. Eukaryot Cell 5:480–487
39. Chen Z, Newman I, Zhou M, Mendham N, Zhang G, Shabala S (2005) Screening plants for salt tolerance by measuring K^+ flux: a case study for barley. Plant Cell Environ 28: 1230–1246
40. Chen ZH, Zhou MX, Newman IA, Mendham NJ, Zhang GP, Shabala S (2007) Potassium and sodium relations in salinised barley tissues as a basis of differential salt tolerance. Funct Plant Biol 34:150–162
41. Chen ZH, Pottosin II, Cuin TA, Fuglsang AT, Tester M, Jha D, Zepeda-Jazo I, Zhou MX, Palmgren MG, Newman IA, Shabala S (2007) Root plasma membrane transporters controlling K^+/Na^+ homeostasis in salt-stressed barley. Plant Physiol 145:1714–1725
42. Cuin TA, Shabala S (2007) Amino acids regulate salinity-induced potassium efflux in barley root epidermis. Planta 225:753–761
43. Shabala S, Shabala L, Van Volkenburgh E, Newman I (2005) Effect of divalent cations on ion fluxes and leaf photochemistry in salinized barley leaves. J Exp Bot 56:1369–1378
44. Shabala S, Shabala L, Van Volkenburgh E (2003) Effect of calcium on root development and root ion fluxes in salinised barley seedlings. Funct Plant Biol 30:507–514
45. Cuin TA, Shabala S (2005) Exogenously supplied compatible solutes rapidly ameliorate NaCl-induced potassium efflux from barley roots. Plant Cell Physiol 46:1924–1933
46. Shabala S, Cuin TA, Pottosin I (2007) Polyamines prevent NaCl-induced K^+ efflux from pea mesophyll by blocking non-selective cation channels. FEBS Lett 581: 1993–1999
47. Pandolfi C, Pottosin I, Cuin T, Mancuso S, Shabala S (2010) Specificity of polyamine effects on NaCl-induced ion flux kinetics and salt stress amelioration in plants. Plant Cell Physiol 51:422–434
48. Chen ZG, Shabala S, Mendham N, Newman I, Zhang GP, Zhou MX (2008) Combining ability of salinity tolerance on the basis of NaCl-induced K^+ flux from roots of barley. Crop Sci 48:1382–1388
49. Demidchik V, Bowen HC, Maathuis FJM, Shabala SN, Tester MA, White PJ, Davies JM (2002) Arabidopsis thaliana root non-selective cation channels mediate calcium uptake and are involved in growth. Plant J 32: 799–808
50. Demidchik V, Shabala SN, Coutts KB, Tester MA, Davies JM (2003) Free oxygen radicals regulate plasma membrane Ca^{2+} and K^+-permeable channels in plant root cells. J Cell Sci 116:81–88
51. Demidchik V, Cuin TA, Svistunenko D, Smith SJ, Miller AJ, Shabala S, Sokolik A, Yurin V (2010) Arabidopsis root K^+-efflux conductance activated by hydroxyl radicals: single-channel properties, genetic basis and involvement in

stress-induced cell death. J Cell Sci 123: 1468–1479

52. Demidchik V, Shabala SN, Davies JM (2007) Spatial variation in H_2O_2 response of *Arabidopsis thaliana* root epidermal Ca^{2+} flux and plasma membrane Ca^{2+} channels. Plant J 49:377–386
53. Shabala S (2009) Salinity and programmed cell death: unravelling mechanisms for ion specific signalling. J Exp Bot 60:709–711
54. Shabala S, Baekgaard L, Shabala L, Fuglsang A, Babourina O, Palmgren MG, Cuin TA, Rengel Z, Nemchinov LG (2011) Plasma membrane Ca^{2+} transporters mediate virus-induced acquired resistance to oxidative stress. Plant Cell Environ 34:406–417
55. Demidchik V, Shang ZL, Shin R, Colaco R, Laohavisit A, Shabala S, Davies JM (2011) Receptor-like activity evoked by extracellular ADP in Arabidopsis root epidermal plasma membrane. Plant Physiol 156:1375–1385
56. Pang JY, Newman I, Mendham N, Zhou M, Shabala S (2006) Microelectrode ion and O_2 fluxes measurements reveal differential sensitivity of barley root tissues to hypoxia. Plant Cell Environ 29:1107–1121
57. Pang JY, Cuin T, Shabala L, Zhou MX, Mendham N, Shabala S (2007) Effect of secondary metabolites associated with anaerobic soil conditions on ion fluxes and electrophysiology in barley roots. Plant Physiol 145: 266–276

Chapter 12

Sampling and Analysis of Phloem Sap

Sylvie Dinant and Julia Kehr

Abstract

The transport tubes of the phloem are essential for higher plants. They not only provide the route for the distribution of assimilates produced during photosynthesis from source to sink organs but also (re-) distribute mineral nutrients. Additionally, the phloem is essential for sending information between distant plant organs and steering developmental and defense processes. For example, flowering and tuberization time are controlled by phloem-mobile signals and important defense reactions on the whole plant level, like systemic acquired resistance or systemic gene silencing, are spread through the phloem. In addition, recent results demonstrate that also the allocation of mineral nutrients is coordinated by phloem mobile signaling molecules.

However, in many studies the important analysis of phloem sap is neglected, probably because the content of sieve tubes is not easy to access. This chapter will describe the current methods for sampling and analysis of phloem sap in order to encourage researchers to include the analysis of this crucial compartment in their relevant studies.

Key words: Phloem, Sampling, Aphid stylectomy, Exudation, EDTA, Sieve element, Companion cell

1. Introduction

The phloem together with the xylem builds the long-distance transport system that is organized in vascular bundles that pervade the whole body of higher plants. While xylem vessels are mainly responsible for the transport of water and mineral nutrients from the root to the aboveground organs, the phloem can also transport minerals and water, but additionally a wide range of organic compounds from photosynthetic organs to auxotrophic ones. To achieve this, the phloem of vascular plants is built of two kinds of highly specialized cells: companion cells (CC) and sieve elements (SE). CC and SE originate from the unequal division of a single mother cell. While SE differentiate into transport tubes that lose many important cellular components like nuclei, vacuoles, and

Frans J.M. Maathuis (ed.), *Plant Mineral Nutrients: Methods and Protocols*, Methods in Molecular Biology, vol. 953,
DOI 10.1007/978-1-62703-152-3_12,

ribosomes, CC become the support units delivering structural and signaling components and mediating the (un-) loading of the transported substances.

The phloem is important for many physiological processes. For instance, it is the major route for organic compounds synthesized during photosynthesis from sites of production (source organs) to sites of import (sink organs). In addition to this central role in photoassimilate allocation, SE also provide the route for spreading information throughout the plant. This information is important to regulate developmental processes, defense reactions, or to communicate the nutritional status of plant organs. A wide range of molecules are present in phloem sap and could potentially act as nutrients or signal molecules including sugars, amino acids, inorganic ions, proteins, mRNAs, small RNAs proteins and peptides.

2. Materials

2.1. Exudation Buffer

5–50 mM EDTA solution buffered at a near-neutral pH value between 6 and 8 (see below).

2.2. Exudation Cuvette

Homemade device composed of a 10 cm Plexiglas sampling cuvette with a 3 mm wide slit connected both ends to a syringe for washing the chamber and harvesting samples (1).

2.3. Chamber for Exudation

Plexiglas chamber or plastic box that enables maintenance of high relative air humidity (over 60 %).

2.4. Stylectomy

Microcautery device, e.g., HF-microcautery unit CF-50, Syntech, Hilversum, the Netherlands (2–4) or laser device using a ruby or a YAG laser (5–7).

3. Methods

3.1. Plant Growth Conditions and Sampling

Plants used for the collection of phloem sap should normally be well-watered and ideally grown in growth chambers or greenhouses with high air humidity to promote exudation. Also, the developmental stage is crucial for the successful collection of sap and needs to be optimized depending on plant species, sampling site, etc. Huge variations in phloem sap amount and composition have been observed during the diurnal cycle. In addition, biotic and abiotic stresses can modify the phloem sap volume that can be collected, as well as its chemical composition. Therefore day time, developmental stage and growth conditions are important parameters to efficiently collect and optimize yield. Variations in phloem sap composition occur within a plant depending on the position and age of the organs that are used for phloem sap sampling (8). Variations in amino acid concentration can also occur among replicate samples of exudates obtained from insect stylectomy. Conversely, when aphids are used for the collection of phloem sap (see below), the aphid's feeding behavior is influenced by phloem sap composition, which can in turn affect their settlement on preferential, although not necessarily representative, positions on the plant.

3.2. Collection of Phloem Sap

Although the importance of the analysis of sieve element contents for understanding plant nutrition, communication, and stress defense is undisputed, one of the major obstacles hindering comprehensive analyses of this transport fluid is the difficulty in accessing the SE. This is caused by the localization of the phloem tubes well inside the tissue and the extreme sensitivity of SE to wounding. In particular, sieve elements possess different efficient sealing mechanisms to avoid loss of its precious contents after physical injuries. However, depending on the plant species that is the experimental system several different methods that allow the acquisition of phloem samples in different quantities are available.

Over the last decades three main methods have proven most successful to collect phloem sap in sufficient quantities for different kinds of analyses: spontaneous exudation, EDTA-facilitated exudation, and insect stylectomy. There are some other methods that have been used to obtain phloem content or phloem cells, respectively. Their use, however, was restricted to individual studies and they will therefore not be covered in more detail. One option is the use of microcapillaries to sample phloem content under the microscope after SE have been visualized by the phloem-mobile fluorochrome carboxyfluorescein diacetate (9, 10). Other possibilities are the use of laser microdissection (1, 11, 12) or fluorescence-tagged cell sorting (13, 14) to isolate phloem cells.

The three most commonly applied phloem sampling methods mentioned above mainly differ in the amount, purity, and dilution of the resultant samples, but also in their degree of injury imposed and thus the level of potential contaminations and artifacts. Despite the fact that all the different methods can introduce different kinds of artifacts and can therefore result in quite unequal types of samples, phloem extracts are normally summarized under the term "phloem sap" or "phloem sample" disguising the different sampling methods and the possibly different characteristics of the samples. Depending on the collection methods more accurate terms should be used in many cases. The major phloem sampling methods will be introduced in the following paragraphs.

3.2.1. Spontaneous Exudation

This method is only applicable in plants where phloem content spontaneously exudes after wounding. In these species, either complete organs can be cut off, small longitudinal incisions can be made with razor blades, or the phloem can be injured by punctures inflicted by syringe needles. In all cases spontaneous exudation usually provides relatively large volumes of phloem sap that can be collected for analysis. However, phloem content only exudes in a limited number of plants including certain trees (e.g., *Fraxinus americana*, *Robinia pseudoacacia*) (15, 16) and some herbaceous plants (e.g., *Yucca filamentosa*, *Cucurbita maxima*, and other members of the *Cucurbita* genus, *Ricinus communis*, *Brassica napus*, and *Brassica oleracea*) (17–22). From pumpkin (*Cucurbita maxima)*, for example, milliliters of phloem exudate can be collected while Yucca and palms can yield even several liters of sap (23). In most plant species, however, phloem sap does not exude from incisions, probably caused by a rapid and reversible accumulation of P-proteins at the sieve plate pores that can be followed by a slower, irreversible formation of callose deposits that block phloem flow.

Protocol for phloem sap collection by spontaneous exudation

1. Cut the main stem or petiole with a razor blade (see Note 1), or use a razor blade to make a narrow, longitudinal incision (see Note 2).

   Alternatively, hypodermic needles can be used to puncture the plant (see Note 3).

2. Remove the first exudate coming from the cut/incision/puncture with a lintless tissue (see Note 4).
3. Collect the subsequently exuding sap with a pipette into a reaction tube that is kept on ice or liquid nitrogen (see Note 5). Whether the sap is collected pure or into a specific extraction solution is dependent on the downstream application (see Note 6).
4. Check samples for contamination (see Note 7).

3.2.2. EDTA-Facilitated Exudation

Plants not exuding spontaneously can be sampled by chelator-facilitated exudation (24–26). This method is normally based on the use of the calcium-chelator ethylene diamine tetraacetic acid (EDTA) in the collecting solution, but similar calcium chelators may be used. Petioles are cut under water and subsequently placed in a chelator-containing collection solution. The addition of chelator to the collecting solution is thought to prevent the occlusion of sieve plate pores due to callose synthesis and P-protein aggregation, processes that are both induced by Ca^{2+} liberated in response to phloem injuries.

The method of EDTA-facilitated exudation is quick and easy to perform, but has the drawback of not allowing quantitative measurements of component concentrations, because the collected exudate is diluted to an unknown extent. In addition, the sampling method is rather crude and is prone to contamination by destroyed cells and apoplastic fluid, and phloem components may also be degraded due to the usually long collection times. Therefore, the EDTA exudation technique can be useful to answer certain, focused questions, but is not suitable for the reliable identification or quantification of phloem constituents.

Protocol for phloem sap collection by EDTA-facilitated exudation

1. Cut petioles under water.
2. Recut in 5–50 mM EDTA (see Note 8) solution buffered at a near-neutral pH value between 6 and 8 (1, 27) and insert sectioned leaves in a microtube or an exudation cuvette.
3. Place in a chamber with high air humidity to avoid transpiration (see Note 9).
4. Let exude for 2–24 h (see Note 10).
5. Concentrate exudation fluid or desired compound from the exudation fluid if necessary or perform dialysis to reduce the amount of EDTA, dependent on the downstream application.
6. Check samples for contamination (see Note 7).

3.2.3. Insect Stylectomy

A classical, elegant method to obtain phloem sap is the insect stylet technique that can yield phloem sap of high purity from many different plant species, including the model plants Arabidopsis and rice. This method was originally established by Kennedy and Mittler (28). It requires the help of aphids or other phloem-feeding insects such as plant hoppers. This method is mainly used for grasses and is more difficult to apply to dicotyledonous herbs (29). Generally, stylets of phloem-sucking insects that are inserted into a sieve tube while the herbivore is feeding are cut off and the subsequently exuding phloem content is collected. In the initial experiments the stylets of big aphids nourishing on woody species, such as the willow aphid, were severed by razor blades or microscissors (30). To use smaller insect species, laser or radiofrequency microcautery has been established (8, 31, 32). Insect stylectomy is regarded as being minimal-invasive and is thought to have lesser impact on the plant than the exudation-based sampling techniques. However, also aphids manipulate plant physiology by acting as additional sinks (33) or by the initial wounding and injection of saliva into the phloem which can cause a systemic response (34–36).

Protocol for phloem sap collection by insect stylectomy

1. In a room with high air humidity place aphids on the plant in a cage until they have settled (see Note 11).
2. Carefully place the plant part with the aphids under a binocular or microscope.

3. Cut stylets by laser beam or radiofrequency microcautery (see Note 12).
4. Immediately cover the cage containing the cut stylet(s) with oil to avoid evaporation.
5. Collect exuding phloem sap with a glass microcapillary (see Note 6).
6. Check samples for contamination (see Note 7).

3.3. Analysis of Phloem Samples

Pure sieve tube exudates can contain up to 1 M of sugars and up to 200 mM of the most abundant amino acids. Inorganic ions such as potassium, which is the most abundant one, can be found in a range of 10–100 mM. However, many other compounds, such as micronutrients, RNAs, or proteins, are less abundant. For proteins 0.1–1 mg/ml have been found in most plant species with the exception of cucurbits, where up to 100 mg/ml can be obtained. Much smaller quantities of mRNAs and small RNAs can occur, dependent on the species of interest.

If sufficient phloem sap can be obtained, all normal analysis methods can be applied. This includes modern high-throughput techniques like metabolomics, proteomics, array hybridization, or multi-parallel sequencing (20–22, 37–41). Hundreds of metabolites, proteins, mRNAs, and small RNAs have been identified by these approaches from different plant species (8, 42–45). A diversity of mineral nutrients has been detected as well. Thus potassium, sulfur, phosphorus, magnesium, sodium, and chloride have been found as well as small amounts of iron, copper, manganese, zinc, boron, and molybdenum (8, 46). In addition to the direct transport of nutrients it has recently been demonstrated that specific micro RNAs (miRNAs), namely miR395 and miR399, can be translocated through the phloem and probably influence nutrient allocation on a systemic level (38, 47).

Recent advances in solute and macromolecule analyses nowadays also allow analyses of small phloem amounts, as obtainable by insect stylectomy. For instance, amino acid analysis by capillary electrophoresis with laser-induced fluorescence detection allowed the determination of amino acid concentrations in nanoliter volume samples (48, 49). Nanoflow liquid chromatography linked to mass spectrometry enabled the detection of phloem sap proteins from rice that were undetectable after gel electrophoresis (50). Use of real-time PCR enabled the detection of miRNAs from as little as 0.1 μl of phloem sap from pumpkin (38, 39, 51, 52). Finally, sophistication of the microcautery technique itself and employment of large numbers of insects allowed the collection of relatively high amounts of phloem sap even with this sampling method (10 μl within a few hours) and this significantly improved transcript and protein analyses (4). However, the increase in aphid number might potentiate the problems with wound responses, saliva injection, and insects acting as additional sinks, as mentioned earlier.

4. Notes

1. This method works in herbaceous plants, for instance cucumber (*Cucumis sativus*) or pumpkin (*Cucurbita maxima*) (19, 20, 53).
2. This method is used mainly in trees (17, 54).
3. This method is applicable to herbaceous plants like cucurbits or brassicas (20–22).
4. This step is crucial to avoid contamination of phloem exudate with the content of cells destroyed by the incision.
5. Phloem exudates from incisions can be contaminated by cell contents of injured tissues surrounding the phloem (55). Moreover, injury of SEs results in a drop in turgor pressure, which may cause a significant dilution of samples, an influx of components from associated cells, e.g., companion cells, or a release of structural components normally attached to the cytoplasmic lining of SEs (8, 23, 56).
6. Capillaries should be filled with oil to avoid evaporation. Depending on the downstream application, samples can be directly collected into Trizol, protein precipitation solution, buffer for 2-DE, etc., on ice or into a tube placed in liquid nitrogen.
7. Evidence for sample purity can, for example, be obtained by measuring sugar composition: a high content of sucrose and the absence of reducing sugars like glucose and fructose indicate a high amount of SE content. Another option is the detection of mRNAs or proteins that should not occur in SE as, e.g., photosynthesis-related compounds. In this case their absence indicates acceptable low levels of contamination (21, 22, 57, 58). The profile of cytoskeletal elements, such as actin and profilin present in sieve tube exudates, and tubulin absent from such samples, is also an indicator of low contamination by surrounding tissues (59).
8. Other calcium chelators like, for example, 1,2-bis(2-aminophenoxy)ethane-N,N,N',N'-tetraacetic acid (BAPTA) may be used instead.
9. If transpiration is not blocked the collection solution will be sucked into the transpiration stream.
10. EDTA-facilitated exudation can induce artifacts (in comparison to sampling using the insect stylet technique) at the level of metabolites and proteins (60, 61). Such artifacts can be minimized by short exudation times. Where levels of amino acids are concerned it has been shown that changes were low if the duration of EDTA-facilitated exudation did not exceed a few hours (62).

11. Phloem-feeding insects influence the plant by wounding and by emitting saliva into the phloem (34–36). These potential secondary effects can be minimized or abolished by cutting stylets shortly after the aphids started feeding.
12. Insect stylectomy is a time-consuming and work-intensive technique. Stylet exudation is often unpredictable, both in terms of success rate and duration of sap exudation (30, 61, 63). The duration of stylet exudation from herbaceous species can vary from a few minutes to hours and the volume of phloem sap from a few nanoliters to about 100 nl (with an exudation rate of 0.5–2 nl/min). In the model species *Arabidopsis thaliana*, for example, only a few nanoliters can be obtained (48).

References

1. Deeken R, Ache P, Kajahn I, Klinkenberg J, Bringmann G, Hedrich R (2008) Identification of *Arabidopsis thaliana* phloem RNAs provides a search criterion for phloem-based transcripts hidden in complex datasets of microarray experiments. Plant J 55:746–759
2. Pritchard J (1996) Aphid stylectomy reveals an osmotic step between sieve tube and cortical cells in barley roots. J Exp Bot 47:1519–1524
3. Doering-Saad C, Newbury HJ, Bale JS, Pritchard J (2002) Use of aphid stylectomy and RT-PCR for the detection of transporter mRNAs in sieve elements. J Exp Bot 53:631–637
4. Gaupels F, Buhtz A, Knauer T, Deshmukh S, Waller F, van Bel AJE, Kogel KH, Kehr J (2008) Adaptation of aphid stylectomy for analyses of proteins and mRNAs in barley phloem sap. J Exp Bot 59:3297–3306
5. Kawabe S, Bukomorita T, Chino M (1980) Collection of rice phloem sap from stylets of homopterous insect severed by YAG laser. Plant Cell Physiol 21:1319–1327
6. Riens B, Lohaus G, Heineke D, Heldt HW (1991) Amino acid and sucrose content determined in the cytosolic, chloroplastic, and vacuolar compartments and in the phloem sap of spinach leaves. Plant Physiol 97:227–233
7. Lohaus G, Burba M, Heldt HW (1994) Comparison of the contents of sucrose and amino acids in the leaves, phloem sap and taproots of high and low sugar-producing hybrids of sugar beet (*Beta vulgaris* L.). J Exp Bot 45:1097–1101
8. Dinant S, Bonnemain JL, Girousse C, Kehr J (2010) Phloem sap intricacy and interplay with aphid feeding. C R Biologies 333:504–515
9. Brandt S, Kehr J, Walz C, Imlau A, Willmitzer L, Fisahn J (1999) A rapid method for detection of plant gene transcripts from single epidermal, mesophyll and companion cells of intact leaves. Plant J 20:245–250
10. Raps A, Kehr J, Gugerli P, Moar WJ, Bigler F, Hilbeck A (2001) Immunological analysis of phloem sap of *Bacillus thuringiensis* corn and of the nontarget herbivore *Rhopalosiphum padi* (Homoptera: Aphididae) for the presence of Cry1Ab. Mol Ecol 10:525–533
11. Yu Y, Lashbrook C, Hannapel D (2007) Tissue integrity and RNA quality of laser microdissected phloem of potato. Planta 226:797–803
12. Asano T, Masumura T, Kusano H, Kikuchi S, Kurita A, Shimada H, Kadowaki K (2002) Construction of a specialized cDNA library from plant cells isolated by laser capture microdissection: toward comprehensive analysis of the genes expressed in the rice phloem. Plant J 32:401–408
13. Brady SM, Orlando DA, Lee JY, Wang JY, Koch J, Dinneny JR, Mace D, Ohler U, Benfey PN (2007) A high-resolution root spatiotemporal map reveals dominant expression patterns. Science 318:801–806
14. Lyer-Pascuzzi AS, Benfey PN (2010) Fluorescence-activated cell sorting. In: Hennig L, Köhler C (eds) Plant developmental biology. pp 313–319
15. Bose JC (1947) Plants and their autographs. Longman Green & Co., London
16. Zimmermann MH, Ziegler H (1975) List of sugars and sugar alcohols in sieve-tube exudates. In: Zimmermann MH, Milburn JA (eds) Encyclopedia of plant physiology. Springer, Berlin, pp 480–503
17. Milburn JA (1970) Phloem exudation from castor bean: induction by massage. Planta 95:272–276

18. Alosi MC, Melroy DL, Park RB (1988) The regulation of gelation of phloem exudate from *cucurbita* fruit by dilution, glutathione, and glutathione reductase. Plant Physiol 86:1089–1094
19. Yoo B, Kragler F, Varkonyi-Gasic E, Haywood V, Archer-Evans S, Lee YM, Lough TJ, Lucas WJ (2004) A systemic small RNA signaling system in plants. Plant Cell 16:1979–2000
20. Walz C, Giavalisco P, Schad M, Juenger M, Klose J, Kehr J (2004) Proteomics of curcurbit phloem exudate reveals a network of defense proteins. Phytochemistry 65:1795–1804
21. Giavalisco P, Kapitza K, Kolasa A, Buhtz A, Kehr J (2006) Towards the proteome of *Brassica napus* phloem sap. Proteomics 6:896–909
22. Buhtz A, Springer F, Chappell L, Baulcombe DC, Kehr J (2008) Identification and characterization of small RNAs from the phloem of *Brassica napus*. Plant J 53:739–749
23. Turgeon R, Wolf S (2009) Phloem transport: cellular pathways and molecular trafficking. Ann Rev Plant Biol 60:207–221
24. King RW, Zeevaart JAD (1974) Enhancement of phloem exudation from cut petioles by chelating agents. Plant Physiol 53:96–103
25. Marentes E, Grusak MA (1998) Mass determination of low-molecular-weight proteins in phloem sap using matrix assisted laser desorption/ionization time of flight mass spectrometry. J Exp Bot 49:903–911
26. Hoffmann-Benning S, Gage DA, McIntosh L, Kende H, Zeevaart JAD (2002) Comparison of peptides in the phloem sap of flowering and non-flowering Perilla and lupine plants using microbore HPLC followed by matrix-assisted laser desorption/ionisation time-of-flight mass spectrometry. Planta 216:140–147
27. Leggewie G, Kolbe A, Lemoine R, Roessner U, Lytovchenko A, Zuther E, Kehr J, Frommer WB, Riesmeier JW, Willmitzer L, Fernie AR (2003) Overexpression of the sucrose transporter SoSUT1 in potato results in alterations in leaf carbon partitioning and in tuber metabolism but has little impact on tuber morphology. Planta 217:158–167
28. Kennedy JS, Mittler TE (1953) A method of obtaining phloem sap via the mouth parts of aphids. Nature 171:528
29. van Bel AJE, Knoblauch M, Furch ACU, Hafke JB (2011) Question(s) on phloem biology. 1. Electropotential waves, Ca2+ fluxes and cellular cascades along the propagation pathway. Plant Sci 181:210–218
30. Fisher DB, Frame JM (1984) A guide to the use of the exuding-stylet technique in phloem physiology. Planta 161:385–393
31. Downing N, Unwin DM (1977) A new method for cutting the mouthparts of feeding aphids. Physiol Entomol 2:275–277
32. Barlow CA, McCully ME (1972) The ruby laser as an instrument for cutting the stylets of feeding aphids. Can J Zool 50:1497–1498
33. Kehr J (2006) Phloem sap proteins: their identities and potential roles in the interaction between plants and phloem-feeding insects. J Exp Bot 2006:767–774
34. Tjallingii WF (2006) Salivary secretions by aphids interacting with proteins of phloem wound responses. J Exp Bot 57:739–745
35. Will T, Tjallingii WF, Thonnessen A, van Bel AJE (2007) Molecular sabotage of plant defense by aphid saliva. Proc Natl Acad Sci USA 104:10536–10541
36. Divol F, Vilaine F, Thibivilliers S, Amselem J, Palauqui J, Kusiak C, Dinant S (2005) Systemic response to aphid infestation by *Myzus persicae* in the phloem of *Apium graveolens*. Plant Mol Biol 57:517–540
37. Fiehn O (2003) Metabolic networks of *Cucurbita maxima* phloem. Phytochemistry 62:875–886
38. Buhtz A, Pieritz J, Springer F, Kehr J (2010) Phloem small RNAs, nutrient stress responses, and systemic mobility. BMC Plant Biol 10:64
39. Pant BD, Musialak-Lange M, Nuc P, May P, Buhtz A, Kehr J, Walther D, Scheible WR (2009) Identification of nutrient-responsive Arabidopsis and rapeseed microRNAs by comprehensive real-time polymerase chain reaction profiling and small RNA sequencing. Plant Physiol 150:1541–1555
40. Lin MK, Lee YJ, Lough TJ, Phinney BS, Lucas WJ (2009) Analysis of the pumpkin phloem proteome provides insights into angiosperm sieve tube function. Mol Cell Proteomics 8: 343–356
41. Omid A, Keilin T, Glass A, Leshkowitz D, Wolf S (2007) Characterization of phloem-sap transcription profile in melon plants. J Exp Bot 58:3645–3656
42. Kehr J (2009) Long-distance transport of macromolecules through the phloem. F1000 Biol Rep 1:31
43. Kehr J, Buhtz A (2008) Long distance transport and movement of RNA through the phloem. J Exp Bot 59:85–92
44. Kragler F (2010) RNA in the phloem: a crisis or a return on investment? Plant Sci 178:99–104
45. Ruiz-Medrano R, Xoconostle-Cazares B, Kragler F (2004) The plasmodesmatal transport pathway for homeotic proteins, silencing signals and viruses. Curr Opin Plant Biol 7:641–650

46. Peuke AD (2010) Correlations in concentrations, xylem and phloem flows, and partitioning of elements and ions in intact plants. A summary and statistical re-evaluation of modeling experiments in *Ricinus communis*. J Exp Bot 61:635–655
47. Pant BD, Buhtz A, Kehr J, Scheible WR (2008) MicroRNA399 is a long-distance signal for the regulation of plant phosphate homeostasis. Plant J 53:731–738
48. Zhu X, Shaw PN, Pritchard J, Newbury J, Hunt EJ, Barrett DA (2005) Amino acid analysis by micellar electrokinetic chromatography with laser-induced fluorescence detection: application to nanoliter-volume biological samples from *Arabidopsis thaliana* and *Myzus persicae*. Electrophoresis 26:911–919
49. Arlt K, Brandt S, Kehr J (2001) Amino acid analysis in five pooled single plant cell samples using capillary electrophoresis coupled to laser-induced fluorescence detection. J Chromatogr A 926:319–325
50. Aki T, Shigyo M, Nakano R, Yoneyama T, Yanagisawa S (2008) Nano scale proteomics revealed the presence of regulatory proteins including three FT-like proteins in phloem and xylem saps from rice. Plant Cell Physiol 49: 767–790
51. Varkonyi-Gasic E, Wu R, Wood M, Walton EF, Hellens RP (2007) A highly sensitive RT-PCR method for detection and quantification of microRNAs. Plant Methods 3:12
52. Varkonyi-Gasic E, Gould N, Sandanayaka M, Sutherland P, MacDiarmid RM (2010) Characterization of microRNAs from apple (*Malus domestica* 'Royal gala') vascular tissue and phloem sap. BMC Plant Biol 10:159
53. Walz C, Juenger M, Schad M, Kehr J (2002) Evidence for the presence and activity of a complete antioxidant defense system in mature sieve tubes. Plant J 31:189–197
54. Hocking PJ (1980) The composition of phloem exudate and xylem sap from tree tobacco (*Nicotiana glauca* Grah.). Ann Bot 45:633–643
55. Ruiz-Medrano R, Xoconostle-Cázares B, Lucas WJ (1999) Phloem long-distance transport of *CmNACP* mRNA: implications for supracellular regulation in plants. Development 126:4405–4419
56. Knoblauch M, van Bel AJE (1998) Sieve tubes in action. Plant Cell 10:35–50
57. Doering-Saad C, Newbury HJ, Couldridge CE, Bale JS, Pritchard J (2006) A phloem-enriched cDNA library from *ricinus*: insights into phloem function. J Exp Bot 57:3183–3193
58. Mendoza-Cózatl D, Butko E, Springer F, Torpey J, Komives E, Kehr J, Schroeder JI (2008) Identification of high levels of phytochelatins, glutathione and cadmium in the phloem sap of *Brassica napus* and role for thiol-peptides in long-distance transport of cadmium. Plant J 54:249–259
59. Schobert C, Baker L, Szederkenyi J, Großmann P, Komor E, Hayashi H (1998) Identification of immunologically related proteins in sieve-tube exudate collected from monocotyledonous and dicotyledonous plants. Planta 206:245–252
60. Gaupels F, Knauer T, van Bel AJE (2008) A combinatory approach for analysis of protein sets in barley sieve tube samples using EDTA-facilitated exudation and aphid stylectomy. J Plant Physiol 165:95–103
61. Girousse C, Bonnemain JL, Delrot S, Bournoville R (1991) Sugar and amino acid composition of phloem sap of *Medicago sativa*: a comparative study of two collecting methods. Plant Physiol Biochem 29:41–48
62. Weibull J, Ronquist F, Brishammar S (1990) Free amino acid composition of leaf exudates and phloem sap. Plant Physiol 92:222–226
63. Girousse C, Bournoville R, Bonnemain JL (1996) Water deficit-induced changes in concentrations in proline and some other amino acids in the phloem sap of Alfalfa. Plant Physiol 111:109–113

Chapter 13

Methods for Xylem Sap Collection

M. Alexou and A.D. Peuke

Abstract

Xylem and phloem are essential for the exchange of solutes and signals among organs of land plants. The synergy of both enables the transport and ultimately the partitioning of water, nutrients, metabolic products and signals among the organs of plants. The collection and analysis of xylem sap allow at least qualitative assumptions about bulk transport in the transpiration stream. For quantification of element-, ion-, and compound-flow, the additional estimation of volume flow is necessary. In this chapter we describe methods for collecting xylem sap by (1) root pressure exudate, (2) Scholander-Hammel pressure vessel, (3) root pressurizing method according to Passioura, and (4) (hand/battery) vacuum pump.

Key words: Xylem sap, Root pressure, Scholander-Hammel pressure vessel, "Passioura vessel"

1. Introduction

One of the most important and characteristic features of higher plants is their adaptation to life on land via the evolution of long distance transport systems. The requirement of long distance transport is fulfilled by phloem and xylem in cormophytes, which enable the transport of water, nutrients, and signals among the organs of plants (1). In the xylem water, minerals, products from root metabolism, and signals are transported from the root to transpiring parts of the shoot, particularly the photosynthetically active leaves. Large and especially tall plants must exhibit special features in their xylem. For example, tall trees (up to 100 m in height) must overcome significant gravitational forces in order to lift transport saps to the top of the tree.

The current view of the driving forces of long distance transport is based on gradients in the transport systems, i.e., gradients in hydrostatic pressure, water potential, and chemical potential. Gas exchange (water vapor, CO_2, O_2) and associated processes are central factors in regulating the long distance transport.

Frans J.M. Maathuis (ed.), *Plant Mineral Nutrients: Methods and Protocols*, Methods in Molecular Biology, vol. 953,
DOI 10.1007/978-1-62703-152-3_13, © Springer Science+Business Media, LLC 2013

Regardless of recent criticism on the common hypothesis for long distance transport, there is undisputed agreement that in the xylem there are complex gradients in pressure and potential (2). This results in difficulties in the investigation of xylem function, especially when conducted *ex situ*, since manipulation can lead to significant artifacts, particularly in quantification of fluxes. The knowledge of xylem sap composition is, therefore, a very important step in understanding plant nutrition and stress conditions of plants (1).

In transpiring plants, xylem sap is usually under tension, rather than under pressure, due to the suction of the transpiration stream (2). For collection of xylem sap this negative pressure has to be compensated and overcome. We present four simple methods for xylem sap collection, most of which can also be applied in the field: (1) root pressure exudation, (2) Scholander-Hammel pressure chamber, (3) root system pressurizing chamber according to Passioura (1980), and (4) hand- or battery-operated vacuum pump.

1. The root pressure method can be applied in some cormophytes by removing the transpiring shoot of a plant near the root–shoot interface. Xylem sap will be exuding after some minutes from the cut stem for hours or even days (depending on species and conditions) due to root pressure. The phenomenon of root pressure is the outcome of active transport of mineral ions and passive inflow of water into xylem vessels of the root. Root pressure provides the driving force to transport the xylem sap some distance up the stem of a decapitated plant. Root pressure can also be observed in springtime, short before bud break, and is used for xylem sap collection. Stored starch in pith rays is degraded and the generated sugars are loaded into the xylem vessel resulting in root pressure. This mechanism can be seen in the case of pruned grapevines (see Fig. 1c) or in sugar maple, where it forms the basis for collecting the feedstock for maple syrup.
2. Xylem sap can also be collected using a modification of the "pressure vessel technique" of Scholander et al. (3). According to this method, gas pressure is applied to a plant part/twig to compensate the negative pressure in the xylem vessels and thus cause xylem sap flow in the opposite direction. It is usually used for tree and bush twigs and roots, which once cut from the stem, must be brought to the Scholander device as soon as possible. This device consists principally of a vertical metal cylinder (50 cm long depending on the model) on a bank with its upper end open and its lower part tightly fixed on the bank. The cylinder is supplied with N_2 from a gas cylinder fixed at the side of the device. The cylinder can be tightly sealed with a special cup, in the middle of which the twig (or root) is appropriately placed for the sampling purposes.

3. The Passioura—method for collection of xylem sap (4) involves applying pneumatic pressure to the root system in order to overcome the xylem tension. This method can be used in the lab or the greenhouse only. Xylem sap can be collected from pressurized root systems using a pressure vessel technique, from cut midribs of leaves, flaps of stem tissue or petioles of otherwise intact plants. Plants used for measurement must be grown in special pots, which are sealed into the pressure chamber at the hypocotyl junction with dental silicone impression material. The entire root system is pressurized to 0.2–2.0 MPa, sufficient to cause flow through the xylem system from the roots at rates just exceeding shoot transpiration, thereby causing exudation from various sampling points. This method is nondestructive, since xylem saps are collected from living plants, which do survive the procedure.
4. The vacuum pump (5) is used when higher amounts of xylem sap are assumed in the collected plant material, i.e., in cases where plants are not xerothermic and the environmental moisture is increased. The (hand-) vacuum-creating device is like a small portable pistol and the procedure is simple. However, it takes two persons to operate it. The advantage of this method is the low weight and low cost of the vacuum-creating device without the need for other equipment.

2. Materials

2.1. Materials for "Root Pressure Method"

Gardner scissors, knife, scalpel, distilled water, silicon tubes (diameter depending on experimental plant), silicone grease, plastic vials/eCups, pipettes.

2.2. Materials for "Scholander-Hammel Pressure Vessel Method"

Gardner scissors, scalpel, distilled water, single-use paper towels, plastic vials, pipettes/Pasteur pipettes (150 mm, VWR International), N_2 gas bottle. Scholander devices are offered by a number of companies, for example, UMS GmbH, München, Germany, (http://www.ums-muc.de) or MMM—Mosler Tech Support, Berlin, Germany (http://www.mmm-tech.de).

2.3. Materials for "Passioura Pressure Vessel Method"

Self-made pressure vessel, tool kits (like screwdrivers, screw wrenches, spanners), knife, scalpel, distilled water, Teflon tubes (diameter depending on experimental plant), silicone grease, dental silicone impression material (e.g., hydrophilic vinyl polysiloxane), correction material (blend-a-gum, light N, normal setting: blend-a-med Forschung, Schwalbach, Germany), plastic vials, pipettes, N_2 or pressurized air bottle.

2.4. Materials for "Vacuum Pump Method"

Gardner scissors, scalpel, distilled water, single-use paper towels, plastic/glass vials, (bung, funnel, *for option 1*), rubber tubes, pipettes/Pasteur pipettes (150 mm, VWR International). The vacuum pump can be made of plastic or zinc and can be acquired from companies such as Shreveport Air Tool, Inc. Los Angeles, USA (http://www.tooltopia.com), Fisher Scientific UK Ltd Leicestershire, UK. (http://www.fisher.co.uk), or Environmental Express, Charleston, South Carolina, USA (http://www.envexp.com).

3. Methods

3.1. Root Pressure Method

1. Supply the plant with water or nutrient solution before sampling in order to create less negative water potential in the soil and facilitate water uptake.
2. Cut the plant at the root–shoot interface, to remove all transpiring leaves.
3. Remove 1–2 cm of the bark (if present) below the cut site.
4. Clean the cut surface with deionized water.
5. Fit a silicon tube to the shoot stump (Fig. 1a), if necessary seal with silicone grease.

Fig. 1. Illustrations from the root pressure method. (**a**) A simple experimental setup for xylem sap sampling. (**b**) A grapevine cane with bud, shortly before bud breaking. (**c**) Xylem sap (*arrow*) exudation shortly after cutting the cane in spring time.

6. Wait for the appearance of the first root pressure sap, some few minutes up to several hours after decapitation, depending on the conditions of the plant or the rhizosphere conditions.
7. Remove the very first μLs to avoid contamination from cut tissue/cells.
8. Collect the appearing xylem sap with a pipette and store it in plastic vials on ice until the needed volume is reached; subsequently freeze the samples.

3.2. Scholander-Hammel Pressure Vessel

1. Cut off a twig from the tree/bush with clean gardener's scissors. The most appropriate diameter of a twig is between 0.5 and 1 cm. Make sure the total length of the twig does not exceed the length of the cylinder of the device, so that it will not fold inside the cylinder, since the slightest damage of the twig must be avoided.
2. Remove carefully an adequate extend of the bark (at least 1 cm long) of the twig from the vicinity of the cut surface, to avoid contamination of the xylem sap.
3. Clean up the cut end of the twig with distilled water.
4. Place the twig vertically into the cylinder, so that the cut end protrudes 2–3 cm out of the upper/open end of the cylinder (Fig. 2a "ts" and "cy"; d).
5. Pass the cut end through the cylinder cup (Fig. 2a "sc"), and through its accessory, the appropriate rubber hole, which sits in the middle of it and is arranged to be extremely narrow for the twig under examination (Fig. 2a "rt").
6. The cylinder cup is placed onto the upper end of the cylinder with the twig emerging from the center of it (Fig. 2a "cy"). The cup is screwed gas-tight to the cylinder, while the twig is immobilized on the screw cup with a second, smaller, screw cup placed externally upon the cylinder cup (Fig. 2a "es"). Make sure once more that the cut end of the twig that emerges out of the cylinder cap has adequate length, since this procedure may decrease the length of the twig.
7. Make sure the cylinder cup is tightly screwed on the cylinder and the external screw cup tightly screwed on the cylinder cup, so that the friction between twig, rubber, and cylinder cup makes it impossible for the twig to move and the N_2 to escape from the vessel (Fig. 2a "tc").
8. Safety glasses must be put on and distance must be kept between one's head and the upper part of the cylinder at all times when operating the device.
9. Having assured that the system is sealed and the valve for pressure release is off (Fig. 2b "vr"), turn the pressure inlet valve on the side of the bank (Fig. 2b "vi") gently, just enough to

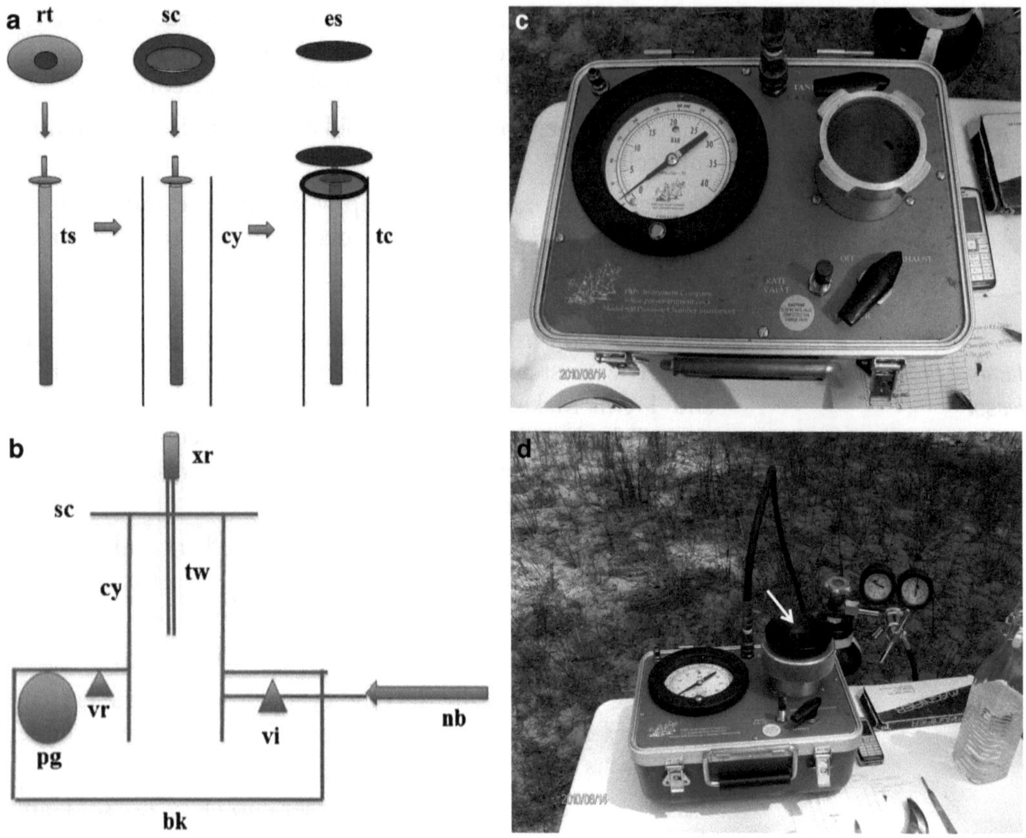

Fig. 2. Illustrations from Scholander pressure vessel method. (**a**) Scheme presenting twig placement. (**b**) Scheme presenting basic traits of Scholander device, bk: bank; cy: cylinder of Scholander device; es: external screw cup securing twig in first screw cup; nb: N_2 bottle; pg: pressure gauge; rt: rubber for twig immobilization; sc: screw cup adjusted to rubber, securing the cylinder closure; tc: twig secured within cylinder; ts: twig during sampling; tw: twig; vi: valve for N_2 inlet; vr: valve for N_2 release; xr: xylem sap concentration rubber. (**c**) Open Scholander device without twig. (**d**) Scholander device connected to N_2 bottle. From the inside of the twig's cut end (*see arrow*) xylem sap emerges.

allow a small amount of N_2 (Fig. 2b "nb") to enter the cylinder. Carefully follow the rise in pressure and then turn off the pressure inlet valve to stabilize the pressure to a low level.

10. The rate of pressure increase should not exceed 50 kPa/s. Do not let N_2 flow into the cylinder for longer than 5 s at a time. Turning it off after short time intervals and monitoring closely the rise of pressure (Fig. 2b "pg") (even after the valve is turned off) is necessary. On the other hand, if the pressure in the cylinder takes long to increase (>10 s N_2 flow), or falls at a rate of more than 50 kPa/s, leakage of the system is likely.
11. The desired pressure is achieved, when xylem sap emerges at the cut surface. The appearance of xylem sap is evident when the color of the cut twig surface darkens and tiny drops with air-bubbles appear.

12. This pressure value is regarded as the apparent shoot water potential of the twig.
13. Discard the first drops of xylem sap to avoid contamination with tissue content. Put a clean, thin, open-ended rubber tube onto the twig (Fig. 2b "xr"), in order to facilitate the accumulation of continuously emerging xylem sap.
14. Raise the pressure above the shoot water potential and maintain this pressure by adding or interrupting the N_2 flow, so that a stable flow of sap is achieved. It is not advisable to let xylem sap flow beyond what is needed for collection. For additional volume of sap one must add N_2 in order to sustain or even increase the pressure within the cylinder.
15. The pressure should not exceed 3.5 MPa because increasing the pressure on the twig tissues will cause cell and tissue damage which might contaminate the xylem sap.
16. Collect the xylem sap using single-use Pasteur pipettes, or automatic pipettes used in biochemical laboratories (most effective are the tips of <20 μl).
17. Transfer the xylem sap to plastic vials and subsequently freeze the samples. The samples must be stored frozen until analysis.
18. After completing the sampling of the xylem sap, open the pressure release valve in order to achieve balance with ambient pressure. Leave enough time for the cylinder to empty, always watching the pressure until it falls to zero, and then unscrew the cylinder cup to replace or discard the twig.

3.3. Passioura "Pressure Vessel": Rhizosphere Pressurizing

1. Construct a pressure chamber from stainless steel (Fig. 3a/b).
2. Grow plants in plastic vessels (Fig. 3a "gv") fitting into the pressure chamber.
3. Supply the plants with water or with nutrient solution.
4. Before transferring the plastic pots to the steel pressure chamber, suck off the excess water, for example, on a tension table, at 5 kPa suction.
5. Place the plant in the middle of the pressure vessel; test the position with the twice-splitted top cover (Fig. 3 "tc") clean and grease the O-sealing ring (Fig. 3 "os").
6. Glue the twice-splitted top cover with dental silicone impression material at the V-formed interface and subsequently screw together the two halves.
7. Screw the merged top cover on the pressure vessel with the fixture (Fig. 3 "tf").
8. Seal the plant stem with dental silicone impression material in a way that a glue plug is formed above and below the top cover to ensure the vessel is gas-tight (Fig. 3 "gp").

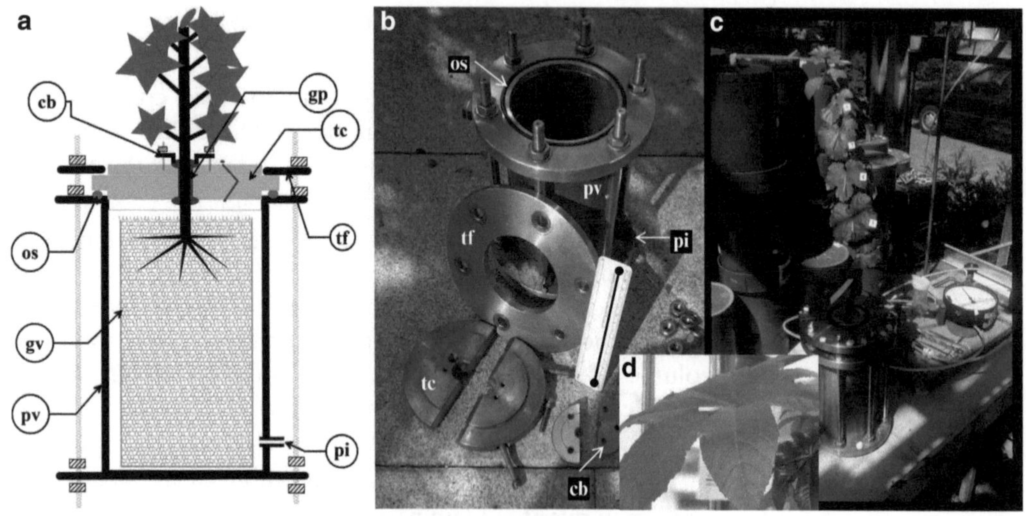

Fig. 3. Illustration from Passioura pressure vessel method. (**a**) Scheme of the Passioura pressure vessel. (**b**) The single components of the vessel. (**c**) The entire system with grapevine, sap from seven individual leaves was collected. (**d**) Teflon tube fitted to the cut midrib of a *Ricinus* leaf. cb: counter bearing; gp: glue (impression material) plug; gv: growing vessel; os: O-ring seal; pi: pressure inlet; pv: pressure vessel; tc: twice-splitted top cover; tf: top cover fixture. The bar in (**b**) indicates 15 cm.

9. Fix the bearing support (Fig. 3 "cb") on top of the stem in order to stabilize the glue plug between the stem and the twice-splitted top cover.
10. Cut potential xylem strands of interest, for example, midrib of leaves in the middle of the laminae (Fig. 3d) or flaps from stems, and fit a rubber to the basal part of the dissected midrib. Several leaves can be cut in the same plant and sap can be sampled simultaneously (Fig. 3c).
11. Tightly fit a tube to the cut xylem strands/midrib, if necessary by additional sealing with silicone grease.
12. Safety glasses must be put on and distance must be kept between one's head and the upper part of the cylinder.
13. Increase the pressure in the root system stepwise at a rate of 0.1 MPa per 10 min, by releasing N_2 into the pressure vessel.
14. The compensation pressure at which the first xylem droplets appear at the cut xylem strands/midrib is recorded as the local apparent water potential.
15. Discard the first μLs of sap to avoid contamination from cut cells and tissues.
16. Increase the pressure to 0.1–0.2 MPa above the compensation point of the sampling threshold. Pressure can be applied in the range of 0.5–2.0 MPa to obtain sufficient flow from all cut sampling points simultaneously.

17. Collect the saps from the fitting rubber tubes (Fig. 3d), store them temporarily on ice, weigh them, and freeze immediately.
18. By determining the volume of collected xylem sap samples (i.e., by weighing the cup before and after sampling) compared with the time for sap collection, the relative apparent volume flow of the different sampling points is estimated.

3.4. Vacuum Pump

1. Prepare twigs as in the description of the Scholander device. There are two options apart from the basic principal (Fig. 4a). First option:
2. After having removed the surrounding bark, tightly place the cut end of the twig (Fig. 4b "tw") in a collection vial (Fig. 4b "cv").

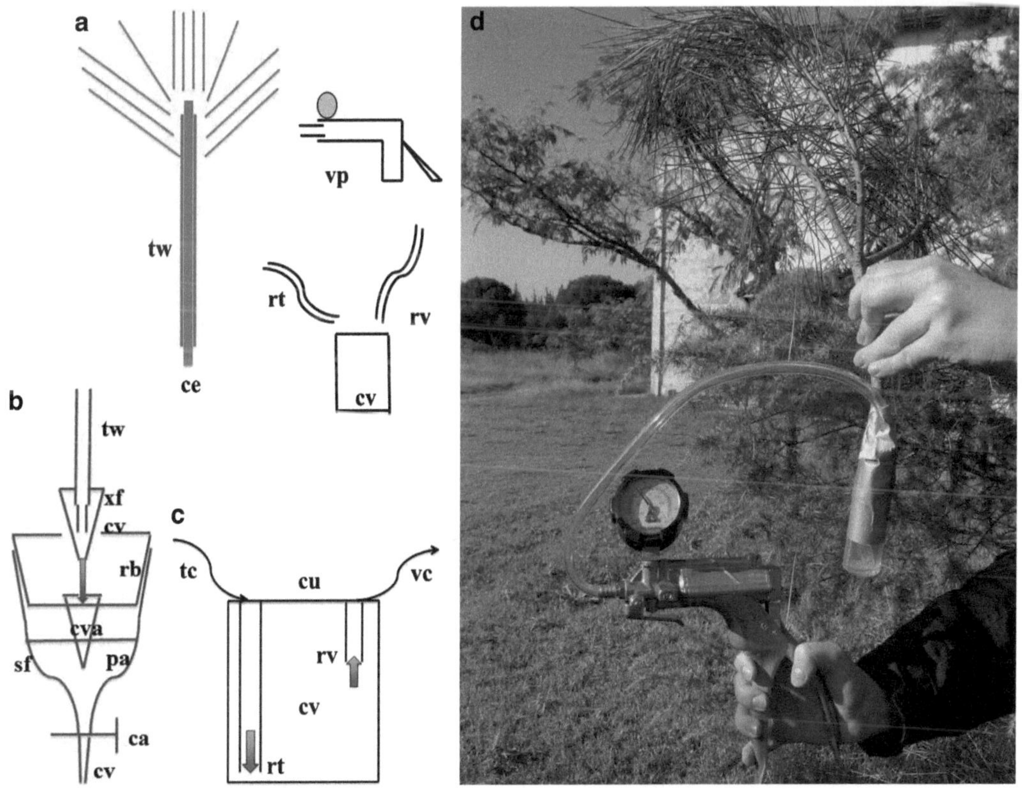

Fig. 4. Illustration from vacuum pump. (**a**) Scheme presenting basic arrangement for sampling. (**b**) Scheme presenting the first option to implement the method. (**c**) Scheme presenting the second option to implement the method. ca: cannula; ce: cut-end prepared; cu: air proof cup of collection vial; cv: collection vial; cva: collection vial, appropriate position; pa: packing holding vial supporting vacuum; rb: rubber bung; rt: ((**c**) ending of-) rubber connecting twig to collection vial; rv: ending of rubber from vacuum pump; rv: rubber connecting vacuum pump to collection vial; sf: separating funnel; tc: twig connection (air proof); tw: twig (progressively trimmed); vc: vacuum pump connection (air proof); vp: vacuum pump; xf: xylem free from bark. (**d**) Photograph of the arrangement (option 2 of the described method) using a glass-vial attached to the vacuum pump and the twig, while being sealed air-tight with packing and duck-tape.

3. Adjust the collection vial for the xylem sap to the rubber bung (Fig. 4b "rb").
4. Fix the vial and the rubber bung on a larger funnel filled with packing to hold the vial and support vacuum (Fig. 4b "sf").
5. Adjust the lower end of the funnel to an opening where the vacuum pump is fixed (Fig. 4b "cv").

Make the arrangement air-tight (Fig. 4b "cva"), and create vacuum with the vacuum pump. Full blank port pressure can be obtained with only two strokes. One stroke is enough to create a vacuum of −15 kPa and enough for the xylem sap to start flowing.

6. As the first drops of sap appear and the first person holds the twig's end and operates the pump (manually or battery functioning), a second person cuts the upper part of the twig with a gardener's scissors in a stepwise manner, so that further drops flow gradually into the vial.
7. After the sampling, the vacuum is dissolved using the cannula (Fig. 4b "ca"), or the appropriate valve on the pump, the arrangement is dismantled, the bung and the funnel are cleaned, and the twig is replaced.

The second option is to use an air-tight trap (a vial) between the twig and the vacuum pump, ideally using a cup with two tube connections:

8. This arrangement leads the xylem sap from the twig flowing through a rubber tube (Fig. 4c "tc") into the vial (Fig. 4c "cv").
9. The vacuum is created by a vacuum pump attached to the same vial by a second tube that connects the cup of the collection vial and the vacuum pump (Fig. 4c "vc"). Full blank port pressure can be obtained with only two strokes. One stroke is enough to create a vacuum of −15 kPa (Fig. 4c "rv") and enough for the xylem sap to start flowing (Fig. 4c "rt"). As the first drops of sap appear and the first person holds the twig's end and operates the pump (manually or battery functioning), a second person cuts the upper part of the twig with a gardener's scissors in a stepwise manner, so that further drops flow gradually into the vial. After the sampling of the xylem sap, the vacuum is released by the appropriate valve of the vacuum pump and both the twig and the vial are replaced, whereas the cap is cleaned.
10. A simpler way for this arrangement to work is to replace the vial with the cup by using a plastic/glass vial for sampling, into which both rubber tubes are inserted (Fig. 4d). The tube leading to the vacuum pump is placed higher into the collection vial than the one connected to the twig. Both are fixed very tightly. The vial with the two tubes is sealed with air-tight duck-tape.

4. Notes

Root pressure method

1. This method works best with plants having big xylem vessel diameters.
2. Do not use on plants under stress, like salinity, nutritional disorder or even drought.
3. Do not collect xylem sap from decapitated transpiring plants for longer than needed for sufficient volume, because recycled compounds derived from the phloem transport will be excluded and energy reserves in the roots for active loading of ions into the xylem vessels only has limited capacity.

Scholander-Hammel pressure vessel

1. In order to complete a sampling task in a preordained amount of time, it is often the case that a number of twigs must be sampled in a short time. It should be considered that the time needed for the sampling of xylem sap from a single twig can vary considerably depending on the experience of the operator, the xylem sap content of the twig and the environmental conditions.
2. In order to ensure stability of compounds in the collected saps, it should be considered that the sap should be aliquoted into multiple vials, since (particularly in the case of amino acid measurements) frozen sap should not be subjected to more than two freeze-thaw cycles.
3. It is risky to try to extract xylem sap from a wet twig in the field. In extreme cases, like on rainy days, first carefully wipe the twig with a clean tissue, while ensuring no mechanical damage which could cause xylem sap to escape from non-visible parts. If the moisture has penetrated the twig, the Scholander device could detach the bark from the woody tissue. The increased pressure could explosively eject the twig out from the pressure vessel and injure the person operating the device seriously.
4. Extreme pressure (>3 MPa) must be avoided due to the existing danger of exceeding the upper pressure limit of the cylinder's capacity, leading to explosive ejection of the cylinder cup.
5. Another case when ejection of the cylinder cap can occur is when very high pressure is applied to collect xylem sap. This can happen when the water potential of the experimental plants is very negative, for example, after a long period of drought or under saline conditions. For this reason always keep an eye on the pressure applied and avoid the maximal capacity of the vessel.

6. The cylinder cap can also be explosively ejected when it is not tightly or carefully screwed in place.
7. The time to put the Scholander device together should always be taken into consideration when planning a field campaign. It is advised to train the experimenter before hand with a similar sized twig. A spare bottle of N_2 should always be available, because the Scholander device can consume high amounts of N_2 when used extensively.

Passioura "pressure vessel": rhizosphere pressurizing

1. Xylem sap can normally not be obtained from the oldest or youngest leaves of a shoot.
2. Plants with aerenchyma are also not appropriate for the Passioura pressure vessel, since the collapse of the aerenchyma may occur under high pressure, creating a direct connection with the apoplastic space.
3. Contamination of the xylem sap with phloem is avoided by this method; no sugars are detected and the pH of the sap is substantially lower than 7.
4. Solute concentrations decrease hyperbolically when pressure increases in the root system and therefore flow rates increase.
5. The highest solute concentration can be recorded under root exudation conditions, while the lowest values were recorded under pressurizing conditions (6).

Vacuum pump

1. In both arrangements the vacuum itself is not easy to sustain.
2. In case the vacuum pump is not battery-powered, it can be quite difficult to operate the vacuum pump for a longer time.
3. One should never try to use the vacuum pump directly connected to the twig in the hope of collecting xylem sap drops from the connection rubber itself. The vacuum pump will be severely damaged.

Additional methods

1. Xylem sap of stem sections of trees can be collected by a decompression method developed for the mechanical drying of timber (7).
2. A continuous, nondestructive method for sampling xylem sap in intact transpiring plants uses the xylem-feeding insect meadow spittlebug (*Philaenus spumarius* L. (Homoptera: Cercopidae)) (8).
3. Small volumes of xylem saps can also be collected with a pressure probe (2).

Storage and analysis of xylem sap

1. Prior to analysis xylem sap should be stored frozen. Depending on the compounds of interest the freezing temperature can be −20°C in case of inorganic ions down to −80°C for amino acids, thiols, hormones, etc.
2. Xylem sap is an aqueous solution and can be easily used after dilution for appropriate analytic methods.
3. Xylem sap in vials often consist of tiny drops which can be difficult to pick up even with the thinnest pipette. It is strongly advised to centrifuge sample vials at 4°C for 2–3 min to coagulate all liquid.
4. Analysis of xylem sap usually involves amino compounds, ionic compounds, thiols, and hormones (e.g., abscisic acid). While necessary precautions must be taken in all cases, proper storage and handling is especially important when organic compounds such as amino acids, thiols, and hormones are of interest.

Acknowledgment

We thank Kalliopi Radoglou and Vassiliki Tsioga for providing us with photos of the Scholander device and vacuum pump.

References

1. Peuke AD (2010) Correlations in concentrations, xylem and phloem flows, and partitioning of elements and ions in intact plants. A summary and statistical re-evaluation of modelling experiments in *Ricinus communis*. J Exp Bot 61:635–655
2. Steudle E (2001) The cohesion-tension mechanism and the acquisition of water by plant roots. Ann Rev Plant Physiol Plant Mol Biol 52:847–875
3. Scholander PF, Hammel HT, Bradstreet ED, Hemmingsen EA (1965) Sap pressure in vascular plants. Science 148:339–346
4. Passioura JB (1980) The transport of water from soil to shoot in wheat seedlings. J Exp Bot 31:333–345
5. Milburn JA, Ranasinghe MS (1996) A comparison of methods for studying pressure and solute potentials in xylem and also in phloem laticifers of *Hevea brasiliensis*. J Exp Bot 47:135–143
6. Schurr U, Schulze E-D (1995) The concentration of xylem sap constituents in root exudate, and in sap from intact, transpiring castor bean plants (*Ricinus communis* L.). Plant Cell Environ 18:409–420
7. Schill V, Hartung H, Orthen B, Weiseel MH (1996) The xylem sap of maple (*Acer platanoides*) trees – sap obtained by a novel method shows changes with season and height. J Exp Bot 47:12–133
8. Malone M, Herron M, Morales M-A (2002) Continuous measurement of macronutrient ions in the transpiration stream of intact plants using the meadow spittlebug coupled with ion chromatography. Plant Phys 130:1436–1442

Chapter 14

Plant Single Cell Sampling

Wieland Fricke

Abstract

Plant single cell sampling and analysis allows the determination of solute concentrations in individual cells and tissues. This is particularly important when studying mineral nutrition, where the cell- and tissue-specific distribution of individual mineral nutrients increases a plant's options to store and mobilize these nutrients in response to a changing external availability. In this chapter, some selected single cell sampling and analysis methods are described in detail, and their advantages and possible pitfalls discussed. These methods include pressure-driven extraction of cell contents (cell sap sampling), and the analysis of extracted cell sap through picoliter osmometry (osmolality), energy-dispersive X-ray (EDX) analysis (concentrations of Na, K, P, S, Cl, Ca), and microfluorometry (concentrations of anions and amino acids). In most cases, the extracted cell sap is mainly vacuolar in origin.

Key words: Microcapillary, Picoliter-sized samples, EDX analysis, Microfluorometry, Picoliter osmometry, Cell contents, Epidermis, Mesophyll, Turgor pressure, Vacuole

1. Introduction

Plant mineral nutrition deals with those chemical components of a plant's body which plants cannot synthesize—mineral nutrients. The availability of mineral nutrients, in particular those ones required at larger quantities, affects the natural distribution of plant species and their productivity in an agricultural setting. Therefore, it is important that plants optimize the way in which mineral nutrients are taken up, transported, stored, and retained or dealt with in any other way should a particular nutrient reach toxic levels. All soil-borne mineral nutrients are taken up as ions. Some mineral nutrients such as nitrogen, phosphorus, and sulfur can be incorporated into larger, organic molecules. However, the majority of mineral nutrients remains in ionic form throughout their residence time in plants, and this has four major implications: (1) the transport of nutrients across membranes is potentially electrogenic; (2) storage of one nutrient

Frans J.M. Maathuis (ed.), *Plant Mineral Nutrients: Methods and Protocols*, Methods in Molecular Biology, vol. 953, DOI 10.1007/978-1-62703-152-3_14, © Springer Science+Business Media, LLC 2013

has to be accompanied by storage of a counterbalancing-charged nutrient (or e.g., organic acid); (3) strong ions (e.g., (1)) such as K^+ affect the pH of the respective compartment; and (4) the three-dimensional conformation and biological activity of larger molecules such as proteins changes with the mineral nutrient microenvironment. A well-known consequence of these implications is that plant cells, as any other eukaryotic cells, have to compartmentalize mineral nutrients between cell organelles in a way that is most compatible with organelle function. Animals have evolved specialized organs through which they discharge waste products or excess of mineral nutrients, at the expense of limited storage capacity of these nutrients. In contrast, plants do not possess any specialized discharge organs (an exception being, e.g., salt glands in some halophytic plants). Instead, each mature living cell has a large central vacuole, where nutrients can be retained and stored. Furthermore, the presence of a cell wall allows nutrients to be accumulated at much higher concentrations and total osmotic pressure compared to animal cells. This in turn provides each cell with the ability to cope with a large range of mineral nutrient concentrations and each plant with the ability to cope with varying external supply in the root environment, both features supporting a sessile evolutionary life strategy. In addition, the larger the concentration of mineral nutrients is within plant cells as compared to the outside environment, the larger is cell turgor pressure and the more leeway plants have to adjust to decreases in external water potential. A prerequisite to understand the role of a particular mineral nutrient in these processes is knowledge about their compartmentation.

Solute compartmentation can occur between organs (root, shoot, flower), growing and nongrowing regions, tissues (e.g., epidermis, mesophyll, vascular bundle), and cell compartments (cytosol, organelles, vacuole) (e.g., (2–9)). In this chapter, we will be looking at methods that make it possible to study compartmentation of solutes between cells and tissues.

A number of different ways is available for sampling the contents of individual cells, and so gain information about compartmentation of solutes between tissues. Protoplasts can be isolated from different tissues (10, 11), or leaf tissues can be fractionated according to their partitioning in nonaqueous media (12). Outlaw and colleagues developed techniques for manual *microdissection*, weighing and analyzing individual cells, more than 40 years ago (13). More recently, manual dissection of cells has been replaced by laser microdissection of tissues, followed by chemical or molecular (PCR) analyses (14, 15). An alternative to dissection of tissue is sampling of single-cell contents in vivo, by inserting a silicon-oil filled microcapillary directly into cells (16). Since living plant cells have turgor pressure and an exponential relation between cell turgor and volume (compare inflating a bicycle tube and tire, where the pressure increases rapidly as the final tube and tire volume is reached), between 10 and 20 % of cell contents shoots rapidly into the capillary when the cell is punctured. This single-cell sampling technique, which will be described here, is suited ideally for the analysis of highly vacuolated cells (too high a proportion of cytoplasm can clog the capillary) and surface tissues such as the leaf or root epidermis, since these can be easily accessed (17–19). Deeper-lying tissues such as mesophyll and bundle sheath or root cortex and stele can also be sampled, albeit, with some modification of the sampling approach, i.e., inserting the microcapillary through a stoma (20). The extracted cell sap consists of almost 100 % vacuolar sap in the case of an epidermal cell, but can contain significant portions of cytoplasm in the case of mesophyll or bundle sheath cells (20). The volume of extracted sap is in the lower (mesophyll) or higher (epidermis) picoliter range. Due to the small volume, the extracted sap must be handled under conditions that suppress evaporative loss. All sample manipulations have to be carried out under a stereomicroscope with the aid of micromanipulators.

The extracted cell sap can be analyzed for osmolality using picoliter-osmometry (for details, see below). In short, a droplet (5–20 pl) of sap is placed under a drop of liquid paraffin onto a small (1 cm^2) copper stage on which a cover slip with a black-and-white background is attached using heat-conducting paste. Standards (NaCl) of known osmolality are placed nearby, and the stage is cooled to –40 °C. Thereafter, the stage is reheated slowly and melting of ice crystals is observed under the stereomicroscope. The temperature at which the last ice crystal in a particular droplet melts is recorded and used, together with values of standards, to calculate the osmolality of the sample. Up to 25 droplets can be analyzed in parallel.

Solutes in extracted cell sap can be analyzed using three types of techniques: EDX analysis, micro-fluorometry, and capillary-zone electrophoresis. The first two techniques will be described here in detail. In short, during energy-dispersive X-ray (EDX) analysis, sample droplets (10–20 pl) are pipetted onto an electro-microscope copper folding grid (100 and 200 mesh), which is coated with a film, ideally Pioloform, and placed under liquid paraffin. The same glass constriction pipette is used for samples and the internal standard (typically $RbNO_3$), which is placed in a 1:1 ratio together with sample droplets. Following pipetting, the liquid paraffin is removed by successive washes in hexane and isopentane, the latter having also a freeze-drying effect on droplets. Grids can then be analyzed with a scanning electron microscope (SEM) equipped with an X-ray analyzer (see Note 1). Solutes such as nitrate, sugars, or malic acid can be analyzed by microfluorometry. The enzymatic assays employed in this method are essentially the same as used in conventional spectro- or fluorimetric assays of metabolites, except that a set of constriction pipettes (10 pl to 5 nl) is used, that all pipetting is done under liquid paraffin and that sample fluorescence is determined under an inverted fluorescence microscope. This technique can also be used to measure enzyme activities in extracted saps (21).

Capillary zone electrophoresis offers potentially the most extensive range of solutes that can be analyzed in single-cell extracts. However, it has been used by only a few laboratories (22, 23), possibly because of difficulties in injecting very small (pl) and reproducible volumes into the capillaries used for this technique.

2. Materials

Apart from a picoliter osmometer and possibly a micromanipulator and capillary puller, no other specialist equipment is required. An inverted fluorescence microscope and an SEM-EDX facility are available in most universities. Unless stated otherwise, all solutions are prepared in distilled water. Most reagents are not toxic, yet general precautions should be taken when dealing with laboratory chemicals. If not stated otherwise, chemicals are supplied by VWR or Sigma. The chemicals that require handling in a fume cupboard are

1. Isopentane.
2. Hexane.
3. 2-Mercaptoethanol (also referred to as "β-mercaptoethanol").
4. Dimethyl-dichlorosilane.

2.1. Cell Sap Sampling and Pipetting

1. Glass capillaries (inner diameter 0.56 mm; e.g., from Clark Electromedical Instruments, Harvard Apparatus).
2. Horizontal or vertical capillary puller (e.g., Harvard Apparatus).
3. Microforge (e.g., Narashige).
4. Oven (80–200 °C).
5. Fume cupboard.
6. Custom-built aluminum capillary holder.
7. Glass beaker (0.5 L; tall version).
8. Dimethyl-dichlorosilane.
9. Silicon oil (AS4 Wacker).
10. Liquid paraffin ("Paraffin oil").
11. Hypodermic needle attached to a 5-ml disposable plastic syringe.
12. Micromanipulator (e.g., Leitz, Narashige) or micropositioner (e.g., Prior, Narahsige, World Precision Instruments).
13. Capillary holder (commercially available or custom-built).

14. Silicone tubing.
15. T-connector (small, as available from any pet shop for connecting water tank tubing).
16. Disposable 50-ml plastic syringe.
17. Aluminum rings (custom-made; 3–4 mm deep; diameter just less than width of standard microscope slide).
18. Two-component glue (e.g., Araldite).
19. Stereomicroscope (standard type, e.g., Leica, Mejri, Olympus, Nikon), with magnification ideally in the range 50–200×.
20. Cold-light source with swan necks (standard type; such as from Volpi or Leica).
21. Petri dish (standard size).
22. Leaf holder (custom-build) to fix leaf.
23. Blu-Tack.

2.2. Picoliter Osmometry

1. Picoliter osmometer (has been available commercially from Bangor University; a Clifton nanoliter osmometer will also do).
2. Cover slips (standard size).
3. Plaster of Paris (also referred to as “Gypsum plaster”). Mix water with $CaSO_4 \times 1/2H_2O$ (calcium sulfate semihydrate) to get a slurry. The slurry should not be too fluid nor too dense either.
4. Permanent-marker pen (black).
5. Acetone.
6. Heat-conducting paste (e.g., RS components).
7. Source of running (tap) water.
8. Wash bottle with silica gel.
9. Water tank pump (standard type as available from pet shops; only smaller versions required).
10. Silicon tubing.
11. NaCl (sodium chloride) standards. Dissolve the following quantities (g) of NaCl in 1 L each of distilled water to reach NaCl concentrations of 0, 100, 200, 300, 400, and 500 mM: 0 mM, 0 g; 100 mM, 5.84 g; 200 mM, 11.69 g; 300 mM, 17.53 g; 400 mM, 23.38 g; 500 mM, 29.22 g. The osmolality of these solutions is 0, 188, 373, 553, 738, and 924 mosmol/kg, respectively.

2.3. EDX Analysis

1. Electron microscopy folding grids (100/200 mesh) coated with Formvar (available commercially; Agar Scientific). It is even better if the grids are bought non-coated and are coated “in-house” with a film of Pioloform (1 %; Agar Scientific),

following standard procedures. The grids can be in copper, which makes them easier to use for pipetting. Since the Cu signal of grids interferes with the Na signal of the EDX analyzer, nickel grids are recommended for analyses which focus on Na (salinity).

2. Internal standard solution containing 0.7 M mannitol and 200 mM rubidium nitrate ($RbNO_3$). Dissolve 12.75 g of mannitol in 90 ml distilled water and make up to 100 ml final volume. Take 2-ml of this and dissolve 59 mg of $RbNO_3$ in it. Store 1-ml aliquots in microcentrifuge tubes at 4 °C for up to 1 month.
3. Elemental standard series A. Pour 400 ml of distilled water into a 1-L glass beaker and add 5.84 g of NaCl and 23.62 g of $Ca(NO_3)_2 \times 4H_2O$ (calcium sulfate tetrahydrate). Dissolve the chemicals with a magnetic stirrer (takes just a few minutes). Make up solution to 500 ml. This solution has concentrations of 200 mM NaCl and 200 mM $Ca(NO_3)_2$ (200 mM Na, 200 mM Cl, and 200 mM Ca, which are relevant for EDX). Take *X* ml of this stock solution and add *Y* ml distilled water to reach final concentrations of each element of 150 mM (*X*/*Y*; 150/50), 100 mM (100/100), 50 mM (50/150), and 25 mM (25/175). Store the stock and dilutions in closed glass containers at 4 °C for up to 6 months. Check periodically for precipitates. If these form, discard the solution.
4. Elemental standard series B. Pour 400 ml of distilled water into a 1-L glass beaker, add 5.1 g of KH_2PO_4 (dipotassium phosphate, also known as “potassium dihydrogen phosphate”), dissolve it and then add to the solution 6.53 g of K_2HPO_4 (monopotassium phosphate, also known as “potassium monohydrogen phosphate”). Dissolve it with a magnetic stirrer (takes just a few minutes). Make up solution to 500 ml. This stock solution contains 75 mM KH_2PO_4 and 75 mM K_2HPO_4. Together this amounts to EDX-relevant elemental concentrations of 225 mM K and 150 mM P. Make the following dilutions: (1) 150 ml stock solution plus 50 ml distilled water (final concentrations of 169 mM K and 112.5 mM P), (2) 100 ml stock solution plus 100 ml distilled water (122.5 mM K and 75 mM P), and (3) 50 ml stock solution plus 150 ml water (61.3 mM K and 37.5 mM P). Store the stock and dilutions in closed glass containers at 4 °C for up to 6 months. Check periodically for precipitates. If these form, discard the solution.
5. Fume cupboard.
6. Tweezers (fine, curved, for electron microscopy; Agar Scientific).
7. Glass beakers (5-ml); three beakers.
8. Isopentane.

9. Hexane.
10. Gelatine capsules to store grids (Agar Scientific).
11. Silica gel (dried) in small (50 ml) plastic container.
12. Carbon stub (Agar Scientific) to mount grid in scanning electron microscope.
13. Scanning electron microscope (SEM) with EDX analyzer.

2.4. Microfluorometry

1. Micromanipulation, pipetting and viewing equipment as for single-cell sampling.
2. Constriction capillaries of approximately 10 pl, 0.5 nl, and 5 nl constriction volume. Construct these from glass microcapillaries using a microforge.
3. Fluorescence microscope (inverted), ideally with photomultiplier (e.g., MPV Leitz), filter combination to detect NAD(P)H-dependent fluorescence (e.g., Leitz filter block A, excitation 340–380 nm, dichromatic mirror 400 nm, and emission filter 430 nm) and, if required, standard image analysis software such as ImageJ (free download) (see Note 2).
4. Triethanolamine (TEA) hydrochloride buffer (480 mM; pH 7.6). Dissolve 8.91 g of triethanolamine hydrochloride in 90 ml distilled water in a tall 200-ml glass beaker. On a magnetic stirrer, adjust pH to 7.0 using 0.1 M or 1 M KOH. Make up solution to a final volume of 100 ml. Store in a glass bottle at 4 °C for up to 2 months.
5. NADPH solution. Prepare a 1 % (w/v) $NaHCO_3$ solution by dissolving 1 g of $NaHCO_3$ in 90 ml of distilled water and make up to a final volume of 100 ml. This solution can be stored at 4 °C for 6 months. On the day of analysis, take 0.5 ml of the $NaHCO_3$ solution and transfer this into a 1.5-ml microcentrifuge tube ("Eppendorf tube"). Depending whether you use NADH or NADPH, weigh out either 4.45 mg NADH (dipotassium salt; Sigma) or 6.85 mg NADPH (tetracyclohexylammonium salt; Sigma), add to the 0.5 ml $NaHCO_3$ solution and dissolve quickly by inverting the tube. The final concentration of either NADH or NADPH is 12 mM. Wrap tube in tin foil and store on ice. Prepare fresh on the day of analysis.
6. FAD (flavin adenine dinucleotide) solution. Take 100 ml distilled water and dissolve 3.32 mg of FAD (FAD disodium salt; Sigma). The final concentration of FAD is 40 μM. Transfer a 1-ml aliquot into a microcentrifuge tube, wrap in tin foil and store on ice on the day of analysis. The remaining FAD solution can be kept at 4 °C for 1 week.
7. BSA (Bovin serum albumin, fraction V; Sigma) solution. Make a 1 % (w/v) BSA solution by adding 100 mg BSA to 10 ml of distilled water contained in a 20-ml Erlenmeyer. Stir gently

with a magnetic stirrer while avoiding foam formation. Store for up to 1 week at 4 °C.

8. Nitrate reductase (from *Aspergillus niger*; Sigma; 300 units per gram lyophilized powder). Dissolve 1 mg of nitrate reductase lyophilized powder in 0.6 ml TEA buffer (see item 4) in a 1-ml microcentrifuge tube. Keep on ice, wrapped in tin foil, on the day of analysis and store at 4 °C for up to 1 week.
9. Nitrate standards. Prepare a nitrate standard stock solution containing 100 mM nitrate by dissolving 5.06 g KNO_3 (potassium nitrate) in 500 ml distilled water. From this stock make dilutions of (1) 75 ml stock plus 25 ml distilled water (final concentration of 75 mM KNO_3), (2) 50 ml of stock plus 50 distilled water (final concentration of 50 mM KNO_3), and (3) 25 ml stock plus 75 ml distilled water (final concentration of 25 mM KNO_3). Store dilutions and stock in closed glass containers at 4 °C for up to 6 months.
10. Assay cocktail for nitrate analyses: 100 μl of assay cocktail contains 20 μl TEA buffer (see item 4), 15 μl NAD(P)H solution (see item 5), 10 μl FAD solution (see item 6), 10 μl BSA solution (see item 7), and 45 μl H_2O. Prepare fresh on the day of analysis in a 1-ml microcentrifuge tube. Wrap the tube in tinfoil and keep it on ice. Discard any unused solution at the end of the day.
11. O-phtalaldehyde (OPA, Sigma) reagent: use 96 μl of the reagent provided by the supplier and add 4 μl 2-mercaptoethanol (fume cupboard!). Prepare fresh on the day of analysis in a 1-ml microcentrifuge tube. Wrap the tube in tinfoil and keep it on ice. Discard any unused solution at the end of the day.
12. KH_2PO_4/K_2HPO_4, buffer. Add 2.72 g of KH_2PO_4 to 90 ml of distilled water in a 0.5-L glass beaker. Dissolve on a magnetic stirrer and make up to a final volume of 100 ml. This gives a 200 mM KH_2PO_4 solution. Similarly, proceed in the same way by adding 3.48 g of K_2HPO_4 to 90 ml distilled water and making up the volume to 100 ml; this solution has a final concentration of 200 mM K_2HPO_4. Using a tall 200-ml glass beaker, mix 50 ml each of the KH_2PO_4 and K_2HPO_4 solutions and record the pH of the mixture while adding further K_2HPO_4 solution on a magnetic stirred until the pH is between 7.2 and 7.4. Store this buffer at 4 °C for up to 6 months.
13. Amino acid standard (here: glutamic acid; you may also use alanine or aspartic acid). Prepare a 50 mM glutamic acid stock standard by adding 423 mg glutamic acid (monosodium salt; Sigma) to 50 ml of water. Dissolve the glutamic acid on a stirrer. Using this stock to make the following dilutions with distilled water: (1) 8 ml of stock plus 2 ml of water (40 mM glutamic acid final concentration), (2) 6 ml of stock plus 4 ml

of water (30 mM), (3) 4 ml of stock plus 6 ml of water (20 mM), and (4) 2 ml of stock plus 8 ml of water (10 mM). Store stock and dilutions at 4 °C for up to 1 month.

3. Methods

3.1. Preparation of Microcapillaries

1. Pulling of microcapillaries and making constrictions. Glass capillaries are pulled with a horizontal or vertical capillary puller. The load and heat settings have to be adjusted so that the capillary has neither a too narrow nor a too wide tip region. The tip of the capillary can be broken under the stereomicroscope or with a microforge to obtain a tip opening of several micrometers. Pipettes that are used for extraction of cell sap do not have to be modified any further, except that they can be (but do not have to be) silanized (see Note 3). Pipettes used for pipetting small reproducible sample volumes must have a constriction melted into them close to the tip (the proximity to the tip defines the constriction volume for a given tip geometry). A constriction is best made with a microforge, where the capillary is positioned with an x–y–z micropositioner close to a platinum wire. This is heated, whilst viewed between 50 and 200× magnification (see Note 4).
2. Silanization of microcapillaries. Capillaries used for pipetting and for extracting cell samples from wet surface tissue or tissue surrounded by liquid (e.g., roots) have to be silanized. Make sure that the tip of the microcapillary is broken to the desired tip opening diameter before the capillary is silanized. Place an oven (80–200 °C) in a fume cupboard. Place upright capillaries (up to 100) into a capillary holder, which is contained within a glass beaker. Seal the beaker with tin foil and leave it in the oven for 1 h. Next, add about 50 μl of dimethyldichlorosilane to the beaker (still in the fume cupboard) and leave the beaker in the oven for a further 30 min. The capillaries can be stored indefinitely.
3. Filling of microcapillaries. For sampling, the capillaries are best back-filled with silicon oil using a hypodermic needle attached to a syringe. The capillary can also be backfilled with liquid paraffin. Alternatively, apply suction to the microcapillary and suck in a sufficient volume of liquid paraffin through the tip opening. The volume should exceed the expected sample volume about 50–100 times.
4. Pipetting with microcapillaries. The microcapillary is mounted on a micromanipulator or micropositioner and held by a capillary holder. Once cell sap has shot into a microcapillary, the sap has to be expelled. Also, samples have to be sucked into and

expelled from capillaries during pipetting. The easiest way to achieve this is as follows: Attach silicone tubing to the wide (base) end of the capillary and connect the other end of the tubing to a little T-connector. Attach more silicon tubing to each of the two open ends of the T-connector and attach the end of one of the tubes to a 50-ml disposable syringe and the other tube to a solenoid-based device (custom-built) that can be switched on and off so that the tube ending is left open or clamped into a "closed" position. If no such device is available, construct some other mechanical device (e.g., strong paper clips) that allows the reversible clamping of tubing; the simplest mechanical device is your teeth! (Beware of Health & Safety in a lab environment). During sampling, the tube ending (e.g., solenoid) is left in the open position. To expel liquid, the ending is in a closed position; the same applies to generate suction for filling the capillary from the tip with sample.

3.2. Preparation of Microscope Slides for Pipetting

1. Once cell sap is extracted it has to be handled in a microenvironment that prevents evaporative loss of liquid. It also has to be viewed under a (stereo)microscope. To meet both requirements, take some aluminum rings and glue, with two-component glue, one ring in the middle of a microscope slide. Prepare five to ten slides.
2. For use, fill the ring with liquid paraffin. After use, remove the liquid paraffin by washing in hand-hot water, then rub (hands) washing-up liquid onto the slide and remove the detergent with further washes with hot water. Dry the slide in an oven (70–80 °C).

3.3. Preparation of Cover Slips for Picoliter Osmometry (See Note 5)

Take 10–20 cover slips and rub them each for circa 1 min on a glass surface in a slurry of abrasive powder (e.g., Plaster of Paris), which is washed off with water. The cover slips are then dipped briefly in acetone to degrease them and enable later maximum contact between sample and cover slip surface. The dried cover slips are then painted on the non-abraded (shiny) side with a black permanent marker. Once the marker has dried, it is scraped off in lines and crosses with the tip of a scalpel blade such that a black and white (transparent) grid pattern forms.

3.4. Single-Cell Sap Sampling

1. Place a leaf that is still attached to an intact plant in a leaf holder, or in case of roots, mount the root in a Petri dish and then add medium. Place the holder or Petri dish onto a little table, mounted on a micropositioner (see Note 6). Next to the tissue, place a microscope slide with an aluminum ring glued to it and that is filled with liquid paraffin.
2. While viewing the illuminated plant tissue under a stereomicroscope, approach the tissue with a microcapillary that is

mounted on a micromanipulator and connected to a syringe (T-connector in "Open" position) and filled with silicon oil. Approach the tissue slowly. Successful puncturing of a cell is visible as a fast "shooting in" of sap into the microcapillary. Immediately retrieve the capillary from the cell and move the specimen table so that the slide with the aluminum ring comes into the center of the viewing field (see Notes 7 and 8). Slowly lower the capillary into the liquid paraffin contained within the aluminum ring, change from "Open" to "Closed" position (T-connector) at the tubing which is attached to the microcapillary, and press on the syringe to expel the single cell sap onto the surface of the slide beneath the liquid paraffin. The complete emptying of sample becomes visible as silicon oil starts to exit ("streaming") from the capillary. Stop applying pressure, change to "Open" position and repeat the procedure to sample another cell sap.

3. Once you have sampled up to ten samples, remove the sampling capillary and take another set of capillaries (as for sampling). Connect a capillary to the micromanipulator and tubing, lower it into the liquid paraffin and apply suction to fill the capillary from the tip with liquid paraffin (about 50–100 times the sample volume). Then "suck in" a sample, followed by more liquid paraffin (about 5–10 sample volumes), followed by another sample, more liquid paraffin, a third sample and a final filling with liquid paraffin (see Note 9). Move the capillary out of the liquid paraffin, put it upright into a closed container and store the samples for 1–2 weeks in the fridge until analysis (see Note 10). Proceed in the same way with the remaining samples. If you do not want to store samples but analyze them immediately, for example, through picoliter osmometry, then place the osmometer stage next to your leaf holder or Petri dish and expel the extracted cell sap directly onto the prepared osmometer stage.

3.5. Picoliter Osmometry

1. Put a small quantity (size of matchstick head) of heat conducting paste onto the stage of the osmometer.
2. Take an abraded and painted cover slip and gently press, with the painted side down, onto the heat conducting past so that the white paste is evenly squashed. You may also opt to not use an entire cover slip but a quarter of it (see Note 11).
3. Place a droplet of liquid paraffin onto the cover slip so that it covers a large portion of it, yet does not run off at the sides (see Note 12).
4. Switch on the cooling water supply (e.g., running tap water) for the osmometer (see Note 13). Switch on the osmometer (with maximum stage temperature setting) and adjust the airflow so that it is at the minimum flow rate required to remove any condensation droplets from the liquid paraffin. The air flow is

generated by a small water tank pump that pumps air through a "washing bottle" filled with dried silica gel (to dry the air), and connected through silicon tubing to the lower and upper inlet on the osmometer stage. The lower inlet guides air directly above the surface of the liquid paraffin, the upper inlet guides air above a cover slip that is used later to "seal" the osmometer stage during a freezing cycle (see step 7). This prevents condensation from interfering with the viewing of samples (see Note 14).

5. Place standard droplets (up to 1 μl each) onto the microscope slide with the paraffin-filled aluminum ring. You can use a Hamilton syringe (10–25 μl) to pipette the standards.
6. Suck in aliquots of standards into a microcapillary, attached to a micromanipulator, and place the sample beneath the liquid paraffin onto the cover slip. Deposit two droplets for each standard concentration. Thereafter, place up to ten single cell extracts onto the cover slip (see Note 15)
7. Once all samples and standards have been pipetted, cover the sample stage of the osmometer with a larger, normal cover slip and start a cooling cycle of the osmometer. Adjust the manual temperature regulator of the osmometer to a setting that is below the anticipated melting temperature (once the osmometer has cooled down the sample stage to below –20 °C and reached the lowest possible temperature, it automatically reheats the stage up to the temperature set by the manual regulator).
8. View samples through a stereomicroscope while the stage cools down to below –20 °C. Observe any formation of condensation droplets and increase, if necessary, the rate of air flow.
9. When the osmometer reaches its lowest temperature and automatically returned to the temperature set by the manual regulator, start to slowly increase the temperature of the stage while viewing the sample droplets and ice crystals contained in them. Write down the temperature at which the last ice crystal of a sample melts and use this information, together with readings for standards, to calculate the osmolality of standards.
10. You may use the same cover slip again for the analysis of more samples. If it is full, remove it carefully with a pair of curved tweezers, trying not to spill liquid paraffin on the osmometer surface (if it spills, it does not damage the osmometer, it just makes it messy). Then, take a cotton swab ("Q-tip") to remove the heat paste. Take care that only the surface of the sample stage is cleaned and that the surrounding area that contains fine wiring is not touched. The osmometer is now ready for another analysis or can be switched off.

3.6. EDX Analysis of Picoliter-Sized Droplets

1. Take a microscope slide with a paraffin-filled aluminum ring. Place the slide onto a stage mounted onto a micropositioner that can be viewed with a stereomicroscope. Using a pair of fine, curved tweezers, take an electron microscope double-folding grid (100/200 mesh) coated with, ideally, Pioloform and place it into the center of the well with the shiny surface down.
2. Using a Hamilton syringe (10–25 μl volume), pipette droplets (0.5–1 μl) of internal standard and elemental standards into the same well, but place these droplets at the periphery next to the aluminum ring. Make sure that none of the droplets touches the grid.
3. Connect a microcapillary (connected through silicon tubing to a suction/pressure device such as a syringe) to a micromanipulator. Lower the capillary into the oil well, apply suction to suck in some liquid paraffin from the tip and then suck in some solution of internal standard. While keeping the filled capillary under the liquid paraffin, move the stage so that the grid is now in the center of the field of view. Lower the capillary and empty a droplet on one large square on the 100 mesh half of the grid. About half of the square should be covered by this "mini-reservoir" droplet (see Note 16). Now, move the stage back so that you have the internal standard stock again in your field of view and empty the standard that remained in the capillary back into this stock (alternatively, you can also expel the excess pipette filling anywhere else on the slide, just make sure that it does not get in contact with the grid).
4. Repeat step 3 for all other standard solutions.
5. Now, it is time to get your single-cell sap samples. If these are stored in microcapillaries, dispel the stored samples, one each, on a large square of the 100-mesh part of grid. If you do not have stored samples, but are harvesting these "fresh," place a leaf holder or Petri dish on the stage next to the slide holding the grid and start sampling (see Note 17). Once you have punctured a cell and sampled some sap, move the stage quickly so that the grid appears now in the field of view. Lower the capillary quickly into the liquid paraffin. Once this is done, you can be sure that the sample contained in the capillary does not evaporate. Now, you can take your time to slowly lower the capillary and expel the sample onto a large square on the 100-mesh half of grid. Repeat this step with up to ten cell sap samples (see Note 18).
6. You have now all your standard and cell sap samples each on a large square of the grid. Now, change the capillary for a constriction capillary of about 10 pl size. Lower it into the liquid paraffin and start pipetting small droplets (ca. 10 pl) of the internal standard onto the smaller squares, one droplet per

small square (see Note 19). Place droplets in rows of five, these serve as technical replicates. Once you have placed down four to five rows, use the same capillary to pipette the same capillary filling of a different elemental standard in each row, by adding the elemental standard to the internal standard droplet. Proceed in the same way with your cell sap samples. Remember that you must pipette the same volumes of internal standard and elemental standard (1:1) and of internal standard and cell sap samples (1:1). This means that you must use the same capillary for the pipetting of internal standard droplet and whichever sample is added to it. Therefore, it is advisable not to place too many internal standard droplets down in advance in case the capillary tip breaks off (which can happen all too easily!).

7. While you pipette all your samples, make a note on a piece of paper where each sample is positioned on the grid (see Note 20). Once you have pipetted all your samples, remove your 10-pl constriction capillary and keep it for the next experiment (see Note 21). Now, take a spare microcapillary, it does not have to be silanized, and clear off all the "mini-reservoir" droplets on the 100-mesh squares by sucking these into the capillary. After that, discard the capillary. You now have a grid where droplets are only present on the 200-mesh half, each droplet consisting of about 10 pl internal standard and 10 pl of either elemental standard or cell sap sample.
8. Take the slide containing the grid and place it in a fume cupboard (no microscope required). Next to the slide, position three 5–10 ml beakers in a row. Fill the first two beakers with hexane and the third one with isopentane. Also position an empty and opened gel capsule in the fume hood in which you can later place the grid.
9. Using a pair of fine, curved tweezers, grab the grid on its 100-mesh side. Remove the grid slowly from the liquid paraffin. While holding the grid in a vertical position, lower it slowly but steadily in the first hexane beaker. Keep holding the grid with your pair of tweezers throughout. You will now see how the liquid paraffin comes off in the beaker and flows to the bottom of it. While keeping the grid submerged in hexane, move the grid slowly up and down to increase the washing effect. After about 10 s, transfer the grid into the second hexane beaker and proceed the same way. After this, transfer the grid into the beaker containing isopentane. Keep it there for about 3–5 s and then rather rapidly remove it. You will see how the grid "dries" and the grid can be transferred into the gel capsule within 10–20 s after removal from the isopentane. Label the capsule and store it in a closed container over dried silica gel. The hexane and isopentane must be disposed of safely.

10. To analyze your grid, mount it onto a carbon stub, typically at an angle of 45 °, and view it under an SEM equipped with an EDX analyzer. Construct a file for appropriate peaks and keV range for each element (see Note 22). View grids at about 14 keV (see Note 23). Analyze each droplet by fitting an appropriate window, while viewing it at magnifications between 1,000× and 3,000×. Adjust the total count rate for the rubidium signal to about 3,000–5,000 cps. This means that each droplet will be scanned and the cps counted for each elemental peak until the rubidium signal has reached the desired count rate. Save the values and later transfer these into an Excel spreadsheet. Express all your elemental counts as percentage of the rubidium integral count rate and use this information to construct a calibration curve from your standards and calculate the concentrations of elements in your samples (see Note 24).

3.7. Microfluorometry of pl-Sized Droplets (Nitrate Assay)

1. Start in the same way as for EDX analyses, except that you do not place a grid into the aluminum well but directly pipette storage droplets of assay cocktail, nitrate reductase and appropriate standards next to the well beneath the liquid paraffin. Place any stored or freshly sampled cell saps onto the glass surface of the slide inside the well.
2. Take a larger constriction capillary of about 5 nl size and pipette rows of four to five droplets each of assay cocktail (technical replicates) onto the glass surface of the slide inside the well. If you have four different nitrate standards, plus the 0 mM control, and have five cell sap samples, you will require ten rows of four to five droplets each.
3. Exchange the 5-nl constriction capillary for a 10-pl capillary and pipette into each assay cocktail droplet your respective standard or sample. Make sure that you use the same 10-pl capillary for standards and samples, otherwise readings for standards and cell sap samples cannot be related to each other.
4. Once you have pipetted all samples, lift the slide and gently wipe the glass surface beneath with laboratory tissue paper to remove any loose fibers that could give off-the-scale fluorescence readings. Place the slide onto the stage of an inverted fluorescence microscope and select the appropriate filter combination.
5. Adjust the sample diaphragm so that it covers an entire droplet and keep the size of diaphragm the same for all samples. Switch off any room lighting and darken off the windows (see Note 25). Take fluorescence readings for each droplet and also take background readings of the glass surface without droplets at the start and end of each row of droplets (see Note 26). Make sure that your aluminum ring and oil well is not leaking and that no liquid paraffin touches the objective lenses, these are sensitive to liquid paraffin and may be damaged.

6. After you have taken all readings, return the slide to the micromanipulator stage and add nitrate reductase to each droplet using a 0.5 nl constriction pipette to start the reaction. Transfer the slide to the microscope as before and take readings of all droplets at time intervals of 10 min until the fluorescence does not decrease any further and the reaction (nitrate-dependent conversion of NAD(P)H to NAD(P)) is completed.
7. Construct plots of fluorescence decrease with time. At some point, the fluorescence should have leveled off. Take this reading for your standards to make a calibration curve and use the calibration curve to calculate the concentration of nitrate in your cell sap samples.

3.8. Microfluorometry of pl-Sized Droplets (Total Amino Acid Assay) (See Note 27)

1. Proceed as for the nitrate assay except that you place down storage droplets of H_2O, OPA-reagent and buffer into the oil well, followed by your stored or freshly harvested cell-sap samples.
2. Using a 0.5 nl constriction pipette, place down rows of droplets of H_2O. Then, using a 10-pl constriction pipette, add amino acid standard or cell sap sample to the water droplets.
3. Read the fluorescence under the fluorescence microscope.
4. On the micromanipulator stage, add 0.5 nl OPA reagent using a 0.5 nl constriction capillary.
5. After 1–3 min, use the same capillary to add 0.5 nl of buffer.
6. Take fluorescence readings and use these readings to construct a calibration curve and calculate the total amino acid concentration in your cell sap samples.

4. Notes

1. Before embarking on any preparations for EDX analyses, check with your local SEM facility that the EDX equipment includes software that is suitable for quantitative analysis of microdroplets. In the past, I have used repeatedly a system from Link, Oxford, and this works fine. These days, some systems are designed mainly for imaging and not for quantitative analysis of X-ray counts for a specified area.
2. When you do microfluorometry of pl-sized droplets, you ideally want an inverted fluorescence microscope equipped with an additional diaphragm to define the size of the measuring window around a droplet, and a photomultiplier together with the respective software to carry out quantitative analysis of these droplets. In the past, the Leica MPV compact system did

exactly that, but unfortunately the MPV system is no longer manufactured—because these days, most researchers are interested in imaging rather than quantitative fluorescence analyses. There are ways around this. You can either use software that is supplied by the microscope manufacturer to measure the fluorescence intensity of droplets; or you can take pictures of droplets and save them in TIFF format and later use a freely downloadable imaging software such as ImageJ, in which you open the pictures as 8-bit images. This effectively changes the fluorescence intensity into a grayscale and provides quantitative information.

3. Silanizing a micropipette has its advantages and its disadvantages. Silanization stops liquid from entering the capillary through capillary forces since it affectively cancels out any adhesion forces and renders the capillary surface very hydrophobic. Therefore, if you lower your sampling capillary into an aqueous medium, for example to puncture and sample the contents of a root cell, you must silanize the capillary. Otherwise, you will get fluid going into the capillary before you sample the cell and the sampled cell contents will be diluted. If you are using a microcapillary that is connected to a cell pressure probe, you do not have to silanize the microcapillary since you can apply some pressures to stop any unwanted solutions from entering the capillary. A possible disadvantage of silanization is that it makes the inner volume of the capillary, in particular around the tip region, too narrow. This can interfere with getting sufficient cell sample volumes. Also, incoming aqueous solution such as cell sap gets mixed up more easily with the liquid paraffin or silicon oil which was used to preload the capillary. It then becomes difficult to later expel a defined cell sap sample that does not have any oil dispersed within it.

4. The way the constriction works is as follows: when a sample is pulled into the capillary through gentle suction (syringe), the liquid enters the capillary easily up to the constriction. This makes it possible to obtain reproducible sample volumes. If the constriction is too narrow, too high suction forces have to be applied but if the constriction is too wide, sap will go past the constriction. It is impossible to make constriction capillaries of a specified absolute volume, for example 10 pl or 5 nl. Therefore, reference to a 10-pl or 5-nl pipette in the following text specifies the approximate range of volume. An idea of the approximate constriction volume can be obtained by sucking water into a capillary and discharging the water in liquid paraffin and calculating the volume of the droplet from its diameter.

5. The principle of picoliter osmometery is that small (>5 pl) sample droplets are frozen down and then slowly melted so that the temperature can be recorded where the last ice crystal

of the sample melts. The extent of melting point (or, freezing point) depression is related to the total concentration of solutes in a sample. The ice crystals that form during a freezing cycle are rather round in shape and transparent, making it difficult to detect them. Supercooling can be a problem that stops ice crystal formation but is overcome by freezing samples well below –20 °C. Formation of ice crystals can be supported by providing crystallization "seeding centers". For these reasons, the glass cover slips on which the samples are placed have to be modified in two ways by rubbing them in a slurry of Plaster of Paris (provision of seeding centers) and painting a grid-pattern on the back of cover slips (improvement of contrast).

6. Lighting is a key issue in single-cell sampling. A pair of swan necks with a cold-light source are ideal and the swan necks can be arranged so that they shine at the specimen and capillary from almost opposite sides, at an angle of about 30–45 ° each. Also, the background for your sample and your specimen that you puncture affects the contrast. When you sample a cell, you must be able to see clearly the meniscus between the aqueous cell solution and the silicon oil/liquid paraffin in the capillary to show that you actually got some sap. Also, you need to see clearly the surface of the plant organ (e.g., leaf and epidermis) to follow the capillary as it approaches the tissue. When you pipette samples, for example, in a liquid paraffin oil well (aluminum ring glued onto microscope slide), it helps to have paper on which you paint a black and white grid pattern, placed underneath the slide. This improves the contrast of samples. Many pay slips also have a grid pattern that suits this purpose perfectly.
7. Plant cells can have very short half-times of water exchange, in the range of 0.5–1 s. This means that when sampling, you have to remove the capillary almost instantaneously once you have punctured a cell. With a bit of practice, you will be able to judge a successful puncture as a "shadow" that moves up the capillary rather than as a well-defined meniscus. In addition, when you lower your sample capillary, e.g., directly following sampling into a liquid paraffin oil well, successful sampling becomes immediately apparent due to the different refractory index of the oil and aqueous sample. If you do not retrieve the capillary rapidly enough, water will move from the surrounding tissues into the punctured cell (since the puncture reduced turgor and therefore lowered the cell's water potential) and significantly dilute your cell sap sample (24).
8. A good capillary, whether for sampling or pipetting, is a precious item. The best way to extend the life time of a capillary and also to have the best possible view of events under the stereomicroscope, is to position the capillary near the center of field (best focus) and to keep it there. Move the specimen or

slide with samples from y- to x- to z-direction rather than the capillary. The only exception is when you puncture a plant tissue: first get your capillary positioned and focused in the center of view. Next, move it slightly back (e.g., to the right if you are right handed and the specimen is on the left side of field of view), move the specimen into the center, then lower it slightly. Now approach it with the capillary and then, in turn, bring up the specimen and adjust the capillary closer to it until the two are in close contact.

9. You can store up to four to five cell sap samples in one capillary. You may also decide to only store one sample in each capillary. This depends on individual preferences.

10. You can construct any device that allows you to store capillaries in an upright position without the tip touching anything and in a closed compartment. Onken (yogurt) used to produce some great yogurt plastic pots that could be used for this purpose. Cut a strip of corrugated cardboard (ca. 5 cm × 12 cm) from a box and fix it with tape to a cubic (3 cm × 3 cm × 3 cm) piece of firm foam material and glue this construction on the inside of a (yogurt) pot lid. You can store the capillaries in the holes of the corrugated paper and close the box with the main body of the (yogurt) pot.

11. An efficient way to cut a cover slip into four almost evenly sized squares is by putting the cover slip on a piece of dust-free firm rubber material and then press evenly with the single edge razor blade onto the cover slip to induce a cut. Cut cover slips after you prepared them for picoliter osmometry.

12. You can construct disposable "mini-wells" on a piece of cover slip square by cutting teflon rings of appropriate diameter and glue these with superglue or two-component glue onto the cover slip. It is a little bit of extra work, but it allows you to retain the liquid paraffin on the cover slip and provides a depth of liquid paraffin of defined and reproducible size (and this is good for minimizing evaporation of single cell saps—see Note 14).

13. Always remember to turn on the cooling water supply! The small gadget that cools down a stage to −20 to −40 °C would otherwise overheat.

14. You are cooling down a small stage to −20 to −40 °C, so condensation of water from ambient air will always be a problem. You want to flush dry air over the osmometer stage and liquid paraffin containing your sample droplet for as much as required to stop condensation, in particular on the liquid paraffin. Condensation droplets will ultimately sink to the surface of the cover slip and fuse with the much smaller cell sap samples located there. You also want to flush dry air as little as possible, since you do not want to dry out the cell samples, despite being

under liquid paraffin. One means to counter the latter problem is to place a ring of water droplets around your samples on the cover slip or to use water-saturated liquid paraffin rather than "normal" liquid paraffin. Take some liquid paraffin, add an equivalent volume of distilled water, stir the mixture for a day and then let it settle for 1 week. The upper phase (liquid paraffin) should be water-saturated. Finally, the local climate plays a role. Pick a nice, dry day to carry out analyses to minimize condensation. A climatized room with low relative humidity is ideal but often not available. Myself, I learned this technique in the laboratory of Prof. Deri Tomos in Bangor, North Wales, UK and later used it for years in the West of Scotland and at the East Coast of Ireland—all geographic areas that are not renowned for their dry, hot climate!

15. When you displace a sample from a capillary that is backfilled with silicon oil into a larger droplet of liquid paraffin, some of the silicon oil comes in contact with the liquid paraffin. Usually, that does not cause any problems. However, I once encountered a persistent problem in a very dry climate in that the silicon oil literally pushed away the liquid paraffin and the samples on the base of the cover slip got exposed to air and dried out almost immediately. I do not know the reason for this "incompatibility" between silicon oil and liquid paraffin; possibly it is of electrostatic nature, but the best way to deal with it is to backfill your capillary not with silicon oil, but with liquid paraffin.
16. The large squares on the 100 mesh half of grid serve as a platform where you place down your "mini-reservoirs" of your standards and also your cell sap samples. The smaller squares on the 200 mesh half of the grid are the actual sites of final droplet deposition for analyses.
17. The specimen and grid-containing slide has to be positioned so that it is possible to move between the two through rapid micropositioner movement, and both within reach of the center of the field of view.
18. The number of cell sap samples depends on your own preference and expertise. If you have two different standard stocks (e.g., A, B) at five different concentrations, with four replicate droplets each, then this amounts already to 40 droplets. Each cell adds another four droplets, so ten cells would result in a total of 80 droplets to be analyzed.
19. The volume specifications for constriction capillaries are approximate figures. For example, it is not so important whether your 10-pl constriction pipette really pipettes 10 pl or whether it pipettes 5 or 20 pl. What is important is that it allows you to pipette reproducible volumes, since you compare your sample readings with those of the standards and also use an internal standard during EDX analyses.

20. Draw a large version of a 200-mesh grid on an A4-paper, make many photocopies of it and use it as template during experiments to mark down the position of your droplets.
21. With time, constriction capillaries tend to accumulate some "cell sap residue" and this stops the pipette from having a uniform filling that is easy to pipette. If you have a "precious" 10-pl pipette, you can try to wash/flush it in absolute ethanol; if that does not work, silanize it again. I have used some capillaries for more than 10 years.
22. To increase the Na-sensitivity of your instrument, remove the Be-window. This though makes the SEM more susceptible to any damage from particles. Depending on the setting (17), the main peak of Ca can partially overlap with the K window (about 10 %); this has to be corrected for. Typical settings for the keV windows are as follows: NaKa1 (0.947–1.148 keV), RbLa1 (1.587–1.808), PKa (1.908–2.128), SKa (2.207–2.408), ClKa (2.507–2.727), K (3.207–3.428), Ca (3.467–3.828).
23. The mannitol contained within the internal standard serves as matrix for your samples. You will see during measurements that this matrix starts to "boil"—a sign of the high energy (and efficient vacuum) under which the system operates. If you view the sample at too high magnification or too high accelerating voltage, you may burn a hole into the film on your grid square or the droplet may "pop off." Therefore, use of a 1 % pioloform film is recommended. Before you start any pipetting, take an empty grid with a film coat and test it under the SEM by doing some fake analyses.
24. You can assume that Na, Cl, K, and Ca are present as ions (Na^+, Cl^-, K^+, Ca^{2+}) in cell sap. The same may not apply to S and P.
25. Avoid incident light shining at the microscope stage during fluorescence measurements. The darker, the better! Also, remember that you do not have to carry out the pipetting and fluorescence analyses in the same room.
26. In this assay you measure a decrease in fluorescence. Before you place down many droplets, pipette just one or two droplets of 3–4 different concentrations of NAD(P)H and measure their fluorescence intensity to make sure that your initial reading is well within the linear scale of values and not off scale.
27. Although proline is commonly referred to as an "amino acid" it is chemically an "imino acid" and is not detected by OPA. Thus, if you are, e.g., particularly interested in changes in proline content in response to drought or salinity, this is the wrong assay! Unfortunately, I am not aware of any other assay that could be employed through this technique that could detect and measure proline specifically.

Acknowledgment

I would like to thank Prof. Deri Tomos and all his former team members and also Professor Roger A. Leigh for teaching me single-cell sampling and analyzing techniques during my postdoc studies at Bangor University, North Wales, UK. Thanks also to Tamas Visnovitz, UCD, Ireland, for help with ImageJ and to Tracey A. Cuin, Montpellier, for reading through an earlier version of this chapter.

References

1. Gerendás J, Schurr U (1999) Physicochemical aspects of ion relations and pH regulation in plants – a quantitative approach. J Exp Bot 50:1101–1114
2. Fricke W, Akhiyarova G, Wei W et al (2006) The short-term growth response to salt of the developing barley leaf. J Exp Bot 57:1079–1095
3. Fricke W, Peters WS (2002) The biophysics of leaf growth in salt-stressed barley. A study at the cell level. Plant Physiol 129:374–388
4. Fricke W, Leigh RA, Tomos AD (1996) The intercellular distribution of vacuolar solutes in the epidermis and mesophyll of barley leaves changes in response to NaCl. J Exp Bot 47:1413–1426
5. Cuin TA, Miller AJ, Laurie SA et al (2003) Potassium activities in cell compartments of salt-grown barley leaves. J Exp Bot 54:657–661
6. Carden DE, Walker DJ, Flowers TJ et al (2003) Single-cell measurements of the contributions of cytosolic Na^+ and K^+ to salt tolerance. Plant Physiol 131:676–683
7. Karley AJ, Leigh RA, Sanders D (2000) Where do all the ions go? The cellular basis of differential ion accumulation in leaf cells. Trends Plant Sci 5:465–470
8. Leigh RA, Storey R (1993) Intercellular compartmentation of ions in barley leaves in relation to potassium nutrition and salinity. J Exp Bot 44:755–762
9. Delane R, Greenway H, Munns R et al (1982) Ion concentration and carbohydrate status of the elongating leaf tissue of *Hordeum vulgare* growing at high external NaCl. I. Relationship between solute concentration and growth. J Exp Bot 33:557–573
10. Dietz KJ, Schramm M, Lang B et al (1992) Characterization of the epidermis from barley primary leaves. 2. The role of the epidermis in ion compartmentation. Planta 187:431–437
11. Karley AJ, Leigh RA, Sanders D (2000) Differential ion accumulation and ion fluxes in the mesophyll and epidermis of barley. Plant Physiol 122:835–844
12. Winter H, Robinson DG, Heldt HW (1993) Subcellular volumes and metabolite concentrations in barley leaves. Planta 191:180–190
13. Outlaw WH, Zhang SQ (2001) Single-cell dissection and microdroplet chemistry. J Exp Bot 52:605–614
14. Brandt S, Kloska S, Altmann T et al (2002) Using array hybridization to monitor gene expression at the single cell level. J Exp Bot 53:2315–2323
15. Roy SJ, Gilliham M, Berger B et al (2008) Investigating glutamate receptor-like gene co-expression in *Arabidopsis thaliana*. Plant Cell Environ 31:861–871
16. Malone M, Leigh RA, Tomos AD (1991) Concentrations of vacuolar inorganic-ions in individual cells of intact wheat leaf epidermis. J Exp Bot 42:305–309
17. Tomos AD, Hinde P, Richardson P et al (1994) Microsampling and measurements of solutes in single cells. In: Harris N, Oparka KJ (eds) Plant cell biology – a practical approach. IRL Press, Oxford
18. Tomos AD, Leigh RA (1999) The pressure probe: a versatile tool in plant cell physiology. Ann Rev Plant Physiol Plant Mol Biol 50:447–472
19. Tomos AD, Sharrock RA (2001) Cell sampling and analysis (SiCSA): metabolites measured at single cell resolution. J Exp Bot 52: 623–630
20. Fricke W, Leigh RA, Tomos AD (1994) Concentrations of inorganic and organic solutes in extracts from individual epidermal, mesophyll and bundle-sheath cells of barley leaves. Planta 192:310–316

21. Roy SJ, Cuin TA, Leigh RA (2003) Nanolitre-scale assays to determine the activities of enzymes in individual plant cells. Plant J 34:555–564
22. Bazzanella A, Lochmann H, Tomos AD et al (1998) Determination of inorganic cations and anions in single plant cells by capillary zone electrophoresis. J Chrom A 809:231–239
23. Kehr J (2001) High resolution spatial analysis of plant systems. Curr Opin Plant Biol 4: 197–201
24. Malone M, Leigh RA, Tomos AD (1989) Extraction and analysis of sap from individual wheat leaf-cells – the effect of sampling speed on the osmotic-pressure of extracted sap. Plant Cell Environ 12:919–926

Chapter 15

Measurements of Cytosolic Ion Concentrations in Live Cells

Veena S. Anil, Kavitha P.G., Sam Kuruvilla, Pavan Kumar, and M.K. Mathew

Abstract

Here, we describe a series of methods suitable for the measurement of cytosolic ion concentrations in living plant cells using ion selective dyes. We describe procedures for the use of SBFI for the measurement of Na^+ in live cells. The resulting material is suitable for most standard cell biology procedures.

Key words: SBFI, Fluorescence, Ratiometric imaging, Two photon excitation, Confocal microscopy, Spectrofluorometry, Salinity stress

1. Introduction

The measurement of ion concentrations in living cells is now feasible using one of two possible approaches: either using ion selective electrodes or using ion-responsive dyes. In the latter case it is possible to monitor either single cells, several individual cells simultaneously, or an ensemble of a large number of cells. The former two approaches use microscopy to image a small number of cells, while the latter is a cuvette-based approach. In this chapter we present protocols to monitor cytosolic sodium using fluorescence dye.

There are rather few dyes available to measure Na^+ levels. We describe here a dual-wavelength ratiometric approach to measure Na^+ levels in the cytosol of rice cells. SBFI (sodium-binding benzofuran isophthalate, Molecular Probes, Inc., Eugene, OR) is a dual-wavelength ratiometric dye which allows the estimation of Na^+ concentrations directly. We describe an imaging approach to monitor cytosolic Na^+ in single cells and a spectrofluorometer-based assay for ensemble average estimates using SBFI.

Veena S. Anil, Kavitha P.G., and Sam Kuruvilla have contributed equally to this work.

Frans J.M. Maathuis (ed.), *Plant Mineral Nutrients: Methods and Protocols*, Methods in Molecular Biology, vol. 953, DOI 10.1007/978-1-62703-152-3_15, © Springer Science+Business Media, LLC 2013

1.1. Cytosolic Na^+ Measurements Using Two-Photon Microscopy of SBFI-Loaded Rice Cells

High salinity disturbs intracellular ion homeostasis, leading to membrane damage, metabolic inactivation, and secondary effects that ultimately result in cell death. It is widely accepted that limiting cytosolic Na^+ improves the survival of plants subjected to saline stress. Direct, noninvasive measurements of ion activities in the cytosol of living plant cells are technically challenging. We have used two-photon excitation for the ratiometric estimation of cytosolic Na^+ in cultured rice cells using SBFI.

The absorption spectrum of the fluorescent indicator SBFI shifts to shorter wavelengths on binding Na^+ together with an increase in quantum yield. It can thus be used as a ratiometric probe for estimating Na^+ concentrations in solution (1). The spectral shift observed in many animal cells, however, is minimal, restricting its use to a single-wavelength intensity mode (2). Reports of ratiometric estimates of intracellular concentrations have been made in plant cells (3–6), although the one-photon excitation spectra have not been presented. Two-photon excitation of SBFI has allowed noninvasive measurement of cytosolic Na^+ levels in a single wavelength mode in a few studies with animal cells (2, 7). This approach using IR (infra red) photons minimizes the problems of photobleaching and phototoxicity that are encountered with ultraviolet (UV) excitation.

We have determined the parameters for the use of SBFI in a ratiometric mode with two-photon microscopy in plant cells to make reliable Na^+ measurements in the cytosol.

1.2. Spectrofluorometric Determination of Cytoplasmic Sodium in Rice Protoplasts Using SBFI

Plants tend to minimize cytosolic sodium concentrations in order to survive salinity stress. This can be achieved either by restricting the entry of sodium into the cell or by partitioning the excess sodium into the vacuole or a combination of both (5, 8). SBFI (sodium-binding bezofuran isophthalate), being a ratiometric dye, has been extensively used in the noninvasive measurement of cytoplasmic sodium. We have carried out spectrofluorometric analysis of cytoplasmic sodium concentrations in rice protoplasts using SBFI.

The major advantage of spectrofluorometry is that it allows the study of a population of cells, which is complementary to the single cell approach described using a microscope. In order to use this approach, the spectrofluorometer should have a monochromator that can slew rapidly in order to change the observation wavelength within a few seconds so that fluorescence intensity can be measured at both excitation wavelengths almost simultaneously. For dyes that change emission wavelength on binding the ion of interest an alternative approach using T-optics is also possible, where emitted light can be observed in two different light paths allowing simultaneous measurement at two different emission wavelengths. The technique has the advantage that the entire reaction can be followed in real time with good time resolution.

2. Materials

2.1. Materials for Two-Photon Microscopy Measurements of Cytoplasmic Sodium

1. Liquid Murashige and Skoog (MS) medium (basal salts-major and minor salts) with 1 mg/L 2, 4-D, pH 5.8.
2. SBFI dye loading medium: Liquid Murashige and Skoog (MS) medium with 1 mg/L 2,4-D and lacking minor salts and $CaCl_2$, pH 5.8. This is supplemented with 0.04% (v/v) of the nonionic surfactant Pluronic F-127 and 10 μM of acetoxymethyl (AM) ester of SBFI (Molecular Probes, Eugene, OR).
3. SBFI-AM ester, 1 mM stock in dry DMSO.
4. Gramicidin (Sigma-Aldrich, St. Louis, MO), 5 mM stock in dry DMSO.
5. Pluronic 127(Sigma-Aldrich, St. Louis, MO), 4% stock in dry DMSO.

2.2. Materials for Spectrofluorometric Determination of Cytoplasmic Sodium

1. Protoplast isolation:
   (a) Enzyme solution: 1% (w/v) cellulase (Onozuka R-10), 0.5% (w/v) macerozyme (Onozuka R-10), 0.5% (w/v) hemicellulase (Sigma), 0.01% (w/v) bovine serum albumin, 1% (v/v) pectinase (Sigma), 5 mM KCl, 10 mM $CaCl_2$, 1 mM $MgCl_2$, 10 mM MES/Tris, pH 5.7, osmolarity adjusted to 600 mosmol/kg using D-sorbitol.
   (b) Protoplast wash buffer (in mM): 5 KCl, 0.1 $CaCl_2$, 10 MES/Tris, pH 5.7, osmolarity adjusted to 500 mosmol/kg using D-sorbitol.
   (c) Protoplast holding buffer (in mM): 5 KCl, 2 $CaCl_2$, 1 $MgCl_2$, 10 Sucrose, 10 Glucose, and 10 MES/Tris, pH 5.7, osmolarity adjusted to 600 mosmol/kg using D-sorbitol.
2. Dye loading:
   (a) Dye loading buffer (in mM): 4 KCl, 1 $MgCl_2$, 0.1 $CaCl_2$, 10 MES/Tris pH 5.6, osmolarity adjusted to 600 mosmol/kg with D-sorbitol.
3. Spectrofluorometry analysis of dye-loaded protoplasts:
   (a) SBFI, acetoxymethyl (AM) ester, 1 mM stock in DMSO.
   (b) Gramicidin, 1 mM stock in dry DMSO.
   (c) Monensin (Sigma-Aldrich, St. Louis, MO), 1 mM stock in ethanol.
   (d) Pluronic F-127, 4% stock in dry DMSO.
   (e) Eserine (Sigma-Aldrich, St. Louis, MO), 30 mM in dry DMSO.
4. Spectrofluorometer, e.g., FluoroMax 4 (HORIBA Jobin Yvon, Inc., Edison, NJ, USA).

3. Methods

3.1. Methods Relating to Two-Photon Microscopy Measurements of Cytoplasmic Sodium

1. Loading of SBFI into rice cells and preparation for microscopy:
   (a) Suspend rice cells (100 μl packed cell volume) in 200 μl SBFI dye loading medium. This medium contains acetoxymethyl (AM) esters of SBFI at 25 mM and 0.04% (v/v) of the nonionic surfactant Pluronic F-127 (see Note 1).
   (b) Incubate at 22 °C for 1 h with gentle shaking in the dark.
   (c) Wash off the unincorporated dye with MS medium after allowing 20 min for hydrolysis of the loaded dye.
   (d) Allow the dye-loaded cells to settle on a coverslip coated with poly-D-lysine prior to microscopy.
2. Microscopy and imaging:
   (a) Two-photon microscopy is carried out on an LSM 510 meta system (Carl Zeiss, Jena, Germany) using a mode-locked Ti:sapphire laser (Tsunami) pumped by a Millennia solid-state laser (Spectra Physics, Mountain View, CA).
   (b) The laser is operated at 730- or 780-nm center wavelength (unless otherwise mentioned), with an 80-MHz repetition rate and 100-fs pulse width.
   (c) The laser beam is focused onto the sample with a 63 × 1.4 numerical aperture (NA), oil immersion objective using a primary dichroic 650SP and the emitted fluorescence collected through a secondary dichroic 560LP and a 515 nm IR corrected bandpass filter.
   (d) Illumination of the sample is restricted to the time of image acquisition.
   (e) Acquisition time per image is 980 ms. The SBFI fluorescence intensities per cell are estimated using inbuilt software (Zeiss LSM 510), which detects fluorescence intensities from the selected region in the frame. Values obtained from several cells can be averaged to estimate mean cellular Na^+ level.
3. Determination of two-photon excitation wavelengths:
   (a) Two-photon spectra for SBFI are determined by exciting dye-loaded cells across the range 700–810 nm in the presence or absence of 200 mM NaCl and 2 μM gramicidin (see Note 2).
   (b) Fluorescence emission is monitored at 515 plus or minus 15 nm (see Note 3).
   (c) Unloaded cells are also similarly analyzed to determine autofluorescence in the wavelength range tested.

4. Standard curves for cytosolic Na^+:
   (a) Rice cells loaded with SBFI are permeabilized with 2 μM gramicidin and suspended in MS medium containing a range of NaCl (0–200 mM).
   (b) Images are acquired in quick succession using 730 and 780 nm excitation.
   (c) The ratio of fluorescence intensities obtained with excitation wavelengths of 730 and 780 nm ($^{730}F/^{780}F$) at a range of Na^+ concentrations is used to obtain a calibration curve (see Note 4).
5. Estimation of the cytosolic levels of Na^+ in rice cells under salinity stress:
   (a) The dye-loaded cells in MS medium are exposed to a range of saline stress (NaCl 0–250 mM) and incubated for an hour.
   (b) The cells are made to settle on a coverslip coated with poly-D-lysine prior to microscopy.
   (c) The SBFI fluorescence is recorded by exciting at 730 and 780 nm and a ratio is obtained.
   (d) The levels of cytosolic Na^+ are then estimated using the calibration curve (see Note 5).

3.2. Methods Relating to Spectrofluorometric Determination of Cytoplasmic Sodium

The methods described in this section include protoplast isolation from rice seedlings, SBFI loading in the protoplasts, and spectrofluorometric measurement of sodium uptake in the dye-loaded protoplasts.

1. Protoplast isolation:
   (a) Wash 1 week old etiolated rice seedlings (0.5 g) in single distilled water and chop with a sharp razor to about 1 mm in length (see Notes 6 and 7).
   (b) Suspend the chopped pieces in 5 ml enzyme solution, vacuum infiltrate for 10 min and incubate at 30 °C for 3 h in an orbital shaker (see Note 8).
   (c) Check protoplast release under microscope and filter the released protoplasts through 50 μm nylon mesh.
   (d) Dilute the protoplast suspension with 5 ml wash buffer and centrifuge at 500 × *g* for 10 min at room temperature.
   (e) Wash the protoplast pellet two times with 5 ml wash buffer by centrifugation at 500 × *g* for 10 min at room temperature.
2. SBFI loading into protoplasts:
   (a) Resuspend the protoplasts in 485 μl dye loading buffer in a 2 ml centrifuge tube (see Note 9).

(b) In the dark, add 5 μl of 30 mM eserine to the protoplast suspension, add 5 μl each of 4% Pluronic F-127 and 1 mM SBFI AM ester on the wall of the tube, mix by pipetting up and down and invert the tube several times to mix the contents thoroughly (see Note 10).

(c) Incubate for 30 min in the dark at 37 °C in an orbital shaker (see Note 11).

(d) Dilute the protoplast suspension with 5 ml wash buffer and centrifuge at $500 \times g$ for 10 min at room temperature.

(e) Wash the pellet with 5 ml wash buffer by centrifugation at $500 \times g$ for 10 min at room temperature.

(f) Resuspend the protoplasts in holding buffer at a concentration of 2×10^6/ml.

3. Spectrofluorometric analysis of the SBFI-loaded protoplasts:

(a) For the spectrofluorometer analysis, 2 ml of 2×10^6/ml protoplasts was used in a quartz cuvette in FluoroMax 4 (HORIBA Jobin Yvon, Inc., Edison, NJ, USA).

(b) Single wavelength measurements were carried out using "FluorEssence" software and multiple wavelength measurements with "Multigroup" software.

(c) We identified 340 nm as the excitation maximum of the sodium bound dye, 400 nm as the excitation maximum of the free dye and 515 nm as the emission maximum for both.

(d) A calibration curve was plotted by treating the dye-loaded protoplasts with 10 μM Gramicidin and 10 μM Monensin for 10 min and subjecting them to different concentrations of sodium. There was a monotonic increase in the fluorescence of the dye with increase in concentration of sodium up to 150 mM (Fig. 1a). 100 mM sodium was used for further experiments.

(e) Fluorescence intensity was recorded at 340 and 400 nm excitation alternately with emission at 515 nm. An integration time of 0.1 at 2 s intervals was used for a total measuring period of 20 min (see Note 12).

(f) Baseline fluorescence of SBFI without the addition of sodium was acquired for 3 min, and with 100 mM NaCl for 17 min (Fig. 1b). The ratio of fluorescence at 340–400 nm gives a measure of the cytoplasmic sodium concentration (Fig. 1c). The increase in concentration of cytoplasmic sodium over the course of the experiment is depicted for two different rice cultivars in Fig. 1d.

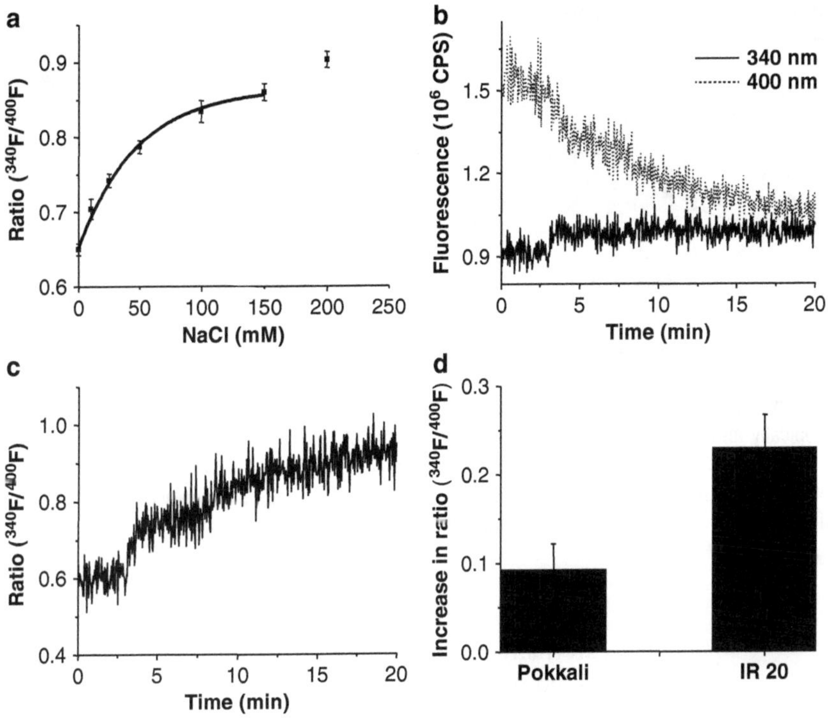

Fig. 1. Estimation of cytosolic sodium in rice shoot protoplasts using SBFI-AM ester. (**a**) Calibration plot of the ratio of fluorescence intensity (^{340}F/^{400}F) with varying external NaCl concentration. (**b**) Fluorescence of SBFI-loaded protoplasts at 340 and 400 nm. 100 mM NaCl added at 3 min, total scan time 20 min. (**c**) Ratio of the fluorescence intensity (^{340}F/^{400}F) for the experiment shown in (**b**). (**d**) Increase in cytosolic sodium concentration calculated from the increase in fluorescence ratio (^{340}F/^{400}F) in shoot protoplasts of two rice cultivars (Pokkali and IR) that vary in salt tolerance, 15 min after exposure to 100 mM NaCl ($n = 3$).

4. Notes

Notes relating to two-photon microscopy measurements of cytoplasmic sodium

1. While loading SBFI into rice cells, it is best to take rice cells (100 µl packed cell volume) in 200 µl of MS (minus minor salts) medium, add required volume of Pluronic F-127 on the wall of the vial and add the dye directly into the detergent droplet and mix using pipette tip. Close the vial and allow the droplet to slide down into the cell suspension and mix by tapping the vial with your finger. This has given better loading efficiency as compared to adding all components to the cell suspension independently.
2. Gramicidin permeabilizes the membrane to Na^+ equilibrating the cytosolic and solution concentrations of Na^+. Spectra recorded with and without gramicidin in the presence of

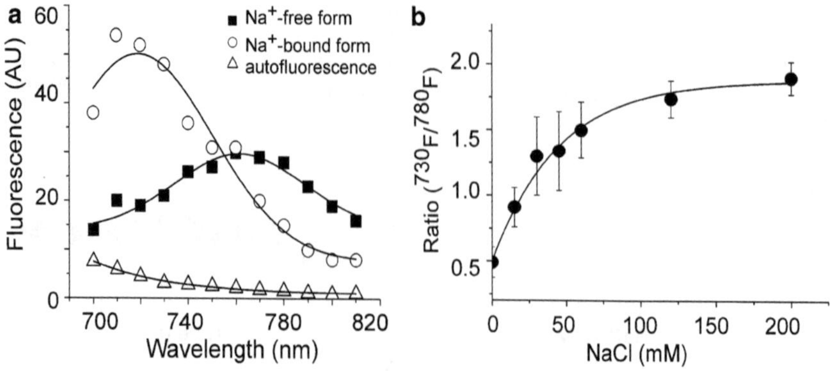

Fig. 2. Sodium-binding benzofuran isophthalate (SBFI) fluorescence in rice cells. SBFI-acetoxymethyl (AM)-loaded cells were microscopically analyzed to identify the subcellular location of SBFI and for ratiometric Na^+ estimations. (**a**) Two-photon excitation spectra of SBFI in rice cells. Na^+-free SBFI (*filled squares*), Na^+-bound SBFI (*open circles*), autofluorescence from unloaded cells (*open triangles*). (**b**) Calibration plot of the ratio of fluorescence intensity ($^{730}F/^{780}F$) with varying external NaCl concentration. Figure reproduced with permission from "John Wiley and Sons" (8).

externally added Na^+ will thus show spectral variation in Na^+ bound dye vs. free dye which can be used to generate a calibration curve.

3. In our experiments, two-photon excitation spectra of SBFI in Jaya (rice cultivar) cells reveal a spectral shift on Na^+ binding with peaks at 780 nm for the free dye and at 730 nm for the Na^+-bound form, together with a significant increase in emission intensity on binding Na^+ (Fig. 2a). Log–log plots of the variation of fluorescence emission intensity vs. excitation power have slopes of approximately 1.8 at both 730 and 780 nm excitation confirming two-photon excitation. The large spectral shift after Na^+ binding facilitates accurate estimations of Na^+ concentrations within these cells (Fig. 2b).
4. In our experiments, a calibration plot generated in rice cells using the ratio of intensities on exciting at 730 and 780 nm, the wavelengths where the two spectra differ most, increases monotonically and plateaus at about 150 mM (Fig. 2b) as has been reported for the ratio $^{340}F/^{380}F$ using UV excitation in *Arabidopsis* cells (4). The incremental increase of the calibration ratio $^{730}F/^{780}F$ with increasing NaCl is significant up to about 100 mM but we have limited our use of the curve to 80 mM.
5. In our experiments, we have demonstrated that the sodium-specific dye SBFI is localized exclusively in the cytosol by confocal microscopy. The two-photon approach is intrinsically confocal and reduces the problems of photobleaching of the dye and phototoxicity to the cell encountered with one-photon UV excitation. The large change in two-photon fluorescence

spectrum on binding of Na^+ permits the use of SBFI as a ratiometric probe for Na^+ in rice cells. In this report, we combine the twin advantages of two-photon excitation with ratiometric analysis of Na^+ concentrations.

Notes relating to spectrofluorometric determination of cytoplasmic sodium

6. Etiolated/dark grown seedlings yield better number and quality of protoplasts than light grown seedlings.
7. Plants older than 2 weeks yield less quantity and quality protoplasts.
8. Adding 300 µg/ml carbenicillin to the enzyme mix eliminates the chances of occasional bacterial contamination.
9. Reaction volume can be scaled down to 200 µl, but larger volume is preferred which minimizes crowding of protoplasts and better movement in the tube which results in efficient loading.
10. Prior mixing of SBFI with the surfactant Pluronic F-127 enhances the solubility of SBFI in the aqueous loading buffer.
11. These conditions yielded 70–80% dye loading efficiency, which is important in the study of a population of cells.
12. The emission slit width was kept at 10 nm and the excitation slit width was adjusted between 1 and 3 nm to keep the emitted fluorescence within the calibration range of the instrument.

References

1. Minta A, Tsien RY (1989) Fluorescent indicators for cytosolic sodium. J Biol Chem 264:19449–19457
2. Despa S, Kockskamper J, Blatter LA, Bers DM (2004) Na/K pump-induced (Na)i gradients in rat ventricular myocytes measured with two-photon microscopy. Biophys J 87: 1360–1368
3. D'Onofrio C, Kader A, Lindberg S (2005) Uptake of sodium in quince, sugar beet, and wheat protoplasts determined by the fluorescent sodium binding dye benzofuran isophthalate. J Plant Physiol 162:421–428
4. Halperin SJ, Lynch JP (2003) Effects of salinity on cytosolic Na^+ and K^+ in root hairs of *Arabidopsis thaliana*: in vivo measurements using the fluorescent dyes SBFI and PBFI. J Exp Bot 54:2035–2043
5. Kader AM, Lindberg S (2005) Uptake of sodium in protoplasts of salt-sensitive and salt-tolerant cultivars of rice, *Oryza sativa* L. determined by the fluorescent dye SBFI. J Exp Bot 56:3149–3158
6. Halperin SJ, Gilroy S, Lynch JP (2003) Sodium chloride reduces growth and cytosolic calcium, but does not affect cytosolic pH, in root hairs of *Arabidopsis thaliana* L. J Exp Bot 54:1269–1280
7. Rose CR, Kovalchuk Y, Eilers J, Konnerth A (1999) Two-photon Na^+ imaging in spines and fine dendrites of central neurons. Pflugers Arch 439:201–207
8. Anil VS, Krishnamurthy H, Mathew MK (2007) Limiting cytosolic Na^+ confers salt tolerance to rice cells in culture: a two-photon microscopy study of SBFI-loaded cells. Physiol Plant 129:607–621

Chapter 16

Real Time Measurement of Cytoplasmic Ions with Ion-Selective Microelectrodes

Anthony J. Miller

Abstract

Ion-selective microelectrodes can be used to report intracellular ion concentrations. The ion-selective barrels of microelectrodes are filled with a sensor cocktail containing several different components including an ion-selective molecule, sensor or exchanger, a solvent or plasticizer, lipophilic cation/anion additives, and a matrix to solidify the membrane. For many ions, the readymade membrane cocktail can be purchased, but the individual chemical components can be bought from suppliers and mixing the cocktail saves money. For commercially available liquid membrane cocktails the membrane matrix is often not included. For plants a matrix is essential for intracellular impalements because without it cell turgor will displace the liquid membrane from the electrode tip, giving decreased or even lost sensitivity. The matrix frequently used is a high molecular weight poly(vinyl chloride). This addition increases the electrical resistance of the electrode, slowing the response time of the electrode. The use of multi-barreled electrodes enables the identification of the cellular compartment. For example, the inclusion of a pH-selective electrode enables the cytoplasm and vacuole to be distinguished.

Key words: Cell compartmentation, Cytoplasm, Ion-selective microelectrodes, Sensor

1. Introduction

Intracellular ion concentrations are important indicators of nutrient status and health. Changes in these concentrations can provide signals that can translate an environmental stimulus into altered gene expression and/or posttranslational modification of proteins. Methods for detecting intracellular ion concentrations suffer from the difficulty that the technique itself can directly alter the parameter that is being measured. Reporter molecules such as fluorescent dyes or proteins function by binding the ion being sensed and therefore will change the concentration of the ion being measured. Inserting glass microelectrodes into a cell measures at a single location and gradients may occur within a cell. No method is perfect

Frans J.M. Maathuis (ed.), *Plant Mineral Nutrients: Methods and Protocols*, Methods in Molecular Biology, vol. 953, DOI 10.1007/978-1-62703-152-3_16, © Springer Science+Business Media, LLC 2013

and the best approach is to use more than one method to check any intracellular ion measurement.

The term *microelectrode* is used to describe a glass micropipette which is pulled into a fine tip at one end and filled with an aqueous salt solution that provides an electrical bridge to a metal wire contact. The junction between the salt solution inside the microelectrode and the input to the electrometer amplifier is provided by a *half-cell*. There are different types of *half-cell*, but usually the metal contact is AgCl-coated silver wire and the salt solution is 0.1 M KCl. The simplest microelectrodes measure voltage and when inserted into cells measure the membrane potential, in mV, between the inside and outside of the cell. An ion-selective microelectrode contains an ion-selective membrane in the tip of the glass micropipette and is responsive both to the membrane potential and the activity (not actually the concentration) of the ion sensed by the selective membrane. The two types of electrode are combined together into a double-barreled tip for intracellular measurements (see Fig. 1). Ion-selective membranes are composed of specific types of chemicals and the roles played by each component have been described in detail (1). Good electrodes have a low detection limit, a near ideal slope, and a small selectivity coefficient for physiologically important interfering ions.

Ion-selective microelectrodes are used to measure ion gradients across membranes and to follow the time course of concentration

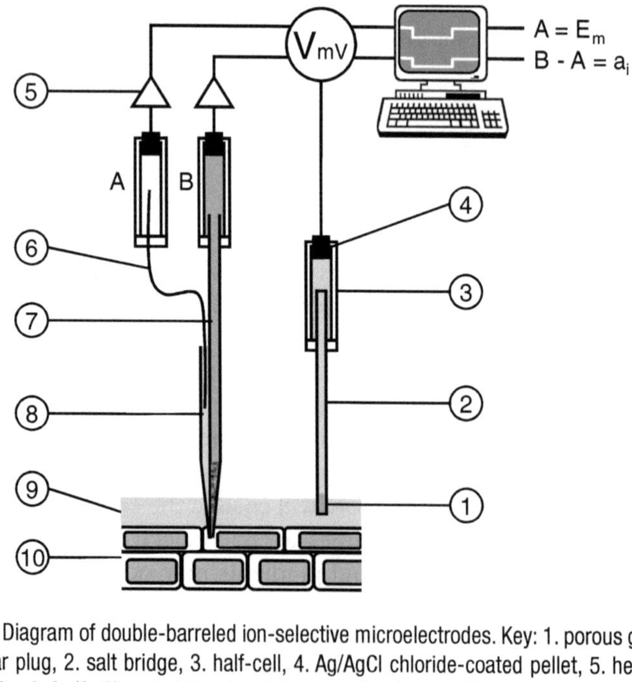

Fig. 1. Diagram of double-barreled ion-selective microelectrodes. Key: 1. porous glass frit or agar plug, 2. salt bridge, 3. half-cell, 4. Ag/AgCl chloride-coated pellet, 5. headstage amplifier, 6. Ag/AgCl coated Ag wire, 7. ion-selective barrel with sensor plug in the tip, 8. cell membrane potential barrel filled with 0.1 M KCl, 9. nutrient solution bathing plant, 10. plant tissue with microelectrode tip in a cell (cytoplasm).

changes. These measurements can be made outside and inside cells. For example, ion fluxes at the surface of roots can be measured by using either conventional ion-selective microelectrodes or by using an ion-selective vibrating probe (2) (also see Chapter 11). Microelectrodes have been used to report the cellular compartmentation of nutrients, dynamics of changes in ion activities (e.g., in intracellular signaling) and transport mechanisms, particularly the energy gradients for cotransported ions (3). For intracellular ion measurement the simultaneous recording of membrane potential is required; this is done by insertion of a second electrode or, for small cells, by combining the ion-selective and voltage-measuring electrodes into a double-barreled microelectrode (see Fig. 1).

The main criticism of intracellular measurements made with microelectrodes is that they report the ion activity at a single point within the cell. This will result in incomplete information if there are significant ion gradients within the cytoplasm of a single cell as may occur in some situations. Other disadvantages of the method are (1) it can only report from single cells that must be accessible close to the tissue surface, (2) specialist equipment and skill is required for the measurements, (3) the chemical response time of the electrodes can be several minutes and therefore too slow to detect faster transient changes.

The main advantages of using ion-selective microelectrodes are (1) it offers a nondestructive method of measuring ions within single cells, (2) it does not change the activity of the ion being measured, and (3) it permits the simultaneous measurement of electrical and chemical gradients across membranes.

Five distinct stages for the manufacture of ion-selective electrodes can be identified.

1. Pulling glass micropipettes
2. Silanization of the inside surface of the designated ion-selective micropipette
3. Backfilling with ion-selective membrane cocktail
4. Backfilling with salt solution
5. Calibration

These stages can be separated in time and are therefore conveniently used to describe the manufacturing processes.

2. Materials

Prepare and store all reagents at room temperature unless specified otherwise. Wear eye protection at all times when pulling and breaking glass. For pulling micropipettes, glass tubing can be purchased of varying diameter and wall thickness according to requirements. An outer diameter of 1 mm is convenient and fits many commercially available half cell holders.

Electrophysiology recordings from plants require very solid anchoring of the plant material while at the same time preserving the normal state of the material as far as possible. It is best to avoid dissecting the plant material as this is likely to lead to local wounding that is known to have major effects on gene expression (4). Plants grown in hydroponic culture can be easily transferred to the microscope stage for electrode impalements of either root or leaf tissue. The hydroponic environment for roots is easily maintained on the stage of a microscope, but leaves are more difficult requiring some wet contact between the tissue and the bathing solution.

Some of the standard equipment required for microelectrode recording is as follows.

1. Voltmeter (also known as an electrometer).
2. Microscope (with long working distance objectives).
3. Micromanipulator.
4. Tissue chamber (for holding and perfusion).
5. Data logging system (e.g., computer or chart recorder).
6. Vibration free table (for microscope and micromanipulator to avoid interference from external vibration sources).
7. Faraday cage (electrical screening around the microscope and micromanipulator especially necessary for high resistance electrodes).
8. Oscilloscope (not essential but useful for identifying and fixing recording noise problems).

Microelectrode impalements are usually made under a microscope using long working distance objectives that allow sufficient space for microelectrode access. Although they are generally used for patch-clamp experiments, inverted microscopes are not so suitable for this type of work. Dissecting microscopes can be used for microelectrode impalements, but they usually do not have sufficient magnification to see individual cells. They can be used for impalements by letting the electrical recording show when tissue contact has been made and a successful impalement can be gauged by the size of the membrane potential measured. Microelectrodes are mounted on micromanipulators for cellular impalement to allow the delicate movement of the tip into a cell. There is a range of different types and hand control of tip movement is achieved by either joystick or rotational manipulation. The size and fine movement axis should be chosen so that the micromanipulator can be conveniently positioned alongside the microscope stage for tissue impalement.

2.1. Pulling Glass Micropipettes

1. Filamented borosilicate glass (see Note 1) is used for membrane potential recording electrodes to make filling the fine tips easier.
2. Non-filamented glass for ion-selective glass electrodes (this is easier to keep dry and treat with a silanizing agent).
3. Multi-barreled glass containing fused bundles of these different types of glass (this can be bought from specialist suppliers, e.g., Hilgenberg GmbH, Malsfeld Germany http://www.hilgenberg-gmbh.com/).
4. Microelectrode puller. Use a specialized puller to obtained microelectrode tips of reproducible dimensions; pullers can operate with the glass held in either horizontal or vertical position. A twisting mechanism is required for multi-barreled tips (see Note 2).
5. Pulled glass tips can be conveniently stored for long periods of time in disposable plastic Petri dishes of an appropriate size and held in place using "*Blu-tack*" (see Note 3).

2.2. Silanization of Glass Micropipettes

1. Prepare a 2% (w/v) silanizing agent in chloroform. A range of different silanizing agents can be used at this concentration, but dimethyldichlorosilane or tributylchlorosilane are most commonly used (see Note 4).
2. Disposable plastic syringe (1 mL) and metal needle (25 G).
3. Heating lamp to give 140 °C on the electrode tips.

2.3. Back-Filling with Ion-Selective Cocktail

1. Ion-selective cocktail dissolved in tetrahydrofuran (THF). This can be purchased direct from a supplier or mixed from the component chemicals (see Subheading 3.3 below).
2. Glass 1 mL syringe and metal needle (30 G is a convenient size).
3. Glass beaker covered with parafilm.
4. After about 48 h the THF has evaporated to leave the sensor cocktail in a plastic membrane embedded in the tip of the glass microelectrode. The microelectrodes are now ready for use or they can be conveniently stored at this stage (see Note 5).

2.4. Back-Filling with Salt Solution, Calibration and Measurements

1. Like backfilling with cocktail (see Subheading 2.3 above), a syringe and needle can be conveniently used to back fill the electrode behind the ion-selective membrane.
2. Back-filling solutions. Typically for plant cell measurements 0.2 M KCl solution is used for back-filling reference barrels. For an ion-selective barrel the choice of back-filling solution depends on the sensor, but it is best to use a high concentration of the ion that is being sensed. For example, pH 4 for H^+-selective microelectrodes or 0.1 M KNO_3 for NO_3^- electrodes.

3. An amplifier voltmeter (or electrometer) with a high input impedance that is at least 1,000 times higher than the impedance of the ion-selective electrode, e.g., $10^{15}\,\Omega$. Furthermore the input leakage current from the electrometer must be low so that no significant offset voltage (>1 mV) is produced across the ion-selective electrode. Other useful electrometer facilities include GΩ range resistance tester and a difference-voltage output so that a direct output equivalent to cell ion activity can be obtained.
4. The backfilled microelectrode can be conveniently connected to a commercially available holder (or "half cell"). A range of holder sizes is available, choose one that fits the outer diameter of the microelectrode glass. Also the holder pin connection size and type (male or female) can be chosen as appropriate for the input on the electrometer. Microelectrode holders can be bought from electrometer suppliers and they are often supplied with the equipment (see Note 6).
5. Standard solutions of known fixed concentrations are required for the calibration of the ion-selective electrode tip. Calibration solutions can be stored at 4 °C for several weeks (see Note 7). Ion-selective microelectrodes can be calibrated using concentration or activity, but the electrodes actually respond to changes in *activity*. Therefore calibrating with ion activity gives a microelectrode output which can be used directly without any assumptions of the intracellular activity coefficient for the ion. Furthermore, activity is actually the important parameter for all biochemical reactions. For these reasons the calibration of microelectrodes generally uses solutions which resemble the intracellular environment in terms of interfering ions, and ionic strength (see Note 8).

3. Methods

3.1. Pulling Glass Micropipettes

1. Break pieces of glass to convenient lengths (approx 10 cm).
2. Pull tips using a puller and following the manufacturers' instructions. Twisting multi-barreled glass before the final stage of the pull helps to ensure that the tips are aligned, see Walker et al. (3) for details.
3. Check that all the microelectrode tips can be used to obtain stable intracellular recordings of membrane potential (see Note 9).
4. For multi-barrelled glass electrodes, after pulling the open glass ends can be broken back to different lengths to enable treatment with a silanzing agent at the end of specific barrels.
5. Electrodes can then be stored dry until they are ready to be used.

3.2. Silanizing Specific Glass Barrels of Micropipettes

1. Pulled glass electrodes are placed under a heating lamp at 140 °C for at least 30 min.
2. Under the lamp the open blunt end of the glass barrel designated to receive ion-selective electrode sensor is treated with a few drops of silanizing agent dispensed using a syringe and needle.
3. The silanizing solution should quickly vaporize giving the ion-selective barrel a hydrophobic coating and leaving no liquid remaining within the glass (see Note 10).
4. After silanizing the microelectrode should remain under the heat for 30 min before directly back-filling with the ion-selective cocktail (Subheading 3.3 below).

3.3. Preparation of the Nitrate-Selective Cocktail

1. Weigh the methyltridodecylammonium nitrate (3 mg), nitrocellulose (2.5 mg), poly(vinyl chloride) (11.5 mg), and a lipophilic cation—methyltriphenyl phosphonium bromide (0.5 mg) into a 1 mL glass screw-topped vial using a balance accurate to 0.1 mg (see Note 11). This cocktail for measuring nitrate was first reported in 1991 (5) and is now sold commercially (Sigma-Aldrich product number 72549).
2. Add the liquid nitrophenyl octyl ether (32.5 mg) to the vial using a glass microcapillary on the balance pan.
3. Dissolve the cocktail in approximately 4 volumes of THF (see Note 12). The cocktail takes at least 30 min to dissolve completely.
4. This cocktail can then be stored at 4 °C for several weeks and is enough to make about 70 nitrate-selective microelectrodes.

3.4. Back Filling with Ion-Selective Cocktail into Silanized Glass

1. Ion-selective cocktail mix dissolved in THF is back-filled into the blunt end of a glass microelectrode using a 1 mL glass syringe and metal needle. The cocktail composition depends on the type of ion-selective microelectrode.
2. The filled microelectrode is placed with the tip down in a glass beaker covered with parafilm.
3. After about 48 h the THF has evaporated to leave the sensor cocktail in a plastic membrane embedded in the tip of the glass microelectrode. The microelectrodes are now ready for calibration and use or they can be conveniently stored at this stage (see Note 13).

3.5. Backfilling with Salt Solution, Calibration and Use for Intracellular Measurements

1. Backfill the electrode barrels with salt solutions. Care must be taken to avoid air bubbles in the fine tip and at the surface of the ion-selective membrane.
2. Newly back-filled ion-selective microelectrodes require "conditioning" for a minimum of 30 min before they become stable

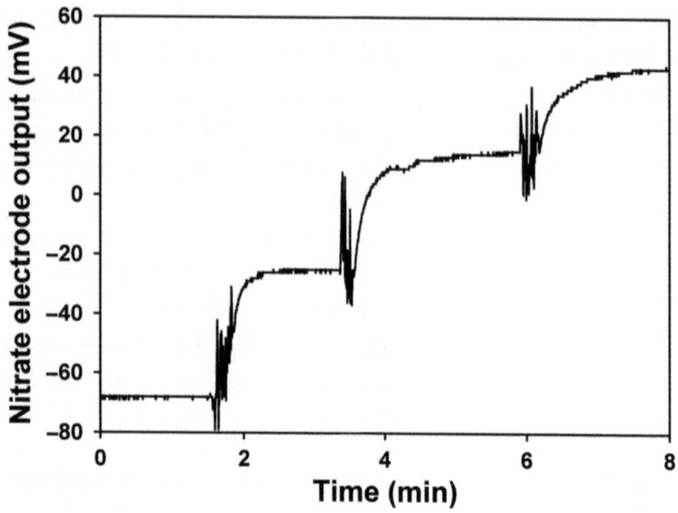

Fig. 2. Calibration of a nitrate-selective microelectrode to show the response time. This recording shows the response times of the electrode as the electrode tip was exposed to decreasing activities of nitrate from 100 mM to 10 mM to 1 mM to 0.1 mM nitrate.

and responsive. This process involves immersing the tip in a solution containing a high concentration (e.g., 0.1 M) of the ion to be measured.

3. Calibration of microelectrodes can be performed in the chamber built to take the plant tissue, or using a U-shaped funnel (see Note 14). Good electrodes should have a low detection limit, a near ideal slope, i.e., around 58 mV per decade for monovalent ions, and a small selectivity coefficient for physiologically important interfering ions (see Note 15).
4. The microelectrodes are now ready for intracellular recordings using a standard electrophysiological equipment rig that includes a high input impedance electrometer and off the shelf data-logging software. If the dynamics of changes in ion activity are particularly of interest then the response time of the electrode must be carefully recorded (see Fig. 2 and Note 16). Usual electrophysiological preparation is needed, ensuring that the plant tissue is well anchored in position under a microscope for the electrode impalement (3). Problem solving can be also followed in the routine way for these types of electrical measurements (see Note 17).
5. In plant cells, the identification of the cell compartment (cytoplasm or vacuole) where the microelectrode tip is located can be a problem for some ions and it may be necessary to grow the plant under conditions in which two populations of measurements can be identified. Alternatively, a triple-barrelled microelectrode can be used where one barrel is pH or Ca^{2+} selective (6). Large gradients of these two ions are known to

exist across the tonoplast, with the cytoplasm maintained at relatively constant values (pH 7.2, Ca^{2+} 100 nM) so compartment identification is possible. Another approach is to use tissues where the two major cell compartments can be identified under the microscope, e.g., root hairs, or cell cultures which have no large vacuole. However, identifying which compartment the electrode is in can still be problematic, particularly if the electrode indents the tonoplast but does not penetrate it.

6. Several criteria for acceptable intracellular measurements can be defined. Firstly, after a recording and retraction of the electrode, the ion-selective microelectrode should be recalibrated and should give a very similar response to that shown before the cell impalement, particularly at activities similar to those measured in vivo. Sometimes the recalibration shows a displacement up or down the Y mV output axis. More often the detection limit of the ion selective microelectrode has changed, but provided the measurement was on the linear response range of the electrode calibration curve this is not usually a reason to disregard the result. Sometimes the performance of the ion-selective microelectrode can even improve with the detection limit actually becoming lower (see Note 18). A comparison between the electrical resistance of the ion-selective microelectrode before and after impalement provides a good indicator of whether the tip will recalibrate. If the resistance decreases below 1 GΩ, the ion-selective membrane has probably been displaced during impalement and the electrode will not recalibrate. Throughout the recording the state of the cell can be assessed by monitoring the membrane potential (which should remain stable unless deliberately perturbed) or by processes like cytoplasmic streaming.

4. Notes

1. Glass can be cleaned before pulling by washing in a solvent like ethanol or acetone, but this is not essential. Various types of glass can be used, but check the melting temperature. Electrode puller heating elements may need changing to reach a higher temperature. Aluminosilicate glass is tougher than borosilicate glass, but it requires a higher heating element temperature.
2. Avoid positioning the microelectrode puller in an air draft as this can cause variable tip shapes.
3. "*Blu-tack*" may not be available, but equivalent products are available in most countries, try a web search. Only lightly press the glass electrode into the Blu-tack, if attached too firmly the glass can snap during removal.

4. Silanizing agents can generate a very toxic vapor (see Note 10) and must be prepared and used in a fume hood. Only small volumes are required, typically a syringe needle drop on each tip. A stock solution of only 10 mL is prepared.
5. For long-term storage, the ion-selective microelectrodes should be stored without backfilling, in a silica-gel dried sealed container in the dark. This can be done in a screw-cap glass jar or parafilm covered beaker containing dry silica gel, with the microelectrodes attached to the inner wall using plasticine or Blu-tack. Ion-selective microelectrodes stored for several years in this way can still give a reasonable performance when back-filled.
6. Commercially available holders have a limited life of a few months, but they can be regenerated by replacing the metal silver/silver chloride pellet electrode in the base of the holder. Microelectrode holders can be made using cut down glass pipettes of an appropriate size and 2 mm jack pins with a melted wax seal onto the electrode glass. The advantage of this method is that a silver wire can be directly soldered to the jack pin and easily re-coated with chloride. The symptoms of deterioration in the silver/silver chloride coating are unstable recordings.
7. Note that the slope of the calibration curve is temperature sensitive and both calibrations and intracellular measurements should be done at the same temperature. If the temperature of the calibration solutions is 4 °C and the cell is at 20 °C, the slope of the electrode calibration for a monovalent ion will be 55 mV per decade change in activity, not the 58 mV expected at 20 °C.
8. The calibration of some microelectrodes for intracellular measurements (e.g., Ca^{2+} and H^+) requires the use of buffering agents such as EGTA for Ca^{2+} because of the very low concentrations being measured.
9. An estimate of the tip geometry of the microelectrode is provided by measuring its electrical resistance when filled with KCl, larger tips having lower resistances. For tips of 2–0.1 μm diameter the electrical resistances of ion-selective microelectrodes are usually in the GΩ range, while the same sized tips filled with 0.1 M KCl have 10^3 times smaller resistances, in the MΩ range. The dimensions of the microelectrodes are usually a compromise between obtaining a stable membrane potential and a good calibration response (detection limit).
10. Care must be taken to ensure that the reagent does not enter the membrane potential-measuring barrel of multi-barrelled microelectrodes. Silanizing agents are corrosive and toxic, protective glasses and gloves must be worn and glass must be treated in a fume hood.

11. These additions are conveniently made one by one on a balance. The cocktail composition is altered to obtain different types of ion-selective electrode. Ion-selectivity is provided by a specific chemical sensor, but in addition most cocktails contain a lipophilic cation or anion, a plasticizer and plastic polymer (often PVC). Some ready mixed cocktails can be obtained from commercial suppliers.
12. No plastic can be used in contact with THF. Glass syringes and metal needles can be conveniently used to dispense small volumes up to 1 mL. Excess THF is required to dissolve the cocktail components for mixing, but as it evaporates to carry the cocktail into the glass tip the volume used need not be measured very precisely.
13. After backfilling and THF evaporation the lifetime of the ion-selective microelectrodes is type dependent. For example, nitrate-selective electrodes can be stored for more than a year in a dry atmosphere without any detrimental decline in their calibration and use, but proton-selective tips are best used within 4–6 weeks after manufacture.
14. When calibrating it is important to make sure that the electrode tip is directly exposed to the complete calibration solution. Flow calibration systems can allow some mixing of the solutions resulting in poor calibration curves. Also, it is best to run through a complete set of calibration solutions and then returning to the first one in the series to check for the stability of the electrode.
15. For statistical analysis and the calculation of means, calculations should be based on data which are distributed normally, i.e., by using the log activity or output voltages, not the actual activities (7). Therefore when mean activity value is used it can only be expressed with 95% confidence limits, whereas -log (activity) can be given standard errors or standard deviations.
16. When measuring changes in intracellular ion concentrations, artifacts can be caused by the differential response times of the two barrels; the ion-selective barrel usually has a slower response time than the membrane potential-sensing barrel. This can be corrected for when the response time of the electrode is known (8). The electrode response time can limit detection of rapid changes in ion activity.
17. The best approach is to try to solve problems by a process of elimination. Firstly, establish whether a problem occurs in the circuitry or is specific to the ion-selective microelectrodes. The circuitry can be tested by putting a broken-tipped KCl-filled microelectrode in place of the ion-selective microelectrode. A broken-tipped electrode should give a stable zero output. It may be necessary to re-coat AgCl-silver contact in the half-cell

or there may be a wiring problem. Noisy recordings can be caused by poor earthing, or air bubbles in backfilling solutions. If the circuitry has no problems then the ion-selective microelectrode must be the cause. When the ion-selective microelectrode does not respond to the calibration solutions then the membrane can be checked by deliberately breaking the tip to expose a larger area of ion-selective membrane. Breaking the tip can displace the ion-selective membrane from the tip so it is important to measure the resistance to check it is still in the GΩ range. If the broken tip gives a good response to changes in ion activity then the problem is independent of the composition of the membrane. When the microelectrode tip diameter becomes too fine the output from the ion-selective electrode will no longer respond to changes in ion activity.

18. For this reason, it can be best to quickly impale a cell with a new tip before calibrating prior to measuring the activity in a cell.

Acknowledgments

John Innes Centre is grant-aided by the Biotechnology and Biological Sciences Research Council (BBSRC) of the UK.

References

1. Ammann D (1986) Ion-selective microelectrodes, principles, design and application. Springer, Berlin
2. Henriksen GH, Bloom AJ, Spanswick RM (1990) Measurement of net fluxes of ammonium and nitrate at the surface of barley roots using ion-selective microelectrodes. Plant Physiol 93:271–280
3. Miller AJ, Cookson SJ, Smith SJ, Wells DM (2001) The use of microelectrodes to investigate compartmentation and the transport of metabolized inorganic ions in plants. J Exp Bot 52:541–549
4. Delessert C, Wilson IW, Van der Straeten D, Dennis ES, Dolferus R (2004) Spatial and temporal analysis of the local response to wounding in Arabidopsis leaves. Plant Mol Biol 55:165–181
5. Miller AJ, Zhen R-G (1991) Measurement of intracellular nitrate concentrations in Chara using nitrate-selective microelectrodes. Planta 184:47–52
6. Walker DJ, Smith SJ, Miller AJ (1995) Simultaneous measurement of intracellular pH and K^+ or NO_3^- in barley root cells using triple-barreled ion-selective microelectrodes. Plant Physiol 108:743–751
7. Fry CH, Hall SK, Blatter LA, McGuigan JAS (1990) Analysis and presentation of intracellular measurements obtained with ion- selective microelectrodes. Exp Physiol 75:187–198
8. Sanders D, Slayman CL (1982) Control of intracellular pH. Predominant role of oxidative metabolism, not proton transport, in the eukaryotic microorganism *Neurospora*. J Gen Physiol 80:377–402

Chapter 17

Large-Scale Plant Ionomics

John M.C. Danku, Brett Lahner, Elena Yakubova, and David E. Salt

Abstract

Large-scale phenotyping methods are at the heart of efficiently deciphering the functions of genes and gene networks in the postgenomic era. In order to obtain meaningful results when comparing natural variants, and mutants and wild-types during large-scale quantitative analyses, necessary precautions must be employed throughout the whole process. Here, we describe large-scale elemental profiling in *Arabidopsis thaliana* and other genetic model organisms using high-throughput analytical methodologies. We also include a description of workflow management and data storage systems.

Key words: Plant ionome, Functional genomics, High-throughput ion profiling, Elemental analysis, Mineral nutrients

1. Introduction

Functional genomics, which endeavors to understand the dynamic relationship between an organism's genome and its phenotype, has become one of the greatest challenges in the life sciences. The approach involves development and employment of high throughput methods because of the genome-wide scale of the problem. One such functional genomics approach is ionomics (1). Ionomics is a high-throughput phenotyping platform that enables one to gain insight into the genes and gene networks involved in mineral-ion accumulation in plants (2–5) and other organisms (6). Applying this approach as a genetic screen is uncovering roles for suberin deposition (7), sphingolipid biosynthesis (8), phloem transport (9), pathogen–response genes (10), and cytokinins (11) in controlling the accumulation of multiple mineral nutrients and trace elements, including Na, Ca, Mg, K, Zn, and Fe. This approach is also helping to identify genes that control natural variation in the accumulation of Na (12), Mo (13), Co (14), Cu (15), and S (16), the ecological role of this natural variation (17), along with the genetic architecture

Frans J.M. Maathuis (ed.), *Plant Mineral Nutrients: Methods and Protocols*, Methods in Molecular Biology, vol. 953, DOI 10.1007/978-1-62703-152-3_17, © Springer Science+Business Media, LLC 2013

of natural variation in the relationship between growth and mineral nutrient homeostasis (18). To be successful this approach requires the application of high-throughput elemental analysis techniques, together with the incorporation of both bioinformatics and genetic tools (1). Mechanisms of mineral nutrient and trace element homeostasis in plants have often been assessed by tracking a single ion or element (19–24). However, mineral nutrient and trace elements usually exist as part of larger multielement regulatory networks (25). It is therefore important to consider the mineral nutrient and trace element composition, or ionome, as a whole. Successful application of ionomics requires rapid and robust multielement quantification in 1,000s of samples with both high sensitivity and precision. The high-throughput nature of this process also necessitates diligent tracking of the data and metadata associated with the multiple samples in the workflow.

Here, we describe the methodologies involved in the implementation of a successful ionomics project, with a focus on *Arabidopsis thaliana*. We detail the necessary conditions and techniques required including growing *A. thaliana*, harvesting of tissue for analysis through to sample analysis workflow management and data storage. As many samples are handled during the course of this undertaking, we combine the physical workflow with an interactive web-based data storage and retrieval system. We also present an overview of similar systems for rice and yeast.

2. Materials

2.1. Growth Room Facility (for A. thaliana and Relatives)

1. A dedicated plant growth room with controlled temperature, lighting and humidity.
2. Sinks supplied with reverse osmosis (RO) water supply and a regular tap water supply.
3. Racks with light fixtures for holding plant growth trays.
4. Storage shelves for trays and other supplies.
5. Refrigerators to store solutions and for seed stratification.
6. Plastic bins with lids for soil storage.
7. Plastic pots/trays of different sizes (20-row trays; 72-pot trays; 6-pot, 12-pot trays; etc.) for plant growth at different conditions.
8. Analytical balance, pH meter, water carboys (reservoirs, e.g., 20-, 40-L for Hoagland stock solutions; also 6-, 10-L), graduated plastic beakers (500, 1,000 mL) with handles.
9. Work desks and carts (2 or 3; could be different sizes).
10. Cryovials or Eppendorf tubes (0.5, 1, 1.5, 2 mL) and yellow envelopes (e.g., 9 × 5.5 cm) for permanent seed storage.

11. Desiccators and card-boxes for seed storage.
12. Labeling tapes and permanent markers.
13. Magnetic stirring device and bars.
14. Trash bins with heavy-duty bags (e.g., 120 L).
15. Photometer.
16. Paper towels and dust pan with brush.
17. Vacuum cleaner for cleaning soil spills and other room cleaning.
18. Plastic ties for light fixtures.

2.2. Soil and Watering

1. Premier PROMIX PGX (Professional plug and germination growing medium, Product of growth medium.
2. Soil mixer (e.g., 140 L capacity. Do not use in Growth Room).
3. Arsenic and selenium group and cadmium group soil element mix solutions. Prepare As group soil mix (consisting of oxides of arsenic (V) ($Na_2HAsO_4 \cdot 7H_2O$) and selenium (VI) (K_2SeO_4)) and Cd group soil mix (consisting of the nitrates of cadmium ($Cd(NO_3)_2 \cdot 4H_2O$), cobalt ($Co(NO_3)_2 \cdot 6H_2O$), lithium ($LiNO_3$), nickel ($Ni(NO_3)_2 \cdot 6H_2O$), strontium ($Sr(NO_3)_2$), and rubidium ($RbNO_3$)). Weigh out the equivalent amount of the elemental concentration (ppb per dry soil weight) of the salts into a 1-L volumetric flask, dissolve, and bring up to the mark with DI water. Stock solutions are usually made for 10 soil batches at a time, where a batch is a bag of soil (dry weight ~5 kg). For instance, the elemental concentrations of the As group are As(V): 5,619.1 ppb and Se(VI): 1,974.0 ppb; the equivalent weights of the salts for 10 batches, given a batch soil dry weight of 4.96 kg, are As: 1,160.9 mg and Se: 274.23 mg. The corresponding elemental and equivalent salt amounts (in parenthesis) for the Cd group are Cd: 202.3 ppb (27.542 mg), Co: 1,178.7 ppb (288.72 mg), Li: 1,388.2 ppb (683.49 mg), Ni: 1,173.9 ppb (288.48 mg), Sr: 8,762.0 ppb (1,071.1 mg for 98 % purity), and Rb: 1,1794.4 ppb (1,009.5 mg). The 500 mL of the elemental mix solutions used in preparing a batch (bag) of soil mix are made by diluting 100 mL of the respective stock solutions (As group and Cd group) to 500 mL and applying to the soil during mixing.
4. Mixed soil storage containers.
5. Plastic or garden scoops (e.g., 0.5–1 kg soil capacity) for soil.
6. RO water (or distilled water).
7. Iron-HBED solution (*N,N'*-di(2-hydroxybenzyl)ethylenediamine-*N,N'*-diacetic acid monohydrochloride hydrate; Strem Chemicals, Inc., Newburyport, MA, USA). Prepare the Fe-HBED from HBED and iron (III) nitrate ($Fe(NO_3)_3 \cdot 9H_2O$, purity 98.1–101.0 %) as follows: Weigh 7.988 g HBED (2 L of

a 10 mM solution) into a container large enough to hold 2 L. Fill to about 75 % (or 1.5 L) with RO water, put in a stir bar and place on a magnetic stir plate. Next, weigh 6.817 g $Fe(NO_3)_3$ (1:1 Fe to HBED molar ratio, and taking into account the purity of the salt) into a 50-mL beaker. Dissolve in sufficient RO water and add to the HBED container. This reaction generates HNO_3 which must be neutralized. Potassium hydroxide which is used for this purpose is prepared by weighing 120.0 g KOH in 2 L (taking the purity into account) to make a 1.0 M solution. Measure out 66.8 mL of the KOH solution using a graduated cylinder and add to the HBED container. Bring the pH of the Fe-HBED solution to 6.0 using the same KOH solution and top with RO water to the mark. Store the Fe-HBED solution in a brown bottle at 4 °C. Caution: KOH is caustic so wear eye protection; it is hygroscopic so weigh quickly. The solution will get hot during dissolution. Store solution in a plastic bottle.

8. 0.25× Hoagland's Mo Type 2 solution (see Subheading 3.1, step 2). Prepare 0.25× Hoagland's Mo Type 2 solution (40 L) as follows. Weigh out these salts for Hoagland's macronutrients into a large beaker, dissolve with RO water, and transfer into a 40-L reservoir: $NH_4H_2PO_4$: 1.15 g, $Ca(NO_3)_2{\cdot}4H_2O$: 6.61 g, $MgSO_4{\cdot}7H_2O$: 4.91 g, and KNO_3: 6.07 g. Add the following Hoagland's micronutrients volumes to the reservoir from the respective stock solutions: $CuSO_4{\cdot}5H_2O$: 10 mL (0.075 g/L stock), $ZnSO_4{\cdot}7H_2O$: 10 mL (0.23 g/L stock), H_3BO_3: 10 mL (2.844 g/L stock), $MnCl_2{\cdot}4H_2O$: 10 mL (1.01 g/L stock), and MoO_3: 12.5 mL (0.016 g/L stock). Fill the reservoir to the 40-L mark with RO water. Caution: Keep the micronutrient stock solutions in the dark at 4 °C when not being used. Also, do not add Fe-HBED to the 40 L 0.25× Hoagland's solution. The iron in Hoagland's solution is replaced by 10 μM Fe-HBED (Make 1 L of 10 μM Fe-HBED by adding 1 mL 10 mM Fe-HBED into a 1-L graduated plastic beaker and dilute to the mark with 0.25× Hoagland's Mo type 2) and is added as part of the watering solution during fertilization of plants.
9. Garden (plastic) sprayers (1,000 mL).
10. Plastic beakers (1,000 mL).

2.3. Harvesting and Drying

1. Harvesting set—Surgical knife (No. 10), plastic forceps (or tweezers), plastic rod (Teflon), beaker (50- or 100-mL filled with DI water), labeled Pyrex digestion tube sets (see Note 1) for plant tissues.
2. Latex gloves.
3. Drying oven.
4. Desiccators.

5. Plastic wrapping films.
6. Storage cabinets for dried plant tissues.

2.4. Analysis of Samples

1. Analytical balance (5 decimal places and connected to computer).
2. Digital dry block heater systems (should have total capacity for 288 sample digestion).
3. Dispensette (1–10 mL) pipettors for dispensing acid and DI (distilled deionized) water.
4. Nitric acid (OmniTrace, VWR Scientific Products or equivalent grade).
5. Distilled deionized water (≥18 MΩ cm) source.
6. Multichannel pipettes (1,000- or 1,200-μL), including adjustable ones, and tips.
7. Deep-well plates (96-well).
8. Pyrex digestion tubes (16 × 100 mm).
9. ICP-MS (inductively coupled plasma-mass spectrometry) instrument.
10. Autosampler (e.g., SC-2, Elemental Scientific, Inc. (ESI), Omaha, NE, USA).
11. Custom-made 60 rod mixing rack (or stir-rod array). We make this from 60-place 5 × 12 plastic racks, plastic rods (polypropylene), rubber stoppers, and tapes.

2.5. Workflow Management and Data Storage

1. Local implementation of the open source web-based Purdue Ionomics Information Management System (PiiMS) to allow scheduling and tracking of samples as well as sample submission to the pipeline, and generation of automated reminders.

2.6. Other Organisms

1. Rice grown in the field and obtained shipments as dehulled grains.
2. Pyrex digestion tube sets (see Subheading 2.4 above for analysis materials).
3. Yeast (*Saccharomyces cerevisiae*) libraries stocks obtained as cultures (in glycerol) in 96-well micro-plates.
4. Freezer (−80 °C) for storage of yeast stocks.
5. Shaker incubators (custom fitted to accommodate 15 deep-well plates) for yeast cultivation.
6. Multichannel pipettes (20-, 300-, and 1,000 μL) and tips.
7. Deep-well plates (96-well, 2 mL) for growing yeast and subsequent processing.
8. Synthetic defined minimal media for yeast cultivation.
9. Adhesive breathable sealing film (e.g., AeroSeal, Dot Scientific, Inc., Burton, MI, USA).

10. Axymat (chemically resistant and flexible lids, Axygen Scientific, Union City, CA, USA).
11. AcroPrep® 96-well filter plates (PVDF (polyvinylidene fluoride) membrane, 0.45 μm, 350 μL, Pall Life Sciences, Ann Arbor, MI, USA).
12. Vacuum filtration manifold (Pall Life Sciences).
13. Clear View micro-plate for optical density measurement.
14. Plate reader (e.g., OpsysMR, DYNEX Technologies, Chantilly, VA, USA).
15. Liquid handling robot (e.g., MultiPROBE II PLUS HT EX, PerkinElmer, Waltham, MA, USA).
16. ICP-MS with autosampler and desolvation sample introduction system (Apex, ESI).

3. Methods

Unless otherwise stated, *A. thaliana* is grown for up to 5 weeks for ICP-MS analysis in a Growth Room facility under these conditions. Temperature: 18–22 °C; Humidity: 15–75 % (Varies depending on season; not controlled but monitored); Day length: 10 h; Light intensity: 90.0 $\mu E/m^2$; and Fertilization: 0.25× Hoagland's Mo Type 2 with 1 mL/L Fe-HBED.

3.1. *A. thaliana* Cultivation and Harvesting

1. Make soil for plant cultivation as follows: Pour a bag of Premier PROMIX PGX soil into the Soil mixer. Cover the opening of the mixer with plastic and tape in place. Cut a 6-cm hole in the center of the cover to spray through. This step is to reduce dust.
2. Add 500 mL element mix solutions to the soil in the mixer and mix thoroughly (A better way to add the solutions is to pour into garden sprayers and then spray onto the soil as the mixing process proceeds).
3. Add more water as necessary during the mixing process to bring the soil to the desired moisture level.
4. Collect soil into storage container, cover with lid and properly label, and store in Growth room.
5. Prepare the pots/trays a day in advance before planting (sowing) as follows: Fill the pots/trays (using plastic scoop) with soil and settle the soil by giving a sharp tap and compress very lightly to give a firm bed (Fig. 1a). Place the pots/tray inside a drip tray (Fig. 1b) and pour 2 L of RO water into this drip tray; also spray water on the soil surface then cover the moistened soil with another drip tray (Fig. 1c). Leave this setup for 24 h.

Fig. 1. (**a**) 20-row seed flat tray filled with soil; (**b**) Drip (display) tray into which the 20-row soil tray is placed; (**c**) Prepared soil tray covered with the drip tray. (**d**) Metal racks with three levels of shelves in the growth room.

6. Dump any residual water from the drip tray on the day of planting. Place stickers with *A. thaliana* line names or order numbers on corresponding pots/rows, and also place tray number labels and other relevant information on tray (see Fig. 2a). Then sow the seeds onto the surface of the soil by scattering them carefully from a piece of paper.
7. Next, pour 1 L of watering solution (0.25× Hoagland's Mo type 2 plus 1 mL/L Fe-HBED) into drip tray as before.
8. Transfer the prepared trays to a cold (4 °C) dark room (or refrigerator) for 2 days for stratification. This improves and synchronizes germination.
9. After stratification, transfer the trays to metal racks with three levels of shelves in the growth room (see Fig. 1d).
10. Water the trays from below (use 1 L per tray) by pouring the watering solution in the drip trays as before (i.e., bottom watering) twice per week (We do this on Monday and Friday). Note that enough solution is added so that it all gets soaked up, so there is none left to dump. And when the room humidity is high (above 70 %) the watering is done once instead of twice per week.
11. Harvest plants 5 weeks after planting following these procedures: Take a picture of the tray to be harvested, ensuring that all tray

Fig. 2. (**a**) *A. thaliana* tray with line names or order numbers on corresponding pots/rows and tray label. (**b**) Harvesting set. Consists of Pyrex digestion tube set, beaker of DI water, surgical knife, plastic tweezers, Teflon rod, and harvest notes.

labels show clearly. Caution: Wear gloves to protect samples from contamination and use only plastic tweezers to ensure that no metal contaminants are propagated.

12. Harvest two healthy leaves from one plant by cutting with a scalpel while holding with tweezers. Do not use metal scissors. Use a harvesting form to record any observations about the plants (see Fig. 2b).
13. Rinse the leaves with DI water in the 50- or 100-mL beaker by gentle shaking.
14. Place rinsed tissue sample in the appropriate labeled glass tube (see Fig. 2b) using the tweezers.
15. Shift sample to the bottom of the glass tube with the Teflon rod, making sure no tissues are left on the inner surface (wall) of the tube.
16. Transfer the tube sets of harvested samples into an oven set at 92 °C for overnight (or 20 h) drying. Note that a tube set comprises 108 samples (corresponding to one *A. thaliana* tray) each in a tube and are placed in two 5 × 12 plastic racks along with ten tubes of calibration standards and blanks. They are kept together and treated under the same conditions (see Note 1).
17. Remove the dried sample tube sets from the oven; let them cool in desiccators for about an hour. Weigh out the appropriate number of sample tubes on the analytical balance (Our laboratory usually weighs seven to ten samples and performs a "weight calculations" for the rest of the samples; see Note 2).
18. Wrap up the tube sets with plastic film and place in storage cabinet for later processing and analysis.

3.2. Sample Processing and Analysis

1. Take out sample tube sets from storage cabinet and add concentrated nitric acid to each tube. Use a dispensette (5 mL)

pipettor fitted to the acid bottle (2.5 L) (or preferably a multichannel pipette for better precision) to dispense the acid into the tubes. The volume of acid used depends on plant tissue type and mass, e.g., 0.7 mL of acid for a 5-mg *A. thaliana* dry leave sample. Note that we usually add the internal standard (usually indium) to the acid bottle and mix thoroughly; thus, the internal standard is included in the digestion stage through to the analysis on the ICP-MS.

2. Transfer the tubes to the dry block heaters. These are set to 110 °C. Start the digestion which should take 4 h. The digestion stage can also be performed overnight. Dry block heaters can be connected to timers that are set so that the digestion will be completed and the tubes cooled before the beginning of the work day the following morning.
3. After cooling dilute the sample digests with DI water following the same procedure used for addition of the acid. The dispensette (10 mL) is attached to a 10-L Nalgene (or similar) bottle (carboy) with the DI water. Dispense the appropriate volume such that the diluted sample solution is 4–12 mL final volume, depending on the sample size and amount of acid used.
4. Mix thoroughly using the custom-made stir-rod array (see Fig. 3). Dip this back into 10 % nitric acid and rinse in between different racks of tubes.
5. The diluted samples can be analyzed on the ICP-MS system in the same tube set as digested in or samples can be transferred into 96-well deep-well plates using an adjustable multichannel pipette. Using 96-well plates allows analysis of up to 384 samples on an ESI SC-2 autosampler without the need to change tube sets. Analyze the samples together with the calibration standards and blanks on the ICP-MS for the suite of elements to be monitored (see Note 4).
6. Import the generated data into Microsoft Excel (or similar spreadsheet) for further processing (see Notes 3 and 4).

3.3. Workflow and Data Management

To allow the management of the ionomics workflow and the formalized collection and storage of metadata and data we developed the Purdue Ionomics Information Management System. This system is accessible at http://www.ionomicshub.org with an in-depth description of the system by Baxter et al. (26). Briefly, the user interface for the system divides functionalities into eLaboratory, eManagement, Data Search/Advance Search, and Order Form. Some of the portals are elaborated on below.

1. Customer Submission: This is where customers wishing to submit lines for ionomics analysis must create an account. Thereafter, use the "Order Form" portal to input critical information about each line to be submitted. Note that the Arabidopsis database accessible from the http://www.ionom

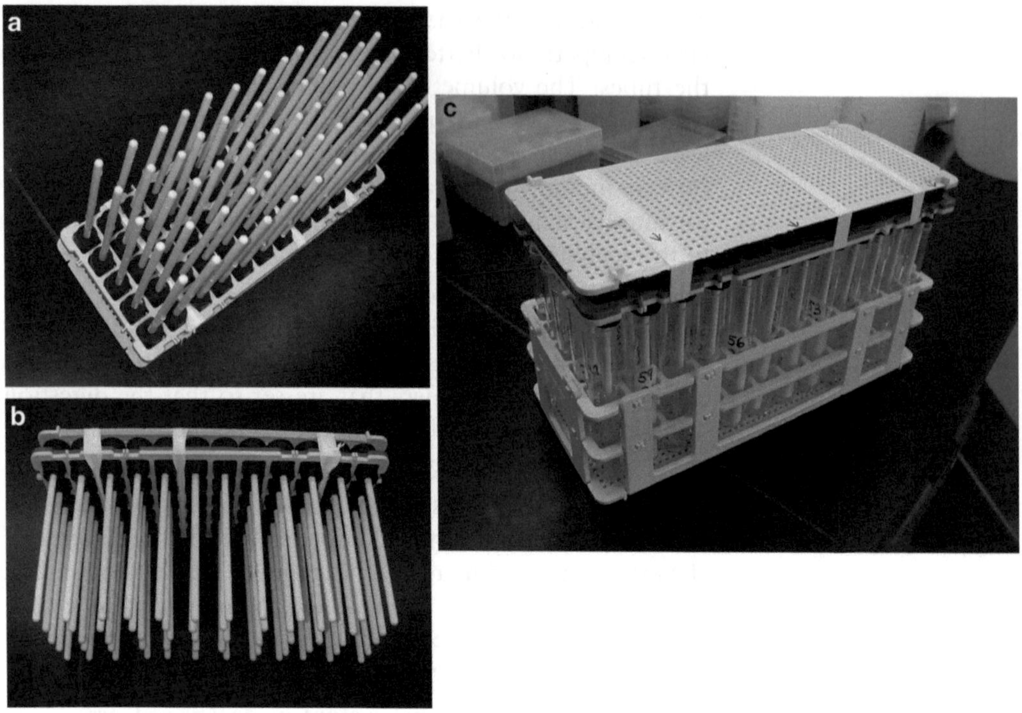

Fig. 3. Custom-made stir-rod array used for thoroughly mixing diluted samples before analysis. (**a**) Top view. (**b**) Side view. (**c**) Stir-rod array dipped in 10 % nitric acid in tube set during storage.

icshub.org welcome page has downloadable PDF tutorial on submitting seed and searching the database.

2. Planting stage: At this stage the type of pot/tray configuration being used in the experiment is chosen, and the placement of each line to be planted is defined. Only lines that have been previously defined in the system can be chosen at this stage. This is also where tray-specific metadata such as soil batch, day length, light intensity, temperature, and humidity, watering solution, date planted, customer identity, and other pertinent planting notes are entered (see Fig. 4b, c).
3. Harvesting Stage: Leaf samples are harvested from plants at the end of the defined growth period into digestion tube sets following a predefined template that maps tubes to specific positions in the pot/tray being harvested (see Fig. 2 for tube sets and Fig. 5 for harvesting stage portal). Metadata about the harvesting stage are also collected, including type of tissue harvested, harvest date, notes on tube set used, (check box if the tube is empty), and general comments about each plant harvested such as its visual appearance. At this stage a photograph of the pot/tray being harvested can also be unloaded and associated with the experiment.
4. Drying Stage: Information pertaining to sample drying is captured into this portal (Fig. 6). This includes oven temperature,

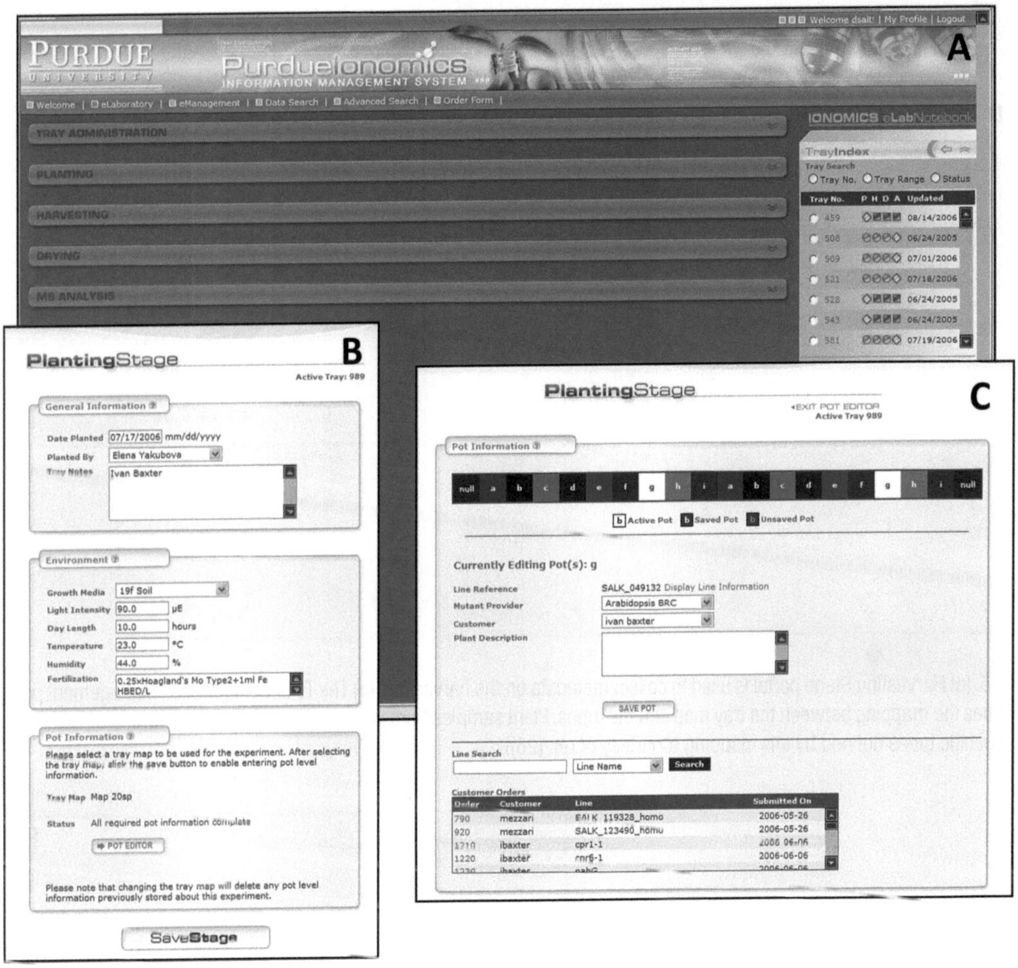

Fig. 4. (**a**) eLaboratory portal used to control workflow in PiiMS with examples of modules accessed through this portal to collect metadata during the planting stage. (**b**) Modules used to collect experimental level information. (**c**) Information that defines each line and its physical location in the planting tray (Courtesy of ref. (26)).

the duration of drying and the type of acid, and volume used in the digestion. Weights on predefined samples are also added at this stage, and this information is used in the ICP-MS analysis stage to calculate weights for the rest of the sample set (see Notes 3 and 4 for weight calculation and connecting different runs).

5. ICP-MS Analysis Stage: This is the stage where data from the ICP-MS analysis are uploaded, processed, and stored in the database. Other relevant metadata such as the date when the samples were analyzed, equipment used, calibration standards used and sample volume are also captured at this stage. Within the eManagement portal tools are provided to define the list of elements that are monitored during the ICP-MS analysis. ICP-MS data can be uploaded as unnormalized solution concentrations and then normalized to weight and background lines using algorithms built into the system (see Fig. 7); alternatively,

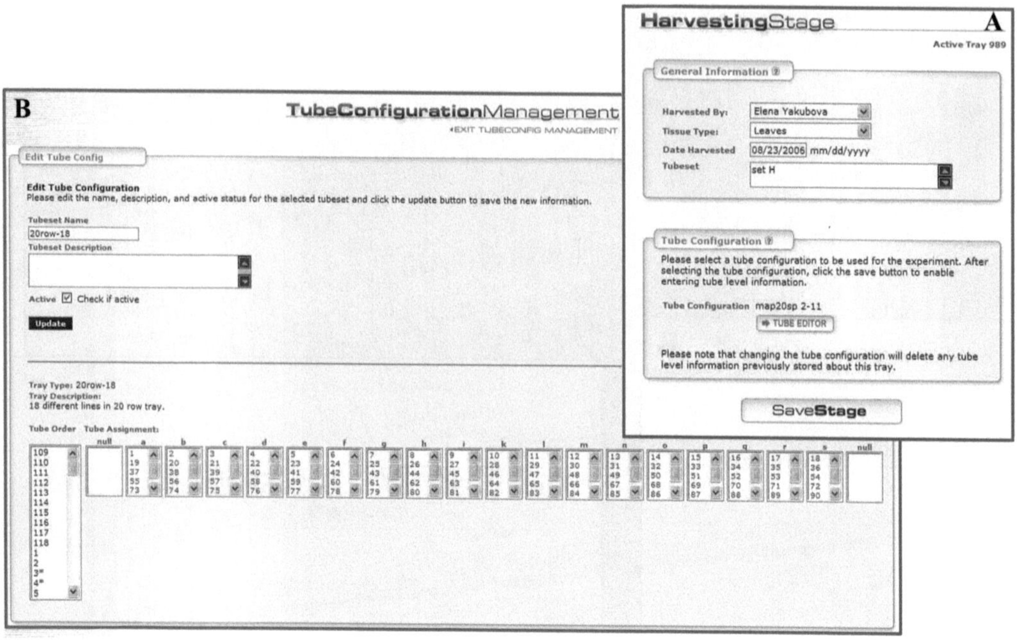

Fig. 5. (**a**) Harvesting Stage portal is used to collect metadata on the harvesting. (**b**) The Tube Configuration Management portal defines the mapping between the tray map and the tubes. Plant samples from each cell in the tray map (a, b, c, etc.) are placed in specific tubes defined by this mapping (Courtesy of ref. (26)).

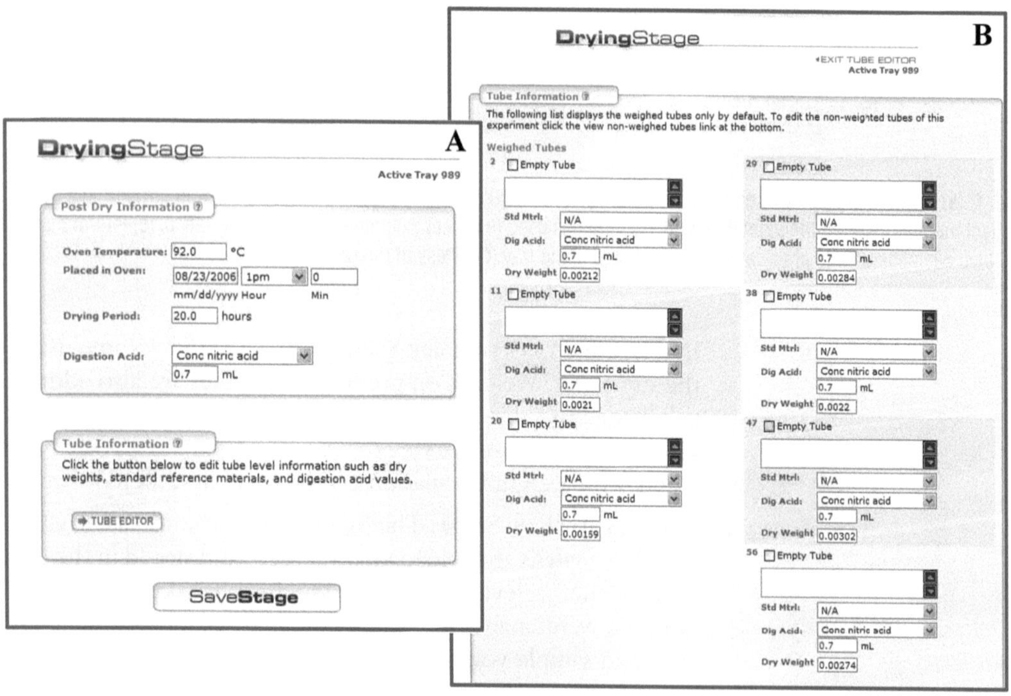

Fig. 6. Drying Stage portal is designed to capture both experiment level (**a**) and sample level (**b**) metadata about the drying process, including sample weight and the type of acid used in the digestion process (Courtesy of ref. (26)).

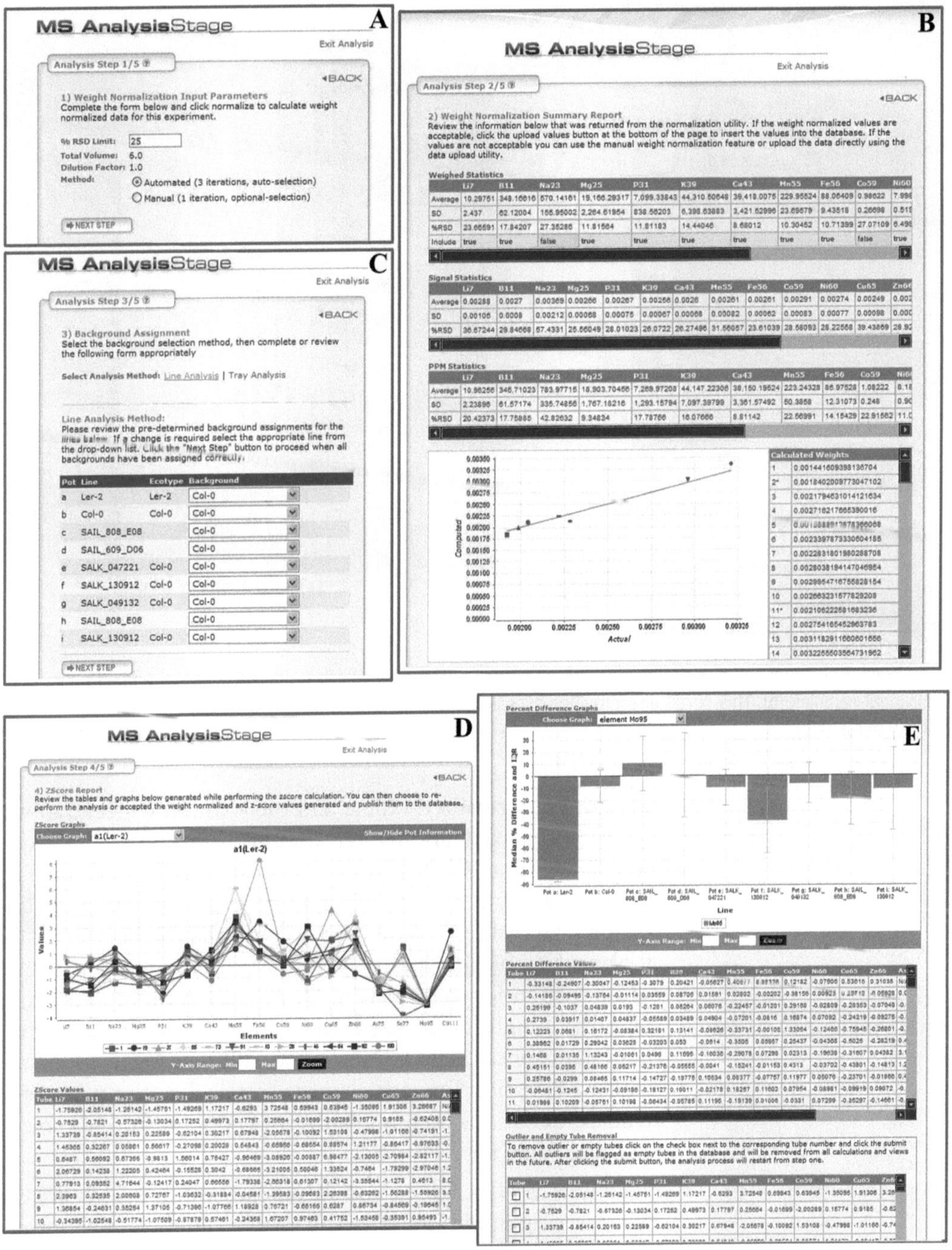

Fig. 7. Data normalization tools are accessed through the MS Analysis Stage. Data normalization is divided into five stages including data upload. These stages include input of the relative standard deviation used for data weight normalization (**a**), a summary report of the output of the weight normalization algorithm (**b**), selection of the background lines for calculation of background-normalized data (**c**), and summary reports of *z*-score (**d**), and percentage change (**e**) calculations (Courtesy of ref. (26)).

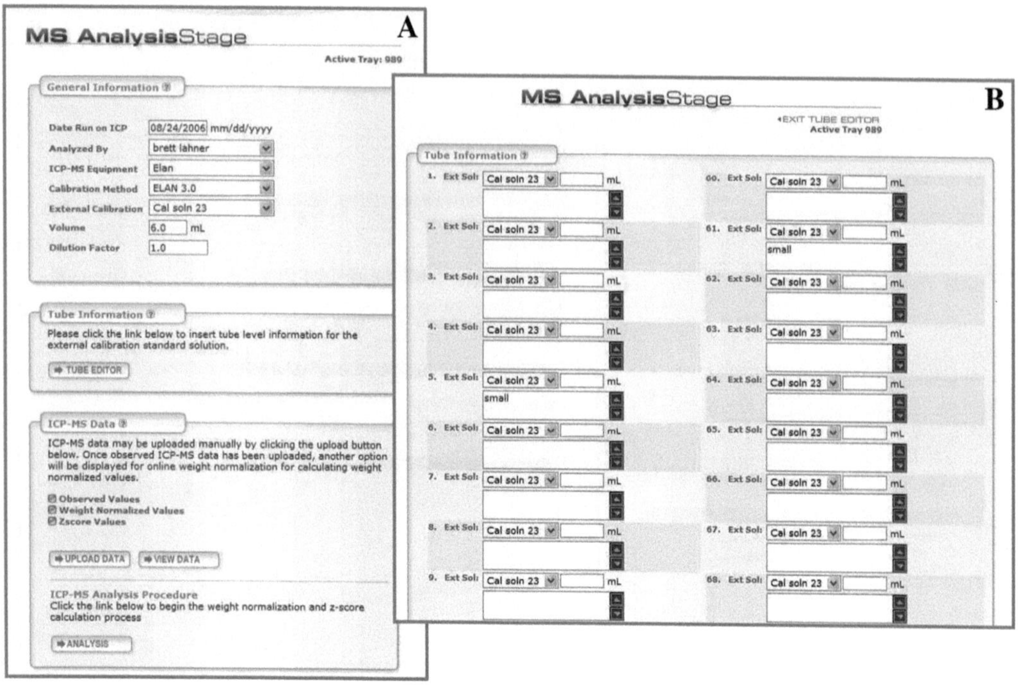

Fig. 8. MS Analysis Stage portal is designed to collect metadata on the ICP-MS analysis processes (**a**) and also comments of individual samples (**b**). This portal also provides access to ICP-MS data upload, viewing and normalization tools (**a**) (Courtesy of ref. (26)).

data can be normalized offline and the solution concentrations, weight-normalized, and background-normalized data uploaded individually (Fig. 8). The weight normalization is performed by dividing the solution concentrations by the sample weights (calculated or measured) taking into account the dilution factor for all plants (samples) in a tray. Next, calculate the mean and standard deviation for each element for all plants (samples) in a tray using the weight-normalized values. The background-normalized values are calculated by using the wild-type plants mean and standard deviation for each element to calculate the number of standard deviations each mutant plant is distant from the mean wild-type value for each element.

6. Data Search: Data Search and Advanced Search portals are the two querying modes the database system supports. The former is accessible to all users to search, view, and download data from their completed sample submissions along with all other publicly accessible data (see Fig. 9). The Advanced Search mode allows for more extensive queries and is accessible to registered users of the system.

7. Display and Download: The system provides various tools to summarize, visualize, and download data (see Fig. 10). These include a summary table for a selected line showing the average

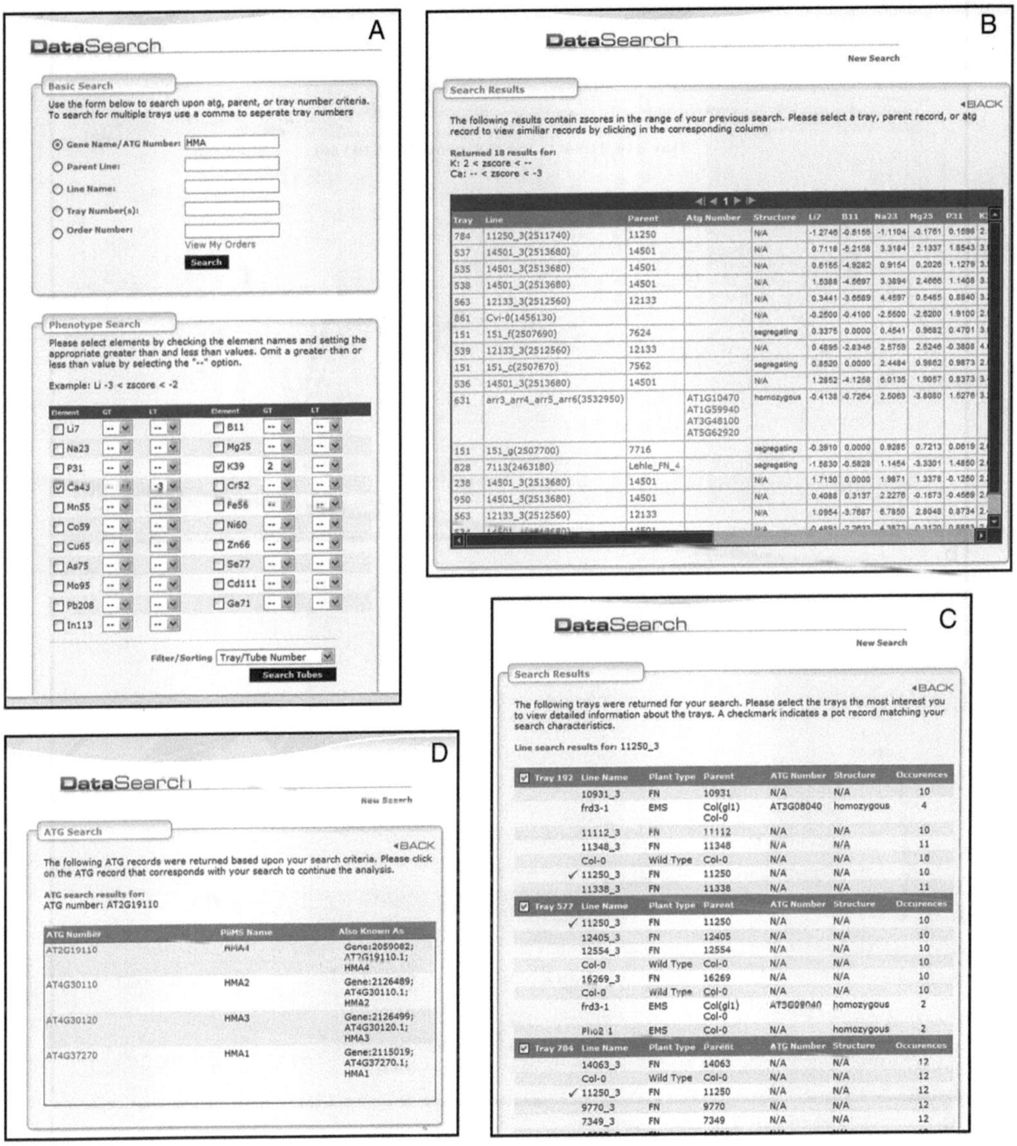

Fig. 9. Modules accessed through the Data Search portal. These include the primary search interface (**a**), the output of searches based on ionomic phenotype (low Ca, high K; **b**) or gene name (NHX; **d**), and the list of experiments the lines returned from the phenotype search (**c**) (Courtesy of ref. (26)).

weight-normalized data together with the line used as a reference (e.g., wild-type background), percentage difference between selected line and reference, and *P* values. Tools also enable the display of results as plots of the *z*-score data (Fig. 10a) or percentage change from the median value of the background line for each element, with error bars for the Interquartile Range (Fig. 10b) for selected lines. Lastly, data formatted as a comma-delimited value (CSV) text file can be downloaded (Fig. 10c); also, a formatted summary report containing data,

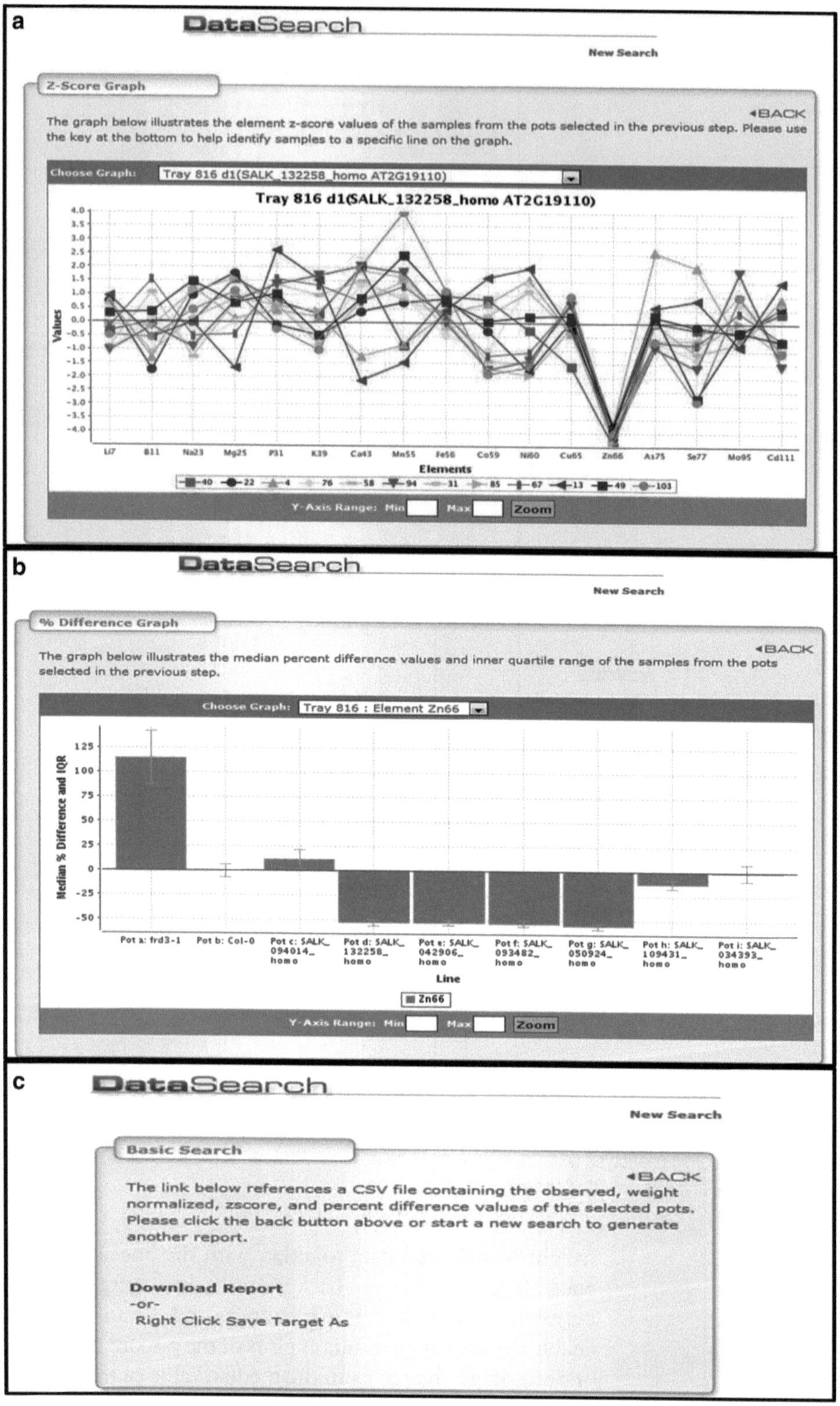

Fig. 10. Visualization of selected ionomics data on the *HMA4* mutants as a plot of *z*-scores (number of standard deviations away from the mean of the wild-type grown in the same experiment) (**a**); and median percentage change from the wild-type for each element analyzed (**b**). Data can also be downloaded as a CSV file for further processing on a local machine or summarized in PDF file (**c**) (Courtesy of ref. (26)).

summary tables, and plots in portable document format (PDF) can be downloaded. A picture of the plants taken at harvest can also be viewed and is included in the summary report.

3.4. Overview of Other Organisms

We have also conducted high-throughput ionomics studies on other organisms. Here we present a brief overview on two of these: rice and yeast. An in-depth description of the yeast methodology is presented in Danku et al. (27).

1. *Rice*: We obtain the rice as grains via shipment from collaborators. They are grown in fields with and without flooding, harvested, dehulled, and then sent in packets. Also included in the shipment are files for the various lines and field descriptions.
2. Upload the formatted master file into the Field Planting Portal of the Rice PiiMS (see http://www.ionomicshub.org/rice) whereby label IDs are automatically generated. Add labels to the packets to help track the lines as they go through the ionomics pipeline.
3. Next, split the rice shipment into runs and randomize.
4. Prepare rice for digestion and subsequent analysis by adding grains (three per tube) from packets into the digestion tube sets. A run is usually made up of tube sets of 288 16 × 100 mm Pyrex tubes.
5. Add 1.0 mL concentrated nitric acid (with 20 ppb indium internal standard) to each tube and place the tube sets in the fume hood overnight to predigest at room temperature.
6. Transfer the tubes to the dry block heaters and digest for about 5 h at 110 °C (see Note 5).
7. Remove tubes from the block heaters, dilute to 10 mL with DI water, and mix thoroughly with the stir-rod array (see Fig. 3).
8. Run the samples on the ICP-MS either from the tubes directly or transfer via adjustable multichannel pipette to 96-well deep-well plates (three) and then run together with external calibration standards and blanks.
9. Upload the data to the Rice PiiMS database for normalization to calculated weights (see Notes 3 and 4), statistics, and data sharing.
10. *Yeast*: Three different yeast libraries (Knockout, Overexpression and Essential Heterozygous Diploid collections, Open Biosystems, Huntsville, AL, USA) have been analyzed in our facility and the data deposited in the Yeast PiiMS (see http://www.ionomicshub.org/yeast).
11. Yeast cells are grown in two stages. A pre-growth stage involving the growing (bulking) up of stock cultures to allow the

generation of replicates. The efficiency of this process can be enhanced using a liquid handling robot. Minimal defined growth medium in 96-well deep-well plates is inoculated (5 μL inoculum per 500 μL medium per well) and incubated for 48 h at 30 °C with shaking at 400 rpm. Plates are sealed with adhesive breathable film. Next, replicates are prepared using a liquid handling robot (10 or 20 μL inoculum per 750 μL medium per well). Three 96-well plates of replicated cultures are then cultured for 36 h as above (see Note 6).

12. Yeast cultures are prepared for ICP-MS analysis using AcroPrep® 96 PVDF filter membrane micro-plates. Using multichannel pipettes, yeast cultures are transferred from the 96-well deep-well plate (200 μL/well) to the AcroPrep plate whose membrane had been wetted with methanol and rinsed (300 μL/well and 400 μL/well, respectively). Concurrently, the same amount of yeast cultures is transferred into clear view microtiter plate from the same deep-well plate for optical density (OD) measurement using a plate reader. Yeast cells in the filter plate are washed with EDTA solution (350 μL/well, 1 μM pH 8) and rinsed with DI water (350 μL/well). The wash and rinse step is repeated four times. A vacuum filtration manifold is used. The filter membrane with the cleaned yeast cells is dried in an oven at 88 °C for 150 min. Thereafter, concentrated nitric acid with gallium internal standard (100 μL/well) is added to the filter plate, covered with a polypropylene lid and the samples digested using block heaters (88 °C for 40 min; this plate is not placed directly in contact with the block floor but rather on top of another (used) plate). Next, the digested samples are transferred into a 96-well deep-well collection plate using vacuum manifold. The final dilution volume is 1.6 mL including 0.005 % Triton X-100 (see Note 7). The plate is covered with a chemically resistant lid (Axymat) and the sample solutions thoroughly mixed.
13. Prepared samples in 96-well deep-well plates are then analyzed by ICP-MS (using external calibration method) using an ESI autosampler and Apex sample introduction system.
14. ICP-MS generated solution concentration data are imported into spreadsheet software for further processing. OD values can be used as a normalization factor, and the data uploaded into yeast PiiMS. Further statistical analysis of the complete data set can be performed to extract relevant biological information, for example, following the linear mixed-effect model approach of Yu et al. (28). After appropriate statistical treatment, complete datasets can be re-uploaded into yeast PiiMS for storage and dissemination to the broader community.

4. Notes

1. A tube set consists of 108 sample tubes and ten calibration standards and blank tubes. The Pyrex digestion tubes (16×100 mm) are labeled from 1 through 108 using permanent marker (We use PYREX® Brand Culture Tubes, Reusable, Borosil, 16×100). Alphabet designation is used to differentiate tube sets. The tubes are put into two 60-place 5×12 plastic racks (e.g., No Wire Rack, Polypropylene, 16 mm test tube, Bel-Art Products, Pequannock, NJ, USA). Tubes are conditioned by performing nitric acid digestion using the same procedure for sample digestion. This is followed by washing with DI water and drying. The process is performed three times. The treatment reduces considerably leaching of sodium and boron during subsequent usage. The tube set is then ready for use.
2. For further discussion of the Arabidopsis high-throughput method including the set of elements monitored and explanation of the "weight calculations" see Lahner et al. (2). The weight calculation explanation and demonstration can also be accessed (i.e., downloaded) from the ionomics home page (http://www.ionomicshub.org) under the Education drop-down menu. Select the "How-To" option to get there.
3. Weight Calculation: Primarily as a time-saving measure, and also to reduce indeterminate errors, sample weights can be calculated based on the combined concentrations of all the measured elements. As a basis, seven to nine samples are weighed, and their averaged dry weight concentrations become the standard which is used to calculate all the other weights. In practice only the best elements, those for which the relative standard deviation of the basis samples is below a set number are used. Elements may also be excluded if they are known to change a lot in the given sample set, even if the basis samples do not show this variance. Indeterminate (random) error may be reduced using this technique. This advantage arises when the sample weights are small and the weighing error exceeds the weight calculation error. On our 5-place Mettler Toledo balance, this crossover point lies between about 3–5 mg. A useful extension of this idea is applicable to samples that are too small to be readily weighed, by pooling enough of these samples to make one or more basis samples. Care must be taken when using this weight calculation technique. The basic assumption is that all the samples are identical in composition, which is only at best approximately true. In cases where a group of samples varies systematically from the basis samples (and are not represented among the basis samples) determinate error may occur. In some situations, such as when the medium is

different between the calculated-weight samples and the basis samples or different organs from a single organism are compared, this technique is clearly inappropriate. At other times, as when comparing mutants and wild-types, the error introduced is well below the signal and the method works quite well.

4. Connecting Groups of Samples That Have Been Run at Different Times: Sample sets that have been run at different times can be connected as an extension of the single-run drift correction. The way we have chosen to deal with both drift correction and run-to-run connection is by pooling a number of the digested/diluted samples and then running this "pooled standard" periodically throughout the group of runs we wish to connect. In our case we run the pooled standard right after the calibration, and then after every nine samples. Linear interpolation between each pair of pooled standards is used to generate a theoretical standard for each sample, which is used to correct for drift by simple proportion to (typically) the first pooled standard in the first run. The reason for using a pool of samples as a standard, rather than say an NIST (National Institute of Standards and Technology) calibration standard, is that the pooled standard matches the chemical matrix of the samples as closely as possible, thereby emulating the sample drift. Note that by using a proportional correction we are making an adjustment for the slope of the calibration curve but not correcting for an actual baseline shift.
5. We generally ramp up the temperature in stages instead of just letting it rise to the final set point without any interruption. For example, we set the temperature to 70 °C to start and let it settle there for about 30 min. Next, it is raised to 90 °C for another 30 min and finally to 110 °C for the rest of the duration. This way the digestion doesn't run out of control.
6. The cultivation conditions are slightly different for the overexpression collection. The first growth stage (I) involves normal growth of yeast for 48 h with SC-ura/Raffinose (500 μL/well) medium. Thereafter, 3× SC-ura/Galactose (250 μL/well) medium is added and grown for 12 h to induce protein expression (growth stage II) (29). Note that all the growth media are supplemented with all or some of these elements: Na, Co, Ni, Mo, Cd, As, Se.
7. The PVDF filter membrane is hydrophobic; wetting with methanol transforms it to hydrophilic status enabling filtration of yeast culture. Drying the membrane restores the hydrophobicity which is required for subsequent nitric acid digestion of yeast cells. Addition of Triton X-100 to the final sample solutions as well as the calibration standards reduces surface tension and enables smoother self aspiration of the PFA micro-nebulizer of the Apex sample introduction system.

Acknowledgment

Work in the Salt lab is support by the National Institutes of Health and the National Science Foundation.

References

1. Salt DE, Baxter I, Lahner B (2008) Ionomics and the study of the plant ionome. Annu Rev Plant Biol 59:709–733
2. Lahner B, Gong J, Mahmoudian M, Smith EL, Abid KB, Rogers EE, Guerinot ML, Harper JF, Ward JM, McIntyre L, Schroeder JI, Salt DE (2003) Genomic scale profiling of nutrient and trace elements in *Arabidopsis thaliana*. Nat Biotechnol 21:1215–1221
3. Chen Z, Watanabe T, Shinano T, Okazaki K, Osaki M (2009) Rapid characterization of plant nutrients with an altered ion-profile: a case study using *Lotus japonicus*. New Phytol 181:795–801
4. Norton GJ, Deacon CM, Xiong L, Huang S, Meharg AA, Price AH (2010) Genetic mapping of the rice ionome in leaves and grain: identification of QTLs for 17 elements including arsenic, cadmium, iron and selenium. Plant Soil 329:139–153
5. Sánchez-Rodríguez E, Rubio-Wilhelmi MM, Cervilla LM, Blasco B, Rios JJ, Leyva R, Romero L, Ruiz JM (2010) Study of the ionome and uptake fluxes in cherry tomato plants under moderate water stress conditions. Plant Soil 335:339–347
6. Eide DJ, Clark S, Nair TM, Gehl M, Gribskov M, Guerinot ML, Harper JF (2005) Characterization of the yeast ionome: a genome-wide analysis of nutrient mineral and trace element homeostasis in *Saccharomyces cerevisiae*. Genome Biol 6:R77
7. Baxter I, Hosmani PS, Rus A, Lahner B, Borevitz JO, Muthukumar B, Mickelbart MV, Schreiber L, Franke RB, Salt DE (2009) Root suberin forms an extracellular barrier that affects water relations and mineral nutrition in Arabidopsis. PLoS Genet 5:e1000492
8. Chao DY, Gable K, Chen M, Baxter I, Dietrich CR, Cahoon EB, Guerinot ML, Lahner B, Lü S, Markham JE, Morrissey J, Han G, Gupta SD, Harmon JM, Jaworski JG, Dunn TM, Salt DE (2011) Sphingolipids in the root play an important role in regulating the leaf ionome in *Arabidopsis thaliana*. Plant Cell 23:1061–1081
9. Tian H, Baxter IR, Lahner B, Reinders A, Salt DE, Ward JM (2010) Arabidopsis NPCC6/NaKR1 is a phloem mobile metal binding protein necessary for phloem function and root meristem maintenance. Plant Cell 22:3963–3979
10. Borghi M, Rus A, Salt DE (2011) Loss-of-function of constitutive expresser of pathogenesis related genes5 affects potassium homeostasis in *Arabidopsis thaliana*. PLoS One. 6:e26360
11. Mason MG, Jha D, Salt DE, Tester M, Hill K, Kieber JJ, Schaller GE (2010) Type-B response regulators ARR1 and ARR12 regulate expression of AtHKT1;1 and accumulation of sodium in Arabidopsis shoots. Plant J 64:753–763
12. Rus A, Baxter I, Muthukumar B, Gustin J, Lahner B, Yakubova E, Salt DE (2006) Natural variants of AtHKT1 enhance Na^+ accumulation in two wild populations of Arabidopsis. PLoS Genet 2:e210
13. Baxter I, Muthukumar B, Park HC, Buchner P, Lahner B, Danku J, Zhao K, Lee J, Hawkesford MJ, Guerinot ML, Salt DE (2008) Variation in molybdenum content across broadly distributed populations of *Arabidopsis thaliana* is controlled by a mitochondrial molybdenum transporter (MOT1). PLoS Genet 4:e1000004
14. Morrissey J, Baxter IR, Lee J, Li L, Lahner B, Grotz N, Kaplan J, Salt DE, Guerinot ML (2009) The ferroportin metal efflux proteins function in iron and cobalt homeostasis in Arabidopsis. Plant Cell 21:3326–3338
15. Kobayashi Y, Kuroda K, Kimura K, Southron-Francis LJ, Furuzawa A, Kimura K, Iuchi S, Kobayashi M, Taylor GJ, Koyama H (2008) Amino acid polymorphisms in strictly conserved domains of a P-type ATPase HMA5 are involved in the mechanism of copper tolerance variation in Arabidopsis. Plant Physiol 148:969–980
16. Loudet O, Saliba-Colombani V, Camilleri C, Calenge F, Gaudon V, Koprivova A, North KA, Kopriva S, Daniel-Vedele F (2007) Natural variation for sulfate content in *Arabidopsis thaliana* is highly controlled by APR2. Nat Genet 39:896–900
17. Baxter I, Brazelton JN, Yu D, Huang YS, Lahner B, Yakubova E, Li Y, Bergelson J, Borevitz JO, Nordborg M, Vitek O, Salt DE

(2010) A coastal cline in sodium accumulation in *Arabidopsis thaliana* is driven by natural variation of the sodium transporter AtHKT1;1. PLoS Genet 6:e1001193

18. Prinzenberg AE, Barbier H, Salt DE, Stich B, Reymond M (2010) Relationships between growth, growth response to nutrient supply, and ion content using a recombinant inbred line population in Arabidopsis. Plant Physiol 154:1361–1371
19. Marschner H (1995) Mineral nutrition of higher plants, 2nd edn. Academic, London
20. Rausch C, Bucher M (2002) Molecular mechanisms of phosphate transport in plants. Planta 216:23–37
21. Sanders D, Pelloux J, Brownless C, Harper JF (2002) Calcium at the crossroads of signaling. Plant Cell 14:S410–S417
22. Curie C, Briat J-F (2003) Iron transport and signaling in plants. Annu Rev Plant Biol 54: 183–206
23. Véry A-A, Sentenac H (2003) Molecular mechanisms and regulation of K^+ transport in higher plants. Annu Rev Plant Biol 54:575–603
24. Zhu J-K (2003) Regulation of ion homeostasis under salt stress. Curr Opin Plant Biol 6: 441–445
25. Baxter IR, Vitek O, Lahner B, Muthukumar B, Borghi M, Morrissey J, Guerinot ML, Salt DE (2008) The leaf ionome as a multivariable system to detect a plant's physiological status. Proc Natl Acad Sci USA 105: 12081–12086
26. Baxter I, Ouzzani M, Orcun S, Kennedy B, Jandhyala SS, Salt DE (2007) Purdue ionomics information management system. An integrated functional genomics platform. Plant Physiol 143:600–611
27. Danku JMC, Gumaelius L, Baxter I, Salt DE (2009) A high-throughput method for *Saccharomyces cerevisiae* (yeast) ionomics. J Anal At Spectrom 24:103–107
28. Yu D, Danku J, Baxter I, Kim S, Vatamaniuk OK, Salt DE, Vitek O (2011) Noise reduction in genome-wide perturbation screens using linear mixed-effect models. Bioinformatics. 27:2173–2180
29. Gelperin DM, White MA, Wilkinson ML, Kon Y, Kung LA, Wise KJ, Lopez-Hoyo N, Jiang L, Piccirillo S, Yu H, Gerstein M, Dumont ME, Phizicky EM, Snyder M, Grayhack EJ (2005) Biochemical and genetic analysis of the yeast proteome with a movable ORF collection. Genes Dev 19:2816–2826

Chapter 18

Applications of High-Throughput Plant Phenotyping to Study Nutrient Use Efficiency

Bettina Berger, Bas de Regt, and Mark Tester

Abstract

Remote sensing and spectral reflectance measurements of plants has long been used to assess the growth and nutrient status of plants in a noninvasive manner. With improved imaging and computer technologies, these approaches can now be used at high-throughput for more extensive physiological and genetic studies. Here, we present an example of how high-throughput imaging can be used to study the growth of plants exposed to different nutrient levels. In addition, the color of the leaves can be used to estimate leaf chlorophyll and nitrogen status of the plant.

Key words: High-throughput phenotyping, Shoot imaging, Growth analysis, Nutrient use efficiency, RGB, Leaf color

1. Introduction

The contribution of mineral fertilizer to crop yield ranges on average between 30 % and 60 % of total yield (1), making fertilizer application essential for global food production. With ever rising fertilizer prices and a growing population there is an increased pressure on a more economic use of mineral fertilizer. In order to raise yields without the need to also increase fertilizer demand, improved agronomic practices in combination with breeding of more nutrient efficient crops are necessary. Both areas, agronomy and crop breeding, can directly benefit from modern phenotyping technologies to work towards improving nutrient use efficiency.

Remote sensing of crop canopies is already used in precision agriculture to assess the nutrient status of crops and thus inform

Parts of this chapter were adapted from a volume on Plant Salt Tolerance and High-throughput Phenotyping in Plants within the series of Methods in Molecular Biology with the permission of the editors.

Frans J.M. Maathuis (ed.), *Plant Mineral Nutrients: Methods and Protocols*, Methods in Molecular Biology, vol. 953, DOI 10.1007/978-1-62703-152-3_18, © Springer Science+Business Media, LLC 2013

improved agronomic decisions. Aerial photographs or ground-based sensors, such as NDVI (normalized difference vegetation index) meters help to determine the most economic way of fertilizer application, with respect to fertilizer rate, location, and timing. This has been reviewed by Montes et al. (2). The aim of this chapter is to show how modern phenotyping technologies can be used in studies of nutrient use efficiency in controlled environments, focusing on genetic improvements.

In a greenhouse situation, NDVI meters are of limited use since they work best on a closed canopy. Digital color photographs, however, offer the possibility to measure plant size and leaf color of individual plants over time to assess growth rates and greenness in response to different nutrient supplies. This allows the non-destructive evaluation of nutrient use efficiency on the single plant level in a pre-breeding context.

If the phenotyping and experimental setup can be performed at a large scale, this allows the use of a forward genetics approach to identify the genetic basis for higher nutrient use efficiency. With nitrogen being the single biggest nutrient applied to non-legume crops, we will focus on the use of image base high-throughput phenotyping for nitrogen use efficiency in a greenhouse situation.

2. Materials

2.1. Seed Treatment

1. Uniformly sized seeds (see Note 1).
2. 70 % (v/v) Ethanol.
3. 3 % (v/v) Sodium hypochlorite (see Note 2).
4. Alternatively, Thiram or similar fungicides.

2.2. Growth Solutions and Hydroponics Setup

1. Stock solution A: 0.04 M NH_4NO_3, 1 M KNO_3. Add about 2.5 L deionized water to a 5 L measuring beaker. Weigh 12.8 g NH_4NO_3 and 404.4 g KNO_3 and add to the beaker. Mix and adjust to a final volume of 4 L. Store at 4 °C (see Note 3).
2. Stock solution B: 0.4 M $Ca(NO_3)_2$. Add about 2.5 L deionized water to a 5 L measuring beaker. Weigh 377.8 g $Ca(NO_3)_2 \cdot 4H_2O$ and add to water. Mix and adjust to a final volume of 4 L. Store at 4 °C (see Note 3).
3. Stock solution C: 0.4 M $MgSO_4$, 0.02 M KH_2PO_4. Add about 2.5 L deionized water to a 5 L measuring beaker. Weigh 394.4 g $MgSO_4 \cdot 7H_2O$ and 10.8 g KH_2PO_4 and add to water. Mix and adjust to a final volume of 4 L. Store at 4 °C.
4. Liquid $NaSiO_3$ stocks can be purchased and diluted accordingly to reach a final concentration of 0.5 mM (see Note 4). Store at 4 °C.

5. Stock solution E: 0.05 M NaFe(III)EDTA. Dissolve 14.7 g NaFe(III)EDTA in about 0.5 L of reverse osmosis (RO) water, mix, and adjust to 0.8 L. Store at 4 °C.
6. Micronutrient stock F: 50 mM H_3BO_3, 5 mM $MnCl_2$, 10 mM $ZnSO_4$, 0.5 mM $CuSO_4$, 0.1 mM Na_2MoO_4. Dissolve 2.47 g H_3BO_3, 0.79 g $MnCl_2{\cdot}4H_2O$, 2.3 g $ZnSO_4{\cdot}7H_2O$, 0.10 g $CuSO_4{\cdot}5H_2O$, 0.02 g $Na_2MoO_4{\cdot}2H_2O$ in about 0.5 L of RO water and adjust to 0.8 L. Store at 4 °C.
7. Polycarbonate pellets, approximately 3 mm in diameter (see Note 5).
8. PVC tubes (4.5 cm diameter, 28 cm height) fitted with mesh at the bottom to prevent pellets from falling through.
9. 80 L storage tank.
10. Two 50 L growth tubs with grid to hold PVC tubes in upright position.
11. Aquarium pumps that are able to lift water at least 1.2 m at a minimum of 300 L/h.

2.3. Growth in Soil Mix

2.3.1. Measurement of Fiold Capacity of Soil Mix

1. Sintered glass funnel.
2. 1.3 m Silicon or clear plastic tubing with diameter to fit the funnel outlet.
3. Retort stand and clamp.
4. Large beaker or bucket as water reservoir.

2.3.2. Preparation of Soil Mix

1. Low nutrient soil or nutrient-free potting mix (see Note 6).
2. Washed sand (see Note 6).
3. Nutrient solution a (10×): 0.6 M K_2SO_4, 0.2 M $MgCl_2{\cdot}6H_2O$, 12 mM $MnSO_4{\cdot}4H_2O$, 12 mM $ZnSO_4{\cdot}7H_2O$, 15 mM $CuSO_4{\cdot}5H_2O$, 6 mM $CoSO_4{\cdot}7H_2O$. Add 104.58 g K_2SO_4, 40.64 g $MgCl_2{\cdot}6H_2O$, 2.68 g $MnSO_4{\cdot}4H_2O$, 3.45 g $ZnSO_4{\cdot}7H_2O$, 3.75 g $CuSO_4{\cdot}5H_2O$, and 1.69 g $CoSO_4{\cdot}7H_2O$ to about 800 mL water, mix well, and make up to 1 L.
4. Nutrient solution b (10×): 3 mM H_3BO_3, 1.5 mM $Na_2MoO_4{\cdot}2H_2O$. Add 0.19 g H_3BO_3 and 0.36 g $Na_2MoO_4{\cdot}2H_2O$ to 900 mL of water, mix well, and adjust to 1 L.
5. Nutrient solution c (5×): 0.7 M $Ca(NO_3)_2{\cdot}4H_2O$, 1.4 M KNO_3. Add 165.2 g $Ca(NO_3)_2{\cdot}4H_2O$ and 141.54 g KNO_3 to 750 mL of water. Mix well and adjust to 1 L (see Note 3).
6. Nutrient solution d (5×): 16 mM $FeSO_4{\cdot}7H_2O$. Add 4.44 g $FeSO_4{\cdot}7H_2O$ to 900 mL water, mix well, and adjust to 1 L.
7. Nutrient solution e (5×): 2 mM NH_4NO_3. Add 160 g NH_4NO_3 to 800 mL water, mix well, and adjust to 1 L (see Note 7).
8. Nutrient solution f (5×): 0.8 M $CaHPO_4$. Add 108.8 g $CaHPO_4$ to 800 mL water, mix well, and adjust to 1 L.

9. Store all nutrient solutions at 4 °C.
10. Cement mixer for preparation of larger quantities.
11. Plastic pots with a capacity of about 1–3 L (see Note 8).

2.4. Image Acquisition

1. Industry grade digital color camera with automated software control (e.g., LemnaTec 3D Scanalyzer system, LemnaTec GmbH Germany).
2. Automated setup to move plants to the camera or vice versa. If manual systems are used, experiments are usually limited to about 150–200 plants per experiment.
3. Adequate computer hardware for image storage (see Note 9).
4. Adequate illumination equipment.
5. A color reference card for calibration purposes (e.g., Munsell Tissue Color Chart, X-Rite, USA).

2.5. Image Analysis

1. Adequate computer hardware for high-throughput image processing.
2. Image analysis software package, included with imaging system like LemnaGrid (LemnaTec GmbH, Germany) and/or standalone software such as MATLAB (Mathworks, USA), Halcon (MVTec Software GmbH, Germany), or Labview (National Instruments, USA). An open source alternative is ImageJ (http://rsbweb.nih.gov/ij).

3. Methodology

3.1. Seed Treatment

1. Surface sterilize uniformly sized seeds for 1 min in 70 % (v/v) ethanol followed by 5 min in 3 % (v/v) sodium hypochlorite.
2. Rinse the seeds several times in deionized water (see Note 10).
3. OR.
4. Surface coat the seeds with Thiram following the manufacturer's instructions (see Note 10).
5. Germinate the seeds on moist filter paper at room temperature in the dark.

3.2. Growth in Supported Hydroponics

There are numerous ways how to design a supported hydroponics system. The method presented here follows the protocol described by Genc et al. ((3), which also contains an image of the hydroponic setup).

1. Two 50 L opaque plastic tubs are mounted on a trolley and connected to an 80 L storage tank with nutrient solution. Each tub holds 42 PVC tubes filled with polycarbonate pellets.

The pellets should be filled to 1 cm above the maximum height of the nutrient solution during the filling cycle. Each tub is filled and drained with 25 L of nutrient solution every 20 min (see Note 11).

2. Fill the storage tank with about 50 L of RO water then add 400 mL of each stock solution A, B, and C and 80 mL each of stock solution D, E, and F. Adjust the pH to 6.0 with 10 % (v/v) HCl and make up to a final volume of 80 L.
3. To each PVC tube, transplant one, uniformly germinated seedling at a depth of 1 cm, ensuring that the roots reach the nutrient solution during the filling cycle.
4. After 1–2 days, once the seedlings have established, gently fill up the PVC tubes another 1–2 cm with polycarbonate pellets to support the seedlings and reduce the growth of algae at the surface.
5. Monitor the pH of the nutrient solution daily and adjust if necessary.
6. Change the nutrient solution weekly.

3.3. Growth in Soil Mix

3.3.1. Measurement of Field Capacity of Soil Mix

When working in pots, it is important to carefully consider the watering to avoid water-logging and hypoxia (4). Many experiments will adjust watering to "water holding capacity" or "pot capacity", which is the volumetric water content of a free draining pot. However, this value greatly depends on the height of the pot and might often result in hypoxia, especially with fine potting mixes or field soil. In our experiments, we measure "field capacity," defined as the volumetric water content of the soil mix at 1 m suction.

The setup described here to measure this parameter is comparable to the one shown in Fig. 2 of Passioura (4).

1. Attach the silicon tubing to the funnel outlet.
2. Mount the funnel with tubing on a retort stand about 1 m above the water reservoir (see Note 12).
3. Add about 2 L of water to the water reservoir below the funnel.
4. Fill the funnel and silicon tube with water ensuring that all air bubbles are removed.
5. Add the soil/potting mix to be tested into the funnel and let it settle. About half to two-thirds of the funnel should be filled with soil.
6. Once the water has drained to just above the soil level, cover the funnel with clingfilm to avoid evaporation from the surface.
7. To ensure hydraulic conductivity, there should be no air bubbles present between the filter plate, tubing, and water reservoir.

8. Adjust the position of the filter to obtain a height of 1 m from the sintered filter plate down to the water level in the reservoir.
9. Let the soil/potting mix equilibrate for several days up to 1 week, ensuring that no air bubbles form.
10. Take out the wet soil from the funnel and record the wet weight (WW).
11. Dry the soil in an oven at 105 °C until constant weight is reached.
12. Record the dry weight (DW).
13. The volumetric field capacity is given by the equation (WW – DW)/DW.

3.3.2. Preparation of Soil Mix

There is a vast amount of recipes for controlled nutrient treatment in soils and the precise nutrient composition will depend on the plant species, the growth duration, and the residual amount of nutrients present in the soil. Here, we present one example adapted from Murphy et al. (5) suitable for cereals such as wheat and barley. While Murphy et al. used a 1:9 soil sand mix, higher amounts of soil may be necessary to increase the field capacity of the mix and the amount of water available to the plants (see Note 6).

1. Prepare the desired mix of dry soil and washed sand.
2. Fill a pot to about 4 cm below the rim after gentle tapping and then weigh it.
3. Use the same weight to fill up all remaining pots.
4. Include several spare pots to monitor water evaporation from the soil during the experiment.
5. Dilute enough nutrient stock solution to add 10 mL of each solution for each kg of dry soil needed.
6. Add 10 mL each of 1× nutrient solution a, b, c, d, e, and f for each kg of dry soil to the pots and mix thoroughly. For larger quantities, a cement mixer can be used to mix the nutrients into the soil mix.
7. Use the soil dry weight to calculate the target weight of a pot at field capacity as determined by Subheading 3.3.1.
8. Adjust the watering level of each pot to field capacity.
9. Transplant evenly sized seedlings to the pots (see Subheading 3.1).
10. Water to weight daily for soils with low field capacity, every second day for soils with higher field capacity as long as the plants are young.
11. Image the plants daily or every second day during the period important for phenotypic measurements.

3.4. Image Acquisition to Monitor Plant Growth and Leaf Color

How images are acquired will greatly depend on the hard- and software available to the researcher and the trait to be measured. There are complete systems available from LemnaTec (LemnaTec GmbH, Germany) that combine plant handling, imaging hardware, and the control software. Other institutes might have the capability to build their own automated in-house solutions (6, 7) or use a fairly simple camera setup and manual handling of plants. We will therefore only present aspects of image acquisition that are generally applicable and important for any type of setup.

1. The aim of any imaging setup should always be to obtain the best possible image of the plants for measuring the trait of interest. Image acquisition should be done as consistent as possible. This will greatly facilitate the image analysis and ideally allow the generation of automated image analysis algorithms that require minimum user input.
2. In general, there are two methods for image acquisition.
   (a) The plants are stationary and the camera is moved to the plant. This is most commonly used for plants with a simple architecture, such as, for example, Arabidopsis, where a single image from the top often provides sufficient data.
   (b) The plants are moved to a stationary camera setup. This is of advantage for plants with a complex morphology, such as wheat and barley, where images from several angles will greatly increase the quality of data obtained through imaging. In addition, the imaging environment, such as background and illumination, is easier to control.
3. Illumination conditions should be as uniform as possible, both over time and throughout the field of view. It is important to pre-heat the lamps until constant illumination is reached before the first images are taken. Hunter et al. (8) give detailed information on how to achieve optimal lighting and avoid shadows and reflections.
4. Use of a color card and ruler allow calibration of the imaging setup. If both are present in an image, it is possible to normalize the recorded colors and calibrate for the zoom factor used. This allows comparisons between different imaging setups that differ in lighting conditions and the cameras used. Munsell Plant Tissue Color Charts are specifically designed for the use in plant science and most shades of plant leaves for calibration.
5. The imaging background should be chosen carefully to facilitate the identification of the plant in subsequent analysis. Backgrounds, such as white or blue are preferable, since the green of the plant will be easy to differentiate.
6. Green and gray should be avoided as pot colors. White, blue, and black are suitable for most plant types and white has the

advantage of keeping the soil cooler than darker colors. Materials with a flat finish reduce undesired reflections.

7. The soil surface can become challenging in the image analysis, since sandy or drying soils can have very similar colors to senescent leaves. Colored plastic mulch or white gravel on the surface can reduce this problem and have the further advantage of reducing water loss from the soil surface.
8. Many plants, especially wheat and barley, will need some sort of support when grown in pots, such as carnation frames. Again, they should not be green and if they are out of metal it needs to be tested if they can be easily eliminated in the image analysis. In some cases, it might be easier to get color-coated frames to avoid problems in the automated image analysis.
9. The number of images taken per plant will depend on the shoot morphology and the desired throughput. We found that three images (two from the side, one from the top) are sufficient for most plants. Plants like Arabidopsis generally require only a single image from the top.
10. When choosing the exposure for the images, it is generally better to have a lower exposure. Overexposure will lead to white spots and thus a loss of color information that cannot be compensated for by image analysis.
11. The file format for storing the images should not lead to loss of image information, such as done by JPG or BMP. PNG or TIFF are commonly used formats and do not lead to loss of information through compression.

3.5. Image Analysis to Measure Projected Shoot Area

Since plant imaging allows daily recordings, simple image analyses, such as plant size measurement, already yield valuable information about plant growth and performance. Nevertheless, basic image analysis also requires the use of specialized software, computing infrastructure and database management if it is to be performed at high-throughput.

Depending on the software solution used, different levels of prior knowledge in image analysis and programming are necessary to develop image analysis algorithms and collaboration with scientists experienced in that area is advisable.

MATLAB (MathWorks, Massachusetts, USA) is possibly the most commonly used and powerful software to develop image analysis algorithms and offers solutions for automated image acquisition. Halcon (MVTec Software GmbH, Germany) is a fairly comprehensive application for image analysis and it is compatible with common programming languages such as C, C#, and .NET. ImageJ (http://rsbweb.nih.gov/ij) presents a Java-based solution for image analysis that is open source, so it is easily accessible. However, all three softwares require a certain amount of programming skills to use, to write, and to implement analysis algorithms. The in-built image analysis solution of LemnaTec setups, LemnaGrid (LemnaTec GmbH, Germany), is designed to allow researchers without prior programming knowledge to create algorithms for image analysis through drag-and-drop software where individual operators can be connected to create a processing pipeline. Unfortunately, algorithms can only be shared among LemnaTec users and the functionalities are not as comprehensive as those of specialized image analysis software.

Since the specific algorithms will depend on the software used and the imaging setup, we will only discuss general steps common to digital image processing (9) that are necessary to measure the size of the plant and to perform subsequent growth analysis.

1. Image retrieval. Recorded images need to be loaded into the software from a database or storage folder. Images may need to be cropped or a Region Of Interest (ROI) may need to be set to shorten the computing time and/or to remove unnecessary parts of the image that can become a source of noise.
2. Image preprocessing. The application of filters to minimize noise or increase sharpness can improve the outcome of the subsequent analysis steps. However, there is a possibility of losing information that cannot be retrieved in later steps. If thresholding is used to make a binary image in the next step, the color image needs to be converted into a grayscale image by transforming the 3D red, green, and blue (RGB) color information into a single channel.
3. Image segmentation. The next step is the segmentation of the image into objects of interest, parts of the plant, and objects that will later be discarded, such as the background, pot, carnation frame, or soil. Depending on the composition of the image, there are several options to produce a binary image. Classification by color with a supervised nearest neighbor algorithm or thresholding of a grayscale image are commonly used. In both instances, the result is a binary image, where pixels that belong to the object of interest are set to a value of 1, all others to 0.
4. Noise reduction. Morphological operations such as erosion–dilation steps or filling holes can be used to correct for unavoidable imperfections in the binary image, that result from noise from image acquisition or difficulties in distinguishing between parts of the object and background that have similar colors.
5. Image composition. Leaves can often become fragmented in earlier steps due to curling of the leaves and the individual fragments need to be merged to create one single object, the plant.
6. Image description. Features of the identified object, such as area, height, width, convex hull, or compactness are quantified. The features mostly consist of mathematical characteristics calculated from the object.

7. Extraction of color information. The mean R, G, and B values of the identified object, the plant, can now be extracted from the original RGB image. See 3.6 for transformation of RGB values for estimation of leaf greenness.

3.6. Estimating Differences in Leaf N Status Using Leaf Color Extracted from Digital Images

Most digital cameras record color information in the RGB color scheme. It may seem self-evident to use the average green value of the identified plant as a measure for greenness and thus chlorophyll and nitrogen content. However, the value of red and blue will greatly influence how green an image appears. In fact, the green value of RGB images can show negative correlation with leaf chlorophyll and nitrogen status, while the blue value is positively correlated with N content (10, 11). Rather than just using the red, green, or blue value, it is desirable to use all color information available, which can be incorporated into a single index. This can be done using the original RGB color information (12) or after transformation into another color space, such as the CIE L*a*b* color scheme (13) or the HSI (hue, saturation, intensity) or HSB (hue, saturation, brightness) color schemes. Here we will give an example of using the HSB color scheme and an index for greenness derived from the H, S, and B values as described by Karcher and Richardson (14, 15).

1. Using your imaging and camera setup, take example photographs of several color plates of the Munsell Plant Tissue color charts covering the range of colors observed for the plants grown under the chosen growth conditions.
2. Extract the recorded RGB values for the color plates and convert them to percentage. For example, for an 8-bit color image with values of R, G, and B ranging from 0 to 255 use the recorded average value and divide by 255.
3. Use the percentage values of R, G, and B for the conversion to the HSB color scheme by using the following equations (see Note 13).

   Hue (H):

   $$\text{If } \max(R,G,B) = R;\ H = 60\{(G-B)/(R-\min(R,G,B))\}$$

   $$\text{If } \max(R,G,B) = G;\ H = 60(2+\{(B-R)/(G-\min(R,G,B))\})$$

   $$\text{If } \max(R,G,B) = B;\ H = 60(4+\{(R-G)/(B-\min(R,G,B))\}).$$

   Saturation (S):

   $$S = (\max(R,G,B) - \min(R,G,B))/\max(R,G,B)$$

   Brightness (B):

   $$B = \max(R,G,B).$$

4. Establish calibration curves for H, S, and B using the values extracted by image analysis and the actual H, S, and B values of the Munsell color disks (see Note 13).

5. Now, extract the average R, G, and B values of your identified objects or plants and convert them to corrected HSB values following steps 2 and 3 and using the calibration curves established in step 4.
6. Using the corrected HSB values, calculate the greenness index (GI) as $GI = ((H - 60)/60 + (1 - S) + (1 - B))/3$.
7. You will now need to establish a correlation between GI and actual leaf N content by doing destructive harvests of several leaves/plants to establish the range for which GI can be used as nondestructive estimate for leaf N content.

3.7. Basic Plant Growth Analysis

The following protocol describes basic measurements of several growth parameters that cannot only be used in assessing the response to different nutrient supplies and also for other treatments affecting plant growth. For more detailed plant growth analyses refer to the excellent publications by Hunt (16, 17). All steps presented here assume a linear correlation between plant biomass and the projected shoot area measured from the images. This correlation needs to be tested at the beginning of image-based phenotyping experiments for each plant species and stress treatment.

1. Increase in shoot area (A) over time (t). For a first evaluation of the data, plot the shoot area for individual plants or treatment groups over time. This will allow a visual assessment of treatment or genotype effects and the identification of biological outliers (entire growth curve affected) or technical outliers from the imaging process (generally only individual points of the growth series affected). Most plant species have a sigmoid growth curve when imaged from seedling stage to early reproductive stage, consistent with other measuring techniques. Once leaves start to senesce during seed ripening, this will obviously result in a decrease in projected leaf area, which is then no longer a good indicator of plant biomass. It is possible to overcome this technical challenge by using the color information of the leaves to differentiate between green and senescent leaf area, if experiments need to extend over the whole growth cycle. However, this needs to be tested for each plant species.
2. Use the data of shoot area over time to generate a growth model through curve fitting. Growth models, such as higher order polynomials or cubic splines, that make no prior assumption about the data are preferable. Higher order polynomials can be generated with basic spread sheet software, such as Microsoft Excel (Microsoft Cooperation, USA). Spline curves generally need statistical software packages.
3. Use the growth model to compute the absolute growth rate of the plants, which is the first derivative (dA/dt) of the growth model. The absolute growth rate will allow measuring how

much area the plant gained per day at any time during the experiment. If plants were imaged over most of the lifecycle, the absolute growth rate will show an increase during early growth, reaching a maximum, when plants shift from vegetative to reproductive growth and a subsequent decline as plants mature. The time interval for plants to reach maximum absolute growth can be regarded as a trait. Certain stress treatments, such as drought or salinity can alter the length of the interval, indicating altered plant development.

4. Relative growth rate ($dA/dt \cdot 1/A$). In addition to the absolute growth rate, a growth model can be used to calculate the relative growth rate (RGR) at any given time. The RGR is generally highest for young seedlings and then declines gradually. Since RGR is independent on plant size it allows comparison of plants and varieties with fairly different growth habits. Analysis of RGR over time can reveal when genotype or treatment effects become apparent.
5. Leaf area duration (LAD). The expression of leaf area duration was used by Watson in 1947 (18) for the integral of the leaf area over the entire lifecycle and was described as the "whole opportunity for assimilation" of the plant. Using the previously developed growth model it is possible to calculate LAD for the entire experiment or certain intervals relevant to the treatment. LAD will give a measure of the leaf area and its persistency over the chosen period.

4. Notes

1. Since part of the assay presented here is based on growth analysis, it is extremely important that the seeds and seedlings used are as uniform as possible. If sufficient seed is available, one should always plant excess amounts to be able to select for evenly sized seedlings. If it is known that the used lines germinate at different rates, the sowing should be staggered to have evenly sized seedlings at the start of the experiment.
2. The sodium hypochlorite solution can be prepared using a household product such as Domestos®, when taking into account the lower active concentration of Cl^- compared to a lab grade solution.
3. To apply low nitrogen treatments, KNO_3 can be substituted by KCl and $Ca(NO_3)_2$ can be substituted by $CaCl_2$.
4. It is easiest to get a homogenous solution by first adding the water and then the sodium silicate stock due to its high viscosity.

5. Polycarbonate pellets can be obtained from plastic manufacturers that use them as a starting product for molding. If using other types of plastics, ensure their density is higher than water.
6. The choice of soil mix or potting mix will obviously depend on the experiment. If the pots are placed on an automated conveyor system the substrate should not be too lose (such as pure sand) since it might shift through the movement on the belt and damage the root system. Also, a high sand content will lead to a low field capacity and hence a low volume of soil solution available to the plant. However, a certain amount of sand is required when plants are grown on conveyor belts to avoid compaction and root hypoxia. We have had best results with soil:sand or peat:sand mixes.
7. The amount of NH_4NO_3 can be reduced for low nitrogen treatments. However, a total lack of NH_4NO_3 may lead to a pH imbalance in the rhizosphere during plant growth and can result in nutrient deficiencies, mainly iron deficiency.
8. The color of the pot should allow an easy distinction from the plants in the image-processing step, preferably white or blue. Black is possible, but it leads to an increased soil temperature. Standard green nursery pots should not be used.
9. We generally take three images per plant (two from the side at 90 ° rotation and one from the top) at about 15–20 time points throughout an experiment. With a file size of about 4 MB, this amounts to 4 MB × 3 images × 20 time points = 240 MB per plant. Even a smaller scale experiment with 200 plants will therefore need 47 GB of storage.
10. Seed treatment might not be necessary, depending on the source of the seed. However, fungal infections of young seedlings can influence the growth rate and its sensitivity to certain stress treatments.
11. To avoid excessive growth of algae, light exposure to the solution should be kept at a minimum and the space between the PVC tubes should be covered.
12. If no large retort stand is available, a smaller one can be placed on a table with the water reservoir on the ground.
13. Many software packages for image analysis will already include a function for conversion into other color schemes. Alternatively, free online solutions can be used, such as http://www.workwithcolor.com/color-converter-01.htm.

References

1. Stewart WM, Dibb DW, Johnston AE et al (2005) The contribution of commercial fertilizer nutrients to food production. Agron J 97(1):1–6
2. Montes JM, Melchinger AE, Reif JC (2007) Novel throughput phenotyping platforms in plant genetic studies. Trends Plant Sci 12(10):433–436
3. Genc Y, McDonald GK, Tester M (2007) Reassessment of tissue Na^+ concentration as a criterion for salinity tolerance in bread wheat. Plant Cell Environ 30:1486–1498
4. Passioura JB (2006) The perils of pot experiments. Funct Plant Biol 33(12):1075–1079
5. Murphy PJ, Langridge P, Smith SE (1997) Cloning plant genes differentially expressed during colonization of roots of *Hordeum vulgare* by the vesicular-arbuscular mycorrhizal fungus *Glomus intraradices*. New Phytol 135(2):291–301
6. Granier C, Aguirrezabal L, Chenu K et al (2006) PHENOPSIS, an automated platform for reproducible phenotyping of plant responses to soil water deficit in *Arabidopsis thaliana* permitted the identification of an accession with low sensitivity to soil water deficit. New Phytol 169(3):623–635
7. Jansen M, Gilmer F, Biskup B et al (2009) Simultaneous phenotyping of leaf growth and chlorophyll fluorescence via GROWSCREEN FLUORO allows detection of stress tolerance in *Arabidopsis thaliana* and other rosette plants. Funct Plant Biol 36(10–11):902–914
8. Hunter F, Biver S, Fuqua P (2007) Light-science & magic: an introduction to photographic lighting. Focal Press Oxford, UK
9. Gonzalez RC, Woods RE (2006) Digital image processing, 3rd edn. Prentice-Hall, Inc. NJ, USA
10. Yadav SP, Ibaraki Y, Gupta SD (2010) Estimation of the chlorophyll content of micropropagated potato plants using RGB based image analysis. Plant Cell Tissue Organ Cult 100(2):183–188
11. Vollmann J, Walter H, Sato T et al (2011) Digital image analysis and chlorophyll metering for phenotyping the effects of nodulation in soybean. Comput Electron Agric 75(1):190–195
12. Pagola M, Ortiz R, Irigoyen I et al (2009) New method to assess barley nitrogen nutrition status based on image colour analysis: comparison with SPAD-502. Comput Electron Agric 65(2):213–218
13. CIE (1976) Colorimetry, 2nd edn (Publication CIE No, 15.2). Central Bureau of the Commission Internationale de L'Eclairage, Vienna
14. Karcher DE, Richardson MD (2003) Quantifying turfgrass color using digital image analysis. Crop Sci 43(3):943–951
15. Rorie RL, Purcell LC, Karcher DE et al (2011) The assessment of leaf nitrogen in corn from digital images. Crop Sci 51(5):2174–2180
16. Hunt R (1978) Plant growth analysis. Edward Arnold Ltd.
17. Hunt R, Causton DR, Shipley B et al (2002) A modern tool for classical plant growth analysis. Ann Bot 90(4):485–488
18. Watson DJ (1947) Comparative physiological studies on the growth of field crops. Ann Bot 11(41):41–76

INDEX

Frans J.M. Maathuis (ed.), *Plant Mineral Nutrients: Methods and Protocols*, Methods in Molecular Biology, vol. 953,
DOI 10.1007/978-1-62703-152-3, © Springer Science+Business Media, LLC 2013

D

E

F

G

H

I

J

K

L

MIX
Papier aus verantwortungsvollen Quellen
Paper from responsible sources
FSC® C105338

If you have any concerns about our products,
you can contact us on
ProductSafety@springernature.com

In case Publisher is established outside the EU,
the EU authorized representative is:
Springer Nature Customer Service Center GmbH
Europaplatz 3, 69115 Heidelberg, Germany

Printed by Libri Plureos GmbH
in Hamburg, Germany